PEOPLE, LAND AND TIME

PEOPLE, LAND AND TIME

AN HISTORICAL INTRODUCTION TO THE RELATIONS BETWEEN LANDSCAPE, CULTURE AND ENVIRONMENT

Peter Atkins, Ian Simmons and Brian Roberts
Department of Geography, University of Durham

Routledge
Taylor & Francis Group
LONDON AND NEW YORK

First published 1998 by
Hodder Education, an Hachette UK Company

This edition published 2014 by Routledge
2 Park Square, Milton Park, Abingdon, Oxfordshire OX14 4RN
711 Third Avenue, New York, NY 10017

Routledge is an imprint of the Taylor and Francis Group, an informa business

First issued in hardback 2015

British Library Cataloguing in Publication Data
A catalogue entry for this book is available from the British Library

Library of Congress Cataloging-in-Publication Data
A catalog record for this book is available from the Library of Congress

Composition by Phoenix Photosetting,Chatham, Kent

ISBN 978-0-340-67714-8 (pbk)
ISBN 978-1-138-13848-3 (hbk)

CONTENTS

Preface vii
Acknowledgements ix

Introduction xiii

PART 1 THE PRE-INDUSTRIAL WORLD **1**

1 Hunters and Gatherers 3
2 The Origins and Spread of Agriculture 13
3 Early Urbanization and the Hydraulic Environment 27
4 Resources, Population and Sustainability 40
5 Environmental Degradation and the Collapse of Civilizations 53
6 Sustainable Resource Management in Pre-industrial Societies 63
7 Large-scale Landscape Modification: Pays and Pre-industrial Planning 77
8 Clearing the Wood 93
9 The Control of Water 104
10 Landscapes on the Margin: Deserts, Hillslopes, Heath, Moor and Grassland 119

PART 2 THE TRANSITION TO MODERNITY **131**

11 Feudal Landscapes 133
12 Urbanization and Proto-industrialization 148

PART 3 THE MODERN ERA **161**

13 The Impact of Agriculture 165
14 Landscapes of Energy Acquisition: the Getting of Power 176
15 Industrial landscapes 184
16 Modern Urban Landscapes: Modern Cities and City Life 197
17 Postmodern Landscapes 212
18 Metaphors and Meanings in Modern Landscapes: Reading the Landscape 219
19 Landscapes of power and pleasure 231
20 'Other' Landscapes 239

PART 4 THE GLOBAL ERA **253**

21 Globalized landscapes 255
22 Conservation 267
23 Conclusion: the Past, Present and Future of the Study of People, Land and Time 274

Index 280

PREFACE

The inspiration for this book has been the involvement of the authors in a long-standing undergraduate course at Durham entitled *Land, People and Time*. Although taught from the Department of Geography, this is open to all students and has proved very popular as a cross-disciplinary mix of those issues of environment and culture which have influenced the evolution of humanized landscapes. The emphasis in the course and in this text is upon material outcomes but there is of course a recognition, first, that there is a profound complexity of interacting processes which drives the production of visible modifications of the *natural* landscape and, second, that the study of landscape is only one among the several approaches which may be adopted in understanding the relationships between society and environment. Despite the recent revival of interest in landscapes in the literatures of cultural geography and environmental history, there is understandable resistance from those who see landscape study as a naïve interpretation in terms of the purely visual and morphological. In this book we have tried to avoid such a superficial view by relating our discussions of the material world of objects to the human processes responsible and to interpret the outcomes as one form of evidence of the evolving human impact upon the environment.

The text is written in a style that we hope will be accessible, with a minimum of technical jargon. The approach is varied, with a mix of what might be called environmental history and historical geography, along with a leavening of cultural studies in Chapters 16–20. This range of voices will illustrate the fascinating range of methodologies and perspectives which are now being deployed to study the culture/nature connexion.

The structure of the book is in four parts, arranged by techno-social stages from the pre-industrial to the global era. Each part is subdivided into chapters which are thematic, and these are supplemented with case studies that accompany the descriptions and explanations of a general nature. They elaborate points of fact and some (for instance, the case study of the Maya in Chapter 5) are full-length location-specific illustrations of a chosen theme.

ACKNOWLEDGEMENTS

Our first debt is to the thousands of Durham students who have taken our course over the years. Their questions and sometimes scepticism, expressed in lectures, tutorials and workshops have helped to sharpen our presentation. We have also benefited from the advice, experience and knowledge of the colleagues who have contributed to the course but are no longer part of the teaching team.

Steven Allan and David Hume worked quickly and efficiently to produce professional artwork from our scribbles and Michèle Johnson did the photographic work. Anne Billen and Gwyneth Thomas from the Inter-Library Loan section of the University Library worked their customary miracles in finding obscure materials for us. A particular thank you is due to Laura McKelvie of Arnold for her infectious enthusiasm and also for her understanding.

The authors and publishers would like to thank the following for permission to use copyright material in this book:

Thames and Hudson for A. Leroi-Gourhan (1968) *The art of prehistoric man in western Europe*; University of Arizona Press for P.S. Martin (1984) Prehistoric overkill: the global model, pp. 354–403 in P.S. Martin and R.G. Klein (eds) *Quaternary extinctions: a prehistoric revolution*; Professor D.R. Harris and Professor R. Ellen for (1996) Domesticatory relationships of people, plants and animals, pp. 437–63 in R. Ellen and K. Fukui, (eds) *Redefining nature: ecology, culture and domestication* Oxford: Berg; Princeton University Press and Professor L. Cavalli-Sforza for L.L. Cavalli-Sforza, P. Menozzi and A. Piazza (1994) *The history and geography of human genes*; The Royal Geographical Society and Dr Dai Morgan for W.T.W. Morgan (1988) Tamilnadu and eastern Tanzania: comparative regional geography and the historical development process, *Geographical Journal* 154, 69–86; Thames & Hudson for S. Lloyd (1984) *The archaeology of Mesopotamia*; Dr J.N. Postgate for *idem* (1992) *Early Mesopotamia: society and economy at the dawn of history* and *Bulletin on Sumerian Agriculture* 5, 147–86; The Royal Geographical Society and Professor P. Beaumont for P. Beaumont (1968) Qanats in the Varamin Plain, *Transactions of the Institute of British Geographers* 45, 169–80; American Philosophical Society for P.W. English (1968) The origin and spread of qanats in the old world, *Proceedings of the American Philosophical Society* 112, 170–81; Routledge for O.H.K. Spate (1967) *India, Pakistan and Ceylon: the regions* 3rd edn. London: Methuen; Blackwell Publishers and Professor K. Butzer for K.W. Butzer, J.F. Mateu, E.K. Butzer and P. Kraus (1985) Irrigation agrosystems in eastern Spain: Roman or Islamic origin?, *Annals of the Association of American Geographers* 75, 479–509; the University of Lund for T. Carlstein (1982) *Time resources, society and ecology. On the capacity for human interaction in space and time. Vol. 1: preindustrial societies*; The Food and Agriculture Organization of the United Nations for B.N. Okigbo (1984) Improved permanent production systems as an alternative to shifting intermittent cultivation, pp. 1–100 in Food and Agriculture Organization of the United Nations, *FAO Soils Bulletin* No. 53; Dr T.T. Chang for *idem* (1989) Domestication and spread of the cultivated rices, pp. 408–17 in D.R. Harris and G.C. Hillman, (eds) *Foraging and farming: the evolution of plant exploitation* London: Unwin Hyman; Stanford University Press and Dr T.H. van Andel for T.H. Van Andel and C. Runnels (1987) *Beyond the Acropolis: a rural Greek past*; The American School of Classics at Athens for T.H. Van Andel, C. Runnels and K.O. Pope (1986) 5000 years of land use and abuse in the southern Argolid, Greece, *Hesperia* 55, 103–28; Academic Press Limited, London and Dr D. Eisma for (1978) Stream deposition and erosion by the eastern shore of the Aegean, in W.C. Brice (ed.) *The environmental history of the Near and Middle East since the last Ice Age*; A.A. Balkema and Neil Roberts for S. Bottema, G. Entjes-Nieborg and W. van Zeist (eds) *Man's role in shaping the eastern Mediterranean landscape: Proceedings of the symposium on the impact of ancient man on the landscape of the eastern Mediterranean region and the Near East, Groningen, 6–9 March 1989*, 1990. 352 pp. A.A. Balkema 1675, Rotterdam, Netherlands; Cambridge University Press for J.A. Tainter (1988) *The collapse of complex societies*; The University Museum, Philadelphia for Figure 5.2; The University of New Mexico Press for F.M. Wiseman (1978) Agricultural and historical ecology of the Maya lowlands, pp. 63–116 in P.D. Harrison and B.L. Turner (eds) *Pre-hispanic Maya agriculture*; The American Association for the Advancement of Science and Professor B.L. Turner for B.L. Turner and P.D. Harrison (1981) Prehistoric raised field agriculture in the Maya lowlands, *Science* 213, 399–405; Cambridge University Press and Lady Darby for H.C. Darby (1983) *The changing fenland* ; Dr O. Rackham for (1976) *Trees and woodland in the British landscape* London: Dent; Committee for Aerial Photography, University of Cambridge for Figures 6.6, 6.9, 7.2, 7.10, 9.7, 9.8, 10.6, 11.5, 11.7, 11.10, 11.13, 13.5, 15.13 and 16.3; English Nature for the data upon which Figure 6.8 is based; The Royal Geographical Society and Dr P. Cloke for P. Cloke, P. Milbourne and C. Thomas (1996) The English National Forest: local reactions to plans for renegotiated nature–society relations in the countryside, *Transactions of the Institute of British Geographers* n.s. 21, 552–71; Mrs M.S. Dilke for O.A.W. Dilke (1971) *The Roman land surveyors* Newton Abbot: David & Charles; Dr Douglas Lockhart for Figure B7.12; Rand McNally for N.J.W. Thrower (1966) *Original survey and land subdivision: a comparative study of the form and effect of contrasting cadastral surveys* Chicago: Rand McNally; the University of Chicago Press for H.C. Darby (1956) The clearing of the woodland in Europe, pp. 183–216 in W.L. Thomas (ed.) *Man's role in changing the face of the earth* Chicago: University of Chicago Press; NASA Landsat Pathfinder Humid Tropical Forest Project, University of New Hampshire for Figure 8.4; Cambridge University Press and Dr G. Peterken for (1996) *Natural woodland: ecology and conservation in northern temperate regions*; Dr Michael Williams for *idem* (1990) The clearing of the forests, pp. 146–68 in M.P. Conzen (ed.) *The making of the American landscape* Boston: Unwin Hyman; Academic Press Limited and Dr Michael Williams for M. Williams (1982) The clearing of the United States forests: the pivotal years, 1810–1860, *Journal of Historical Geography* 8, 12–28; The Royal Geographical Society and Dr A.S. Mather for A.S. Mather (1992) The forest transition, *Area* 24, 367–79; W.J. Mitsch, R.H. Mitsch and R.E. Turner (1994) Wetlands of the Old and New Worlds: ecology and management, pp. 3–56 in W.J. Mitsch (ed.) *Global wetlands: old world and new* with kind permission from Elsevier Science, Amsterdam and from Professor Mitsch; The Royal Geographical Society and Dr Michael Williams for M. Williams (1990) Agricultural impacts in temperate wetlands, pp. 181–216 in M. Williams (ed.) *Wetlands: a threatened landscape* Oxford: Blackwell; KLM Aerocarto for Figure 9.5; Blackwell Publishers for W.M. Denevan (1992) The pristine myth: the landscape of the Americas in 1492, *Annals of the Association of American Geographers* 82, 369–85; Professor W.M. Denevan for Figure 9.10; Mr Eric H. Mose for Figures 9.11 and 9.12; P. Armillas for (1971) Gardens on swamps, *Science* 174, 653–61. *Scientific American* for M.E. Coe (1964) The chinampas of Mexico, *Scientific American* 211, 1, 90–8; Cambridge University Press and Dr K. Ruddle for K. Ruddle and G.F. Zhong (1988) *Integrated agriculture-aquaculture in south China: the dike-pond system of the Zhuijiang delta*; Nebraska State Historical Society for Figure 10.1; the Editor of *Pacific Viewpoint* for J.E. Spencer and G.A. Hale (1961) The origin, nature and distribution of agricultural terracing, *Pacific Viewpoint* 2, 1–40; N.R. Webb (1989) Studies on the invertebrate fauna of fragmented heathland in Dorset, UK, and the implications for conservation, *Biological Conservation* 47, 153–65 with kind permission from Elsevier Science Amsterdam and from Dr Webb; Figure 10.9 is reproduced courtesy of Dr K.R. Olwig and the Geographical Institute, University of Copenhagen; Figure 10.10 is the copyright of the Scottish Office, source: J. Miles (1994) The soil resource and problems today: an ecologist's perspective, pp. 145–58 in S. Foster and T.C. Smout (eds) *The history of soils and field systems*, Scottish Cultural Press; Cambridge University Press and Lady Darby for H.C. Darby (1973) *A new historical geography of England*; The Royal Geographical Society for M.J. Harrison, W.R. Mead and D.J. Pannett (1965) A Midland ridge-and-furrow map, *Geographical Journal* 131, 365–69; C.T. Smith (1967) *An historical geography of western Europe before 1800* London: Longman by permission of Addison Wesley Longman Ltd; Dr R. Shirley for the data upon which Figure 11.11 is based; Academic Press Limited, London and Dr Tim Unwin for P.T.H. Unwin (1981) Rural marketing in medieval Nottinghamshire, *Journal of Historical Geography* 7, 231–51; Addison Wesley Longman Ltd for A.E.J. Morris (1979) *History of urban form before the industrial revolutions* 2nd edn. London: Godwin; Academic Press Limited and Professor D. Cosgrove for D. Cosgrove (1982) The myth and the stones of Venice: an historical geography of a symbolic landscape, *Journal of Historical Geography* 8, 145–69; British Nuclear Fuels for Figure 14.5; Mick Dunford and Diane Perrons for (1983) *The arena of capital* London: Macmillan; Chris Freeman for M.N. Cleary and G.D. Hobbs (1984) The fifty year cycle: a look at the empirical evidence, pp. 164–82 in C. Freeman (ed.) *Long waves in the world economy* London: Pinter; The Tate Gallery for Figure 15.5; Dr Rosemary Bromley for (1991) The Lower Swansea Valley, pp. 33–56 in G. Humphrys (ed.) (1991) *Geographical excursions from Swansea. Volume 2: human landscapes* Swansea: University of Wales; Birmingham Central Library for Figure 15.8; Addison Wesley Longman Ltd for K.J. Hilton (ed.) (1966) *The Lower Swansea Valley Project*; Durham County Council for Figures 15.12, 16.1, and 19.3; Nissan for Figure 15.14; Blackwell Publishers and Professor D. Ward for D. Ward (1964) A comparative historical geography of street car suburbs in Boston, Massachusetts and Leeds, England: 1850–1920, *Annals of the Association of American Geographers* 54, 477–89; The Alfred Stieglitz Collection, The Carl Van Vechten Gallery of Fine Arts, Fisk University and the Georgia O'Keeffe Foundation for Figure 16.4; Consolidated Edison for Figure 16.5; Cambridge University Press and Dr R. Dennis for R. Dennis (1984) *English industrial cities of the nineteenth century: a social geography*; Harper & Row and Professor L.M. Sommers for Figure 16.7, source L.M. Sommers (1983) Cities of western Europe, pp. 85–122 in S.D. Brunn and J.F. Williams (eds) *Cities of the world: world regional urban development*; Professor Dr D. Denecke for (1992) Ideology in the planned order upon the land: the example of Germany, pp. 303–29 in A.R.H. Baker and G. Biger (eds) *Ideology and landscape in historical perspective* Cambridge: Cambridge University Press; Edward Arnold and Professor A. Sutcliffe for A. Sutcliffe (1970) *The autumn of central Paris: the defeat of town planning 1850–1970*; The Disney Corporation for Figure 17.3; Cambridge University Press and Professor J. Duncan for J.S. Duncan (1990) *The city as text: the politics of landscape interpretation in the Kandyan kingdom*; Blackwell Publishers for J.T. Kenny (1995) Climate, race, and imperial authority – the symbolic landscape of the British hill station in India, *Annals of the Association of American Geographers* 85, 694–714; The Trustees of the National Gallery for Figure 18.6; Professor H. Clout for Figure 19.4; Dr Mike Heffernan for Figure 19.5; The Toledo Museum of Art for Figure 20.1; The Academic Press Limited, London and Peter Howard for P. Howard (1985) Painters' preferred places, *Journal of Historical Geography* 11, 138–54; The Museum of the City of New York, the Harry T. Peters Collection for Figure 20.4; Blackwell Publishers and Professor D. Meinig for D.W. Meinig (1965) The Mormon culture region, *Annals of the Association of American Geographers* 55, 191–220; Dr Gwyn Rowley for Figure 20.7; Blackwell Publishers and Dr G. Falah for G. Falah (1996) The 1948 Israeli-Palestinian war and its aftermath: the transformation and de-signification of Palestine cultural landscape, *Annals of the Association of American Geographers* 86, 256–85; Routledge and Professor Janet Abu-Lughod for J. Abu-Lughod (1993) Discontinuities and persistence: one world system or a succession of systems?, pp. 278–91 in A.G. Frank and B.K. Gills (eds) *The world system: five hundred years or five thousand?*; The Royal Geographical Society and Professor A.J. Christopher for A.J. Christopher (1985) Patterns of overseas investment in land, 1885–1913, *Transactions of the Institute of British Geographers* N.S. 10, 452–66; Dr P.J. Atkins for Figures 2.9, 2.10, 3.1, 4.10, 4.11, 6.4, 9.2, 10.3, 10.4,

10.5, 11.2, 11.3, 12.7, 13.1, 13.2, 13.3, 14.1, 14.2, 14.6, 15.4, 15.10, 15.11, 15.15, 16.9, 16.10, 16.11, 17.1, 17.2, 17.4, 17.5, 18.4, 18.8, 18.9, 18.10, 19.1, 19.6, 19.7, 21.3, 21.5, 21.6, 21.7, 21.8, and 21.9; Dr B.K. Roberts for Figures 7.4, 7.5, 7.11, 8.2, 8.3, 8.5, 11.9, 11.11, 11.12, and 11.14; Professor I.G. Simmons for Figures 1.4 and 1.5.

Every effort has been made to trace the copyright holders of quoted material. If, however, there are inadvertent omissions, these can be rectified in any future editions.

The book is dedicated with love and thanks to Liz, Carol and Jan.

INTRODUCTION

Even if the scope of this book is limited to the last 10 000 years and places its emphasis on the later centuries of that era, there is still a lot of time and space to be dealt with. If the globe is covered with a grid 1 lat × 1 long and we assume 40 per cent of it is land and adjoining seas, and then divide each 1 cell vertically into 100-year sections, then there are 162 000 cells to be filled with information about the development of their landscapes under the influence of both natural and human-directed factors. Either there have to be some very gross generalizations made or the account has to be very selective. In fact, of course, both will occur. But there are also very broad statements to be explained about the terms used together with the ideas underlying the selections made and the explanations given.

Society, culture, environment

The terms *society, culture,* and *environment* are used often. Society refers to the fact that humans live in groups and seem always to have done so: even hermits need someone to bring them water and a little food. One balance in societies which will be reflected in this book is that between humans acting according to their own wishes, as agents, and the constraints exerted by the rest of their society and, indeed, by their non-human surroundings, the structures of their total environment. The possession of culture is one of the distinctive qualities of the human animal and exhibits two distinct forms. The first is *material* and refers to basic needs such as food, water, shelter, clothing (outside the tropics), and tools for subsistence. The second is *non-material* and refers to factors such as love and affection, esteem, fears and hopes, education and meaningful leisure, and cooperation with others. It can also include a transcendent or spiritual dimension such as religion. A culture may only satisfy the basic needs which are just above those required to keep an individual alive and capable of reproduction or it may supply the abundances of food, clean water and material goods to which westerners have become habituated. Environment is taken here to mean the non-human surroundings on all scales from the local to the global and impinges in two main ways. In the first, it is *deterministic* in the sense that the features of the planet allow or prevent certain human activities. Deserts are still unpopular places for growing lettuces in spite of the ameliorations made possible by technology; bananas could be grown at the South Pole but they would prove to be expensive on the world market. On the other hand, many environments have undergone human impact and are no longer *natural* environments but culturally changed environments, of the kinds which form the main material of this book. Some are deliberately formed, perhaps even planned years ahead, whereas others are accidental by-products of some other set of processes.

Approaches to the past

In English, the word *history* has two meanings. The first refers to past events and processes: the chronology of what happened when, who was involved and what traces are left of those times. The second is the way later generations *represent* the past. This involves, inevitably, the selection of partial survivals (maps, statistical records, even people's memories) which have been winnowed by time and thus it is incomplete. In some ways, history is to society what remembered experience is to the individual, thus it is usually selective in its representations. The state, for example, may want to use history to promote a sense of national unity and cohesiveness (Chapter 22). So the calamities and mistakes made during a war are forgotten in favour of the glorious victory which was eventually achieved: the Britain–Argentine conflict of 1982 in the Falklands/Malvinas is an example. Opposition groups and radical individuals will then try to bring out a history of suppressed groups, such as peasants or women.

History, too, has always been valued as a guide to conduct in practical affairs. This is especially so for the management of the state, a practice which in the West dates from classical times and was made most famous by the Italian renaissance courtier Niccolò Machiavelli. The broad pattern of the future can be inferred from the direction of history up to the present. So a modern historical consciousness holds in tension (a) the fact that there is a difference between all previous ages and our own; and (b) the recognition that our world has grown from those past stages. A further use of history may be added: as an inventory of assets. The achievements of humanity are recorded and stored in historical traces and accounts and are present therefore as a store of ideas; this cache may also help to show us what is durable in our present condition and what is transient or subject to chance, i.e. is *contingent*.

For this work, the lessons are reasonably obvious. No piece of historical representation can be complete, nor can it ever be totally objective as, say, an experiment in physics is supposed to be. History is situated in being mostly written (though it may also be in pictures or as drama, for instance) by an individual or a group (like this book) and has to be validated. You, the reader, accept much of what is said in these pages because of the names, titles and institutional affiliations of the authors. If those were replaced with text *written* by the members of Class 2 at Coketown Junior Mixed Infants School, you would doubt the information. Further, we can only fill in a few of the 162 000 cells which makes it more tempting to have an agenda: we might, for example, dedicate all the material to showing that the present cultural landscapes of the world are the expressions of unsustainable economies; or to show that capitalism is the only system under which people compose landscapes which are both productive and beautiful. Careful selection of material could achieve either of these. In fact, neither is true: an inventory of past times (incomplete though it must necessarily be) and the plurality of human achievement are what we have set out to display.

The last 10 000 years

No account by geographers can fail to mention the diversity of natural conditions over the surface of the planet, nor the fact that they are in a constant state of change. No time will be spent here in a potted natural geography but some of the natural transformations of the last 10 000 years (known geologically as the Holocene) form an essential element of the processes we shall describe. We then set out a simple classification of human cultures which can encompass all their diversity during this era and which forms the skeleton of the book's layout.

Natural changes

In humanity's current concerns with matters like the possibility of global warming, it is often forgotten that the last major withdrawal of ice from temperate latitudes was only 10 000 years ago and that many natural systems have spent much of that time adjusting to changed natural conditions. Climate, for example, warmed very rapidly in the temperate zones during the period 9000–7000 Before Present (B.P.), with a period of greatest warmth at about 5500 B.P., before beginning to cool; periods of considerable coolness such as the Little Ice Age (*c.* AD 1450–1800) can be detected in most mid-latitude countries. The fluctuations in the tropics during these millennia were less, and in arid zones it was the rainfall rather than the temperature which varied most. We cannot assume, then, that in the past the natural background for human actions was the same as that of today: climate, soils, vegetation, slopes, river regimes all might have been different in any one place at any one time. These data might also mean that we are living in an interglacial period and that any day soon (within the next few thousand years at any rate) the ice sheets will start to reclaim much of Canada and Scandinavia, the North Sea basin, Chile, southern New Zealand, and Great Britain north of the Thames.

Periods of global human history

The creation of cultural landscapes involves human societies making changes to the ecological systems into which they come into contact (though not necessarily live among) and the role of technology is clearly crucial in allowing people to effect all kinds of changes to soils, water flow, plants and animals and slopes, and to ameliorate climate, for instance, by building structures which insulate them from the cold or by installing air conditioning. Hence we can erect a classification centred on technology (see Table 0.1).

Note that no dates have been put to the three phases. This is because the change from one to the next is *diachronic*, i.e. at a different time in one place from another. The introduction of agriculture into southern Europe was around perhaps 9000 B.P. (7000 BC) but only in the nineteenth century AD into Australia, since the native populations never developed it. Industrialization came to parts of England in the late eighteenth century AD but to the Inuit populations of Canada mostly with the Second World War.

Technology and society

It is obvious that the possession and use of technology act as a powerful agent of change in producing cultural landscapes. So much so that some scholars have argued that it is now autonomous: it has a life of its own beyond human control. They point to changes which were caused by technology and suggest that since they cannot be reversed then the control of technology by society is highly incomplete. Examples of changes caused by technological innovation might include:

- navigation
- automobiles
- printing
- discovery of the New World by Europeans
- the Reformation
- cotton gin
- oral contraceptives
- American Civil War
- suburbia
- the sexual revolution

By contrast, other commentators are adamant that technology exists only in a matrix of many human

Table 0.1 *The structure of the book*

Period	Characteristics	Part
Pre-industrial	Economies powered by recent solar energy as plants and animals, wood, wind and water. Two sub-divisions: *Food collectors* who garner subsistence by hunting and gathering; *Food producers* who produce food from domesticated plants and animals.	1, 2
Industrial/ modern	Economies continue to utilize solar energy but subsidize its flows through human actions with large amounts of energy derived from fossil fuels (photosynthesis stored on geological time scales) like coal, oil and natural gas.	3
Global/post-industrial	Retains most of the characteristics of the industrial phase but the economy uses electricity as its preferred form of energy and becomes oriented towards the production of services rather than goods. Many processes operate on a world-wide scale.	4

actions and that its development is due to a complex of social, economic, political and other cultural factors. The development of industrial technology in the seventeenth and eighteenth centuries in the West (mainly in Britain and its North American colonies and Western Europe) cannot, they say, be explained simply by the accumulation of capital or the favourable business climate of Protestantism. These are not, after all, the factors in the surge of industrial development in Japan, Taiwan and Singapore in recent times. We have to be wary, therefore, of presenting the role of technology either as an evil genie let out of a non-returnable bottle or as a key agent in a triumphal progress from ignorance and darkness to today's wonders. But that technology is a key instrument in changing nature into cultural landscapes or of making an intensively humanized landscape out of a partly altered one, is not in doubt.

Historical geography/ geographical history

The description and explanation of cultural landscapes have mostly been the concern of historical geography, though fields of study such as landscape history, landscape ecology and environmental history have all overlapped with it. The primary document of description has often been the paper map (now supplemented by Geographical Information Systems) and words, other maps and graphics are used as vehicles of explanation. Most of the factors in the study of geography and of history come into play: the diversity of sources, the scale of the study, the political perspective of the writer, the intended audience, and the likely acceptance of the legitimacy of what is produced, are all relevant. In historical geography there are a number of traditions of study, most of which are atheoretical, i.e. they aim to be objective in the sense of believing that if the data are all collected then successive and progressive analysis of them will yield the truth. Later studies might examine the same body of data from the viewpoint of those under-represented in the main sources or in the light of an over-arching body of theory such as Marxism. Happy the reader who is told by authors where they stand; less sensible the reader who believes them.

What is a cultural landscape?

First, a cultural landscape (Box 0.1) is something to be seen, a visible entity. It may be desirable to try and reconstruct what it looked like in the past. If a photograph could have been taken of Captain Cook landing on Hawaii, what would the landscape have looked like? If there existed aerial photographs of upland England during the Mesolithic, would they show the impact of humans on the ecology of the forests? Second, it integrates in visual form all the influences mentioned above: the ecology of the natural environment or its altered successor states, the culture of the society which inhabits or visits it, in all the diversity (economic, social, political, religious) of that culture. What we see (today) or reconstruct (for the past) is usually a *synoptic* picture, a snapshot at one moment in time. We recognize, nevertheless, that it is almost certainly in the process of change and that any explanation of its condition must be sought in the processes that link the slices in time which form the synoptic pictures. So the landscape is a starting-point which demands explanation which has often to be found in the archive, the laboratory or the local informant.

Box 0.1 *The meanings of 'landscape'*

- **Landscape as scenery:** the visual, tangible aspect of the world. Often this has aesthetic connotations because in popular parlance a landscape refers to a pleasing outlook or panorama, which is either 'unspoilt' by humans or has been 'improved' to enhance its beauty. In both cases the countryside dominates our thinking and we rarely think of industrial or urban vistas as being landscapes in the same way.
- **Landscape as topography, or the 'lie of the land'.** Geologists or geomorphologists study landscapes as regional assemblages of landforms and geological structures.
- **Landscape as nature.** The stresses here are upon physical structures and processes, with human impact seen as an interfering force which disrupts and often damages natural systems. Nature is essentially that part of the world which is beyond humankind, with implications of a pure and unmodified state that is both given, i.e. in most

societies it is assumed to be the handiwork of a God-creator, and is also the essence of existence and a yardstick or norm. The word 'natural' carries these meanings.

- **Landscape as environment or habitat.** The influences surrounding people, including flora, fauna, climate, water and soil. People were seen by scholars in the nineteenth century to be dependent upon their environmental surroundings, and especially the gifts and constraints of available resources. The landscape therefore had implications of uncertainty and danger. In the twentieth century this view was substantially revised as it was realized by 'environmentalists' that the earth needed protection from degradation by human actions.
- **Landscape as artefact,** where humans are perceived as active modifiers and even tamers of nature for their own purposes. This may vary according to cultural context. Landscapes may themselves be symbols and become a focus of group identity. Objectified landscape is raw material for town and country planners and for landscape architects, professionals dedicated to translating our ideas about landscape into physical form.
- **Landscape as place, location or territory.** This is a very common usage in geography, although the study of place has varied from the analysis of geometrical spatial structure and the relationships between its elements, to the humanistic interpretation of place as an existential concept which emphasizes the relationships between people and the physical landscape, and the meanings derived from these settings.

Note: for further discussion see Chapter 23.

In any landscape we may find areas which are still pristine nature. There are not many of these in the world now but some still exist. They are often called wilderness areas, especially if they are large, but collectively such landscapes, no matter what their size may be labelled by the German term *Urlandschaft* (primal landscape). Those landscapes, by far the majority on the world's land surfaces, which exhibit human actions are the *cultural landscapes* on which we focus here. The German term *Kulturlandschaft* is sometimes used but less than its primal equivalent. Any landscape will show current landscape-forming influences but in varying degree will contain elements from the past as well. These may be from the far past as well as of recent origin. The metaphor is sometimes used of the palimpsest. This was a medieval and early modern European document written on vellum. The vellum was so expensive it had to be re-used but the previous writing could never be totally erased from animal skin. Each new layer of writing, thus, showed the relatively faint traces of previous uses. It is quite a good analogy except that today's technology can eradicate yesterday's landscape so effectively that no visible traces remain: the archaeologist needs to be brought in to the study group. This is especially so in cities where constant and complete renewal has recently been the practice: look in vain for traces of the Dutch occupation of Manhattan, let alone any of the aboriginal population from whom they bought it.

FURTHER READING

There is a large and rapidly growing general literature in the area of the present volume. The following will provide further examples and elaborations for the reader to explore.

Goudie, A. and Viles, H. 1997: *The earth transformed: an introduction to human impacts on the environment*. Oxford: Blackwell.

Simmons, I.G. 1991: *Earth, air and water*. London: Arnold.

Simmons, I.G. 1993: *Environmental history: a concise introduction*. Oxford: Blackwell.

Simmons, I.G. 1996: *Changing the face of the earth*. 2nd edn. Oxford: Blackwell.

Simmons, I.G. 1997: *Humanity and environment: a cultural ecology*. London: Longman.

Thomas, W.L. (ed.) 1956: *Man's role in changing the face of the earth*. Chicago: University of Chicago Press.

Turner, B.L. (ed.) 1991: *The earth transformed by human action*. Cambridge: Cambridge University Press.

Part 1

The pre-industrial world

INTRODUCTION

In economic and ecological terms, this phase of human history comprised two types of culture: that of hunter–gatherer–fisher communities, and that of agriculturalists. The spread of these cultural types encompassed two major transitions, namely the occupation of the land surface by anatomically modern humans, and the changes from hunting and gathering to farming.

Both economies had in common a base of solar power. It was the energy of the sun which they harvested as plants and animals, which they harnessed as wind and falling water in mills, and which they released as fire from stored form in vegetation. All of these human communities thus lived off the recent products of photosynthesis, unlike their industrial successors who would tap vast reserves of long-stored products of photosynthesis like oil and coal. So hunter–gatherers and agriculturalists alike relied upon the energy of human bodies to carry out any actions which their cultures suggested (or dictated) that they perform. For them there was no steam shovel, no bulldozer. Instead, a scoop made from the shoulder-blade of a deer, perhaps, and a digging stick.

Food and other materials used in the pre-industrial phase of human culture derived their substance from the very recent fixing of solar energy as plant and animal tissue: the maize cob tastes sweetest straight from the stalk into the boiling water. Wood represents several years of accumulation at least and so it is a repository of energy which can, along with more recent material like grass stems, be released by burning. Controlled fire at the landscape scale can be used in many beneficial ways to improve yields for hunters and agriculturalists and not for nothing has been called 'the first great force employed by humans'.

Hunting–gathering is sometimes classified as *food-collecting* or *foraging* since the humans gather wild species of plants and animals as resources; agriculture and pastoralism, on the other hand, are labelled *food production* because the biota are under a much greater degree of human control, in terms of the characteristics of the varieties used, the sites of cultivation or herding, and the degree of manipulation of the natural scene needed to accommodate this economy. These two relatively simple ecologies have, however, given rise to an immense variety of cultural practice in time and space. On one hand, the climatic and topographic zones of the world have enforced particular patterns: the Inuit of the Arctic gained most of their subsistence from the sea like any other predator, since that is where the high-energy biological resources such as seals and whales were found. In tropical forests, hunters likewise caught mammals when they could, but depended for basic nutrition upon a variety of plant materials. Agriculturalists in the wetter tropics might grow rice but their distant cousins in Scandinavia depended upon rye instead. Both crops are cultivated grasses but their climatic tolerances are markedly different. Herders in the Andes kept flocks of llamas, and in the Himalaya a similar ecology revolved round the yak.

Human inventiveness, however, always sought to transcend the apparent limitations imposed by nature and much of what we call material culture consists of ways of pushing back these limits. Irrigation, for example, allows the surplus water from a wet season or a water-rich neighbourhood to be stored and then released to allow crops to grow at a time of year when otherwise it would be too dry: the spread of Islamic cultures into the Mediterranean basin, for example, brought irrigation practices which added a whole season (the summer, when it rains little) to the agricultural year, as well as some lovely flowers like the lilac and the rose. Another inventiveness which improved food supplies was the ocean-going ship, which might search for richer fishing grounds at a considerable distance and then bring home the catch preserved in salt or by drying and smoking. Thus the Spanish and the Bretons were accustomed to fishing and whaling off Newfoundland as early as the fourteenth century. The addition of large areas of ocean to a culture's resource base has been vividly described as the cultivation of *ghost acreage.*

Wherever there were surpluses of food, then, people could be supported whose contribution to society was not directly economic: the pyramids of Egypt, the universities and cathedrals of Europe, and Angkor Wat are examples, as are the works of Mozart and of Botticelli. These remind us that not all culture is devoted to material ends.

1

HUNTERS AND GATHERERS

> Of the estimated 80,000,000,000 men [sic] who have ever lived out a life span on earth, over 90 per cent have lived as hunters and gatherers, about 6 per cent have lived by agriculture and the remaining few per cent have lived in industrial societies. To date, the hunting way of life has been the most successful and persistent adaptation man has ever achieved.
>
> (*Lee & DeVore 1968, 3*)

INTRODUCTION

Within our chosen theme of cultural landscapes, our interest here is the emergence of the genus *Homo* as the creator of landscapes. At some early stage of our history we must have been no more creators of landscapes than any other animal that builds nests or tears down trees. Indeed, we must have been less inventive than some, like the mound-building species of termites, for example. There are two ways in which hunter–gatherers may place an impress upon the land. The first of these is to make another species extinct, e.g. by culling it until it is no longer reproductively viable or by changing its habitat; the second is to unleash energy upon the landscape at an intensity not possible if just the energy of human bodies (even if augmented by tools) is used. For hunter–gatherers, fire constituted the avenue down which such energy could be loosed.

HUMAN EVOLUTION AND THE ACQUISITION OF FIRE

The story of the stages of human emergence is subject to constant review. When exactly the remnants of something burned which are found in association with human remains or tools constitutes the control of fire is uncertain. There are at least two major views about the control of fire by humans. One group thinks that the species *Homo erectus* gained control of fire in Africa about 0.5 million years ago and took it with them when they radiated out into Europe and Asia. Another group thinks that it was much later, about 100 000 years ago that modern humans of the *Homo sapiens* type acquired control of this tool and likewise spread outwards from their African heartland. Without it, no group would have been able to survive anywhere near the fringes of the great ice sheets of the Pleistocene. If the former group were the initiators, then we need to note that control over and use of fire were present even before our physical evolution was complete; even in the latter case, the role of fire might be said to have been bred into early cultures even if not actually 'bred in the bone'.

The ecology of fire

A fire running through dry vegetation evokes fear in all animal species except humans and even they may get a little scared now and again, especially if surrounded. The heat and the smoke cause most species to flee and so they are vulnerable to being caught: in traps, nets and by dogs as well as by men (usually) with spears, bows and blowpipes. If this process is repeated over many years, then the vegetation shifts in composition. Any species which is killed by the burning and cannot regrow loses its place to a fire-resistant species. Having thicker bark or a seed which is not killed by fire confers a selective advantage and indeed there have come to be species of conifers whose cones will not open and release the seed unless they are burned. So an oft-fired landscape bears the marks of its history in its plant composition and thence in its animal ecology as well. Knowing this, we need not doubt that hunter–gatherer communities throughout time have burned in order to improve the resources available to them: to flush out deer, to encourage lush young shoots on bushes or in the case of one Australian cycad to improve the number of edible seeds per tree. At a material level, therefore, fire has uses beyond the community's blaze where cooking makes a lot of plant material more edible by breaking down tough cellulose plant walls. Non-materially, it must have acted as a powerful social bond, both in the settlement and in the cooperative action needed for successful use in the landscape.

HUNTERS AND SPECIES EXTINCTION

Archaeozoologists and palaeontologists working on the environmental context of the first evidence of tool-making humans on the High Plains of North America, dated to *c.* 14 000–11 000 BC, noticed one remarkable pattern. The introduction of the earliest human cultures into the region coincided with an immense extinction of large mammals. Not only species but entire genera were wiped out within a few thousand years.

The causes of this phenomenon have been debated among scholars, with two main schools of thought in contention. The first suggests that desiccation of climate occurred at that time and so the animals became extinct either because there was no longer the water supply to sustain them or because the increased aridity brought about changes in vegetation which in turn could carry fewer herbivorous mammals. If the herbivores declined, so would the dependent carnivores. The second explanation is more radical and hinges on populations of animals facing a new predator against which they had no evolutionarily acquired defence mechanisms, as they would have had against wolves, for example. The new predators were the palaeo-indian groups working their way southwards from their Alaskan entry point into the North American continent. Climate need not be jettisoned as part of this hypothesis, for if the climate were changing in any direction at all, as was likely at the end of the Pleistocene, then animal populations might well be under some stress.

The credibility of this human-cause model is boosted by evidence from elsewhere in the world (*see* Figure 1.1). Where there is good evidence of changes in animal populations from the times of the very first entry of human populations (*see* Figure 1.2), then it seems that a number of species become extinct and among them the large mammal herbivores are pre-eminent. Where no mammals were present, as in New Zealand, then large flightless birds such as the moa took their place in both the ecology and the path to extinction. Although these surges of extinction took place well into the Holocene period, the event is usually known as *Pleistocene overkill* and is the first of a

FIGURE 1.1 *A silhouette of Palaeolithic age from a cave in southern France (Villars, France). The human figure is throwing up his/her arms in front of a bison: in many ways an early example of European attitudes to nature: 'stop: we are in charge!'*
Source: Leroi-Gourhan, A. 1968: *The art of prehistoric man in western Europe*. London: Thames & Hudson

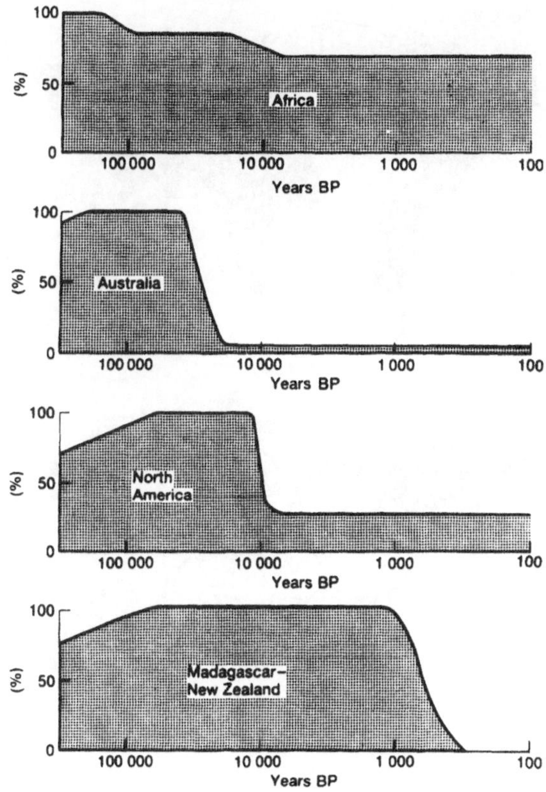

FIGURE 1.2 *In many regions, the advent of humanity is coincident with a sharp drop in the number of species of large animals (including flightless birds); the exception is Africa, where the antiquity of the human presence suggests either that any sharp drop occurred before 100 000 years ago or that there was something resembling a co-evolution of ecosystems which slowly adapted to the membership of humans and other species*

Source: after Martin, P.S. 1984: Prehistoric overkill: the global model. In Martin, P.S. and Klein, R.G. (eds) *Quarternary extinctions: a prehistoric revolution*. Tucson: University of Arizona Press, 354–403

number of periods of accelerated extinction of plant and animal species known to have occurred at human hands.

THE EARLIEST CULTURAL LANDSCAPES IN THE WORLD

If we could invent a miraculous satellite which could reclaim the pictures of the Earth when the first humanized landscapes occurred, what would we see, where and when? We do not have the palaeoecological evidence for all parts of the globe but a few patches present a clear picture.

East and central Africa

These savannas yield the earliest remains of human presence, in the form of bones and of tools. Conservatively, there has been about 2 million years of human presence here, in a set of ecosystems which is clearly adapted to fire. Some of this fire is started by lightning strikes but many other burns are human creations. Ecologists have argued that there has been a co-evolution of vegetation type and its adaptation to fire, a suite of animals closely adapted to the seasonality of the vegetation, and humans. Fittingly, therefore, the region in which the genus *Homo* evolved appears to carry evidence of the landscapes created during that evolution.

Western Europe in the great interglacial

During the interglacial period of the middle Pleistocene which is known as the Hoxnian (*c.* 320 000 B.P.), a similar episode appears in several palaeoecological analyses of lake deposits of the period. During the phase of the interglacial when a semi-tropical climate brought mixed deciduous forest and mammals like the hippo to north-western Europe, the trees are replaced by grasses for an episode of some hundreds of years. At one site in eastern England, this time is also one when flint implements (of the type associated with *Homo erectus*) and charcoal are found in the deposits. Some biologists have argued for a climatic explanation of this phenomenon, but it does not occur in every profile of the period which has been investigated. One palaeoecologist has suggested that the hippos feeding on marginal grasslands have dispersed the grass pollen throughout the deposit in their liquid faeces, which they disperse through the lake water with their tails. Archaeologists writing about the interlude tend to see a human interpretation: that here is evidence of human presence and of fire: hence, the openings in the forest are probably due to human activity. If so, then here is testimony, albeit transient, to an early cultural landscape. The forest closed back during the interglacial and there were then further glacial episodes, so that no trace of the actual landscape of the period remains, as in Africa.

The rise and fall of the hunter–gatherers

By 10 000 B.P., hunter–gatherers occupied nearly all the liveable space on the planet (*see* Figure 1.3). Only a few remote islands remained to be colonized by humans for the first time and there was a short moment when the Pleistocene ice had to a large extent withdrawn polewards and agriculture had yet to be discovered and spread. At that time, for the last time, 100 per cent of the human population were hunters, gatherers and fishers. As later economies developed, they adopted the new ways or were conquered by them, and so now less than 0.0001 per cent of humans live mainly this way; probably none do so entirely without any contact with the industrial world beyond their immediate surroundings.

In their day, the hunter–gatherers were very successful in occupying a very wide range of habitats, with very different biological resources and climatic regimes. At one extreme, the harsh landscapes of the polar coasts were tenanted by a low density of people who largely subsisted off fish and mammals from the sea, with the addition of migrating caribou where possible; but there was little plant material in their diet. There was, however, very little attempt at manipulation of the environment and so the cultural landscape was confined to the environs of the settlements. At the other extreme, hunters successfully colonized the tropical rain forests, even though the edible biomass at ground level is not very high. One adaptation to the presence of many bird and mammal species in the high canopy (monkeys, for instance) was the invention of the blowpipe, which directs the energy of the muscles of the face down the narrow channel of the blowpipe to project the poisoned dart. But given the high growth rates of tropical plants, any cultural features of the landscape were quickly subsumed by the natural ecosystems. In temperate landscapes, the

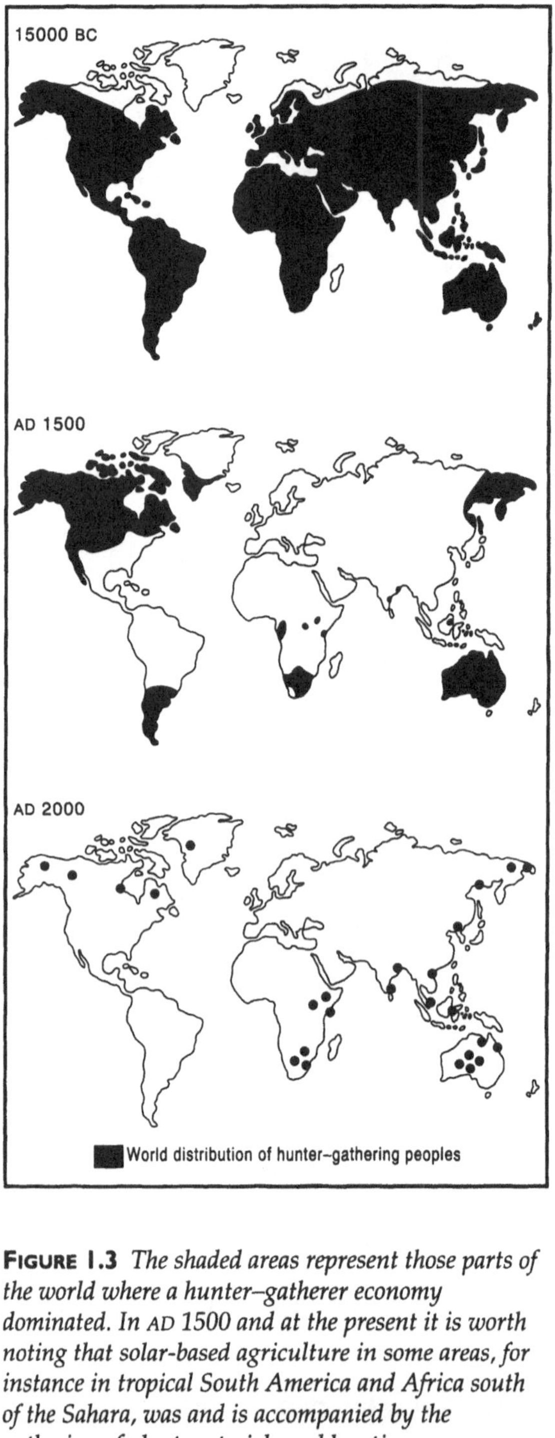

Figure 1.3 *The shaded areas represent those parts of the world where a hunter–gatherer economy dominated. In* AD *1500 and at the present it is worth noting that solar-based agriculture in some areas, for instance in tropical South America and Africa south of the Sahara, was and is accompanied by the gathering of plant materials and hunting*
Source: redrawn after Scarre, C. (ed.) 1988: *Past worlds: the Times atlas of archaeology*. London: Times Books

evidence is for a more lasting impact upon the deciduous forests and the same is probably true for mid-latitude grasslands. The native American occupants of the High Plains certainly burned the grasslands as part of their hunting repertoire, using the fire to drive antelope and buffalo towards concealed hunters or over cliffs. Fire was also used for concealment and for harassing enemies. So a cultural landscape was created, though it was not obviously humanized in the sense that later agricultural landscapes here as elsewhere were to appear. Equally, just as the tenure of the hunter–gatherers was transient, so their landscapes have been superseded by those of agriculture and of industrialization.

Children of nature?

It used to be thought that all hunter–gatherers simply lived off the land, without changing the natural ecosystems in any way. The accounts given above, however, contain evidence that this was not always so. All humans have, of course, altered their immediate surroundings: the trampling of vegetation, cutting of branches, lighting of camp fires and making of tools all contribute to this. Such changes are usually small-scale and transient. But the discussion of animal extinctions and of the use of fire raises the possibility of both large-scale and more permanent changes of a deliberate nature: that hunter–gatherers consciously manipulated nature in order to produce an order which they saw as culturally more desirable. No doubt such desirability included reliable supplies of food but need not have been confined to them.

A survey of the anthropological and archaeological literature reveals no single picture. There are clearly a number of cultures where animals were killed in large numbers without much consideration of the future. On the High Plains, many buffalo were run over cliffs and into blind canyons. So many were killed that much of the meat must have been left to rot *in situ*. Only in a few cases is there any evidence that the killing tried to avoid gravid females. Since the demise of the buffalo came at the time of European colonization of the plains, we must conclude that the density of humans and the intensity of their hunt were never such as to threaten the reproductive capacity of the buffalo herds. By contrast, some of the native populations of the northern forests of the same continent seem to have had a very ecologically aware attitude towards their food sources. It was common for certain animals to be taboo to sub-groups within a settlement, for example, and also for prohibitions to be erected against hunting in certain areas, which had to be *rested* so that e.g. beaver populations might recover. Only in the face of the European fur trade, it is argued, did these customs break down. To show the variety of cultural relations with surrounding landscapes, two differing examples will be given in more detail.

Case study: the later Mesolithic in upland England

The last hunter–gatherers of upland England occupied the land above 300 metres above sea level in the period 9000–5500 B.P. Their remains are mostly confined to flint and charcoal material, together with a few structures and a great number of pollen diagrams showing their probable presence in the form of charcoal and the pollen of plants of open and disturbed ground. Small microliths with narrow blades and the appearance of more geometric shapes (including a small scalene triangle) are diagnostic of the later Mesolithic from *c.* 9000 B.P. onwards. Pure geometric assemblages are especially common on the uplands of northern England.

An analysis of form, function and distribution of Mesolithic stone technology has been undertaken by Myers (1987; 1989). He notes that the frequency of sites from the later Mesolithic suggests a filling up of the landscape, though with smaller sites. The shift to microlith-dominated tools is taken to indicate the demand for reliable projectiles, capable of functioning without the need for time-consuming repair.

Most of the upland spreads of stone tools are near water, either in the form of streams, springheads or standing water. A sheltered aspect not far from the tops of hills and ridges seems also to have been sought: most of the sites are found in the 415–500-metre zone, on sites facing east through south and off the crests. At a regional scale, Mellars (1976) points out that in northern

England a transect from east to west reveals a clustering of sites between 350–480 metres. Only a few places have been deemed to show evidence of structures. Most lithic sites also have evidence of charcoal concentrations which are taken to be the remains of hearths. Some places have a concentration of several of these spreads within a small area and are presumably the result of either a number of groups camped near together or repeated visits by smaller groups. Wood is very scarce, bone totally so. Commonest are the shells, often burnt, of hazel nuts. Hints of worked lengths of wood are given by the layout of microliths: these are generally interpreted as arrowheads, though D. L. Clark (1976) has argued for their consideration as aids to gathering plant material.

All the findings by palaeoecologists suggest that management of the ecology of the uplands was taking place at a small, local scale during the later Mesolithic. Deer were encouraged to visit certain places within the oak forest by the provision of leaf-fodder, which resulted in some canopy lightening, a practice carried on into the Neolithic (*see* Figure 1.4). The production of edges and openings away from natural margins was carried out by ring-barking rather than fire. This process was aided by the barking propensities of concentrations of red deer especially in severe winters. Fire was important for the maintenance and production of edges at the upper limit of forests advancing upslope in response to periods of favourable climatic change. Fire at the hazel scrub-heath interface often resulted in a very acid, greasy, humus that was the precursor of paludification (bog formation). It was also important near water bodies such as lakes and streams, to keep open a grassy meadow in conjunction with grazing wild animals. Where and when dry enough, alder woods and their edges were subject to fire in order to enhance their grass-herb flora (attractive to wild cattle) and their browse content to interest red deer. Natural openings in the forest canopy were maintained by suppressing regeneration with fire: an aim of this practice was to encourage grass-herb ground vegetation as summer food for

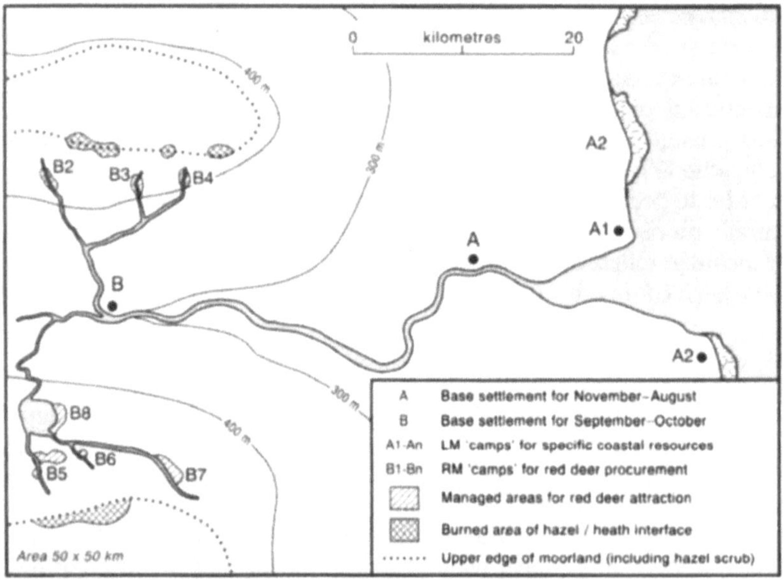

FIGURE 1.4 *A diagram of a possible yearly round of Later Mesolithic groups in Britain. Most of the environmental impact that we know about was exerted in the B1–B8 areas where fire was used to enhance the chances of attracting and killing mammals, especially deer. The lessened impact at lower altitudes may be due to our ignorance rather than its actual absence*
Source: I.G. Simmons

red deer; the edge effect will also increase the amount of browse which is an important food for deer. This burning probably took place in the period of the autumn when conditions might be dry enough for a controlled ground fire. This might not be every year. The incidence of fire might have interacted with gathering bracken rhizomes as a food but this seems less likely at a small extraction camp (unless other resources were scarce) than near a base settlement. Visiting such areas for hunting might, however, have occurred over a longer season: if wild cattle were being hunted, then it might have coincided with the production of ground flora in the streamside woodlands; deer were also likely visitors since this is the season when they are most likely to feed off herbaceous material. Although the same general locality within a territory hosted the extractor groups' camps year after year, the same exact site was not used. Multiple spreads of charcoal and flints are probably evidence of repeated visits but burned areas are not necessarily the same spots used for sleeping, tool repair and food processing.

This essentially small-scale, watershed-based, approach to a set of ecosystems would, if carried out by a number of groups over an upland, result in a fair degree of humanization of the landscape. This notwithstanding, the natural world was still a profound influence in the life of groups of hunter–gatherers. Hence, it is reasonable to generalize a sequence of events in which both natural and human-directed processes have important roles. A possible progression might be:

1. A mixed oak forest on the uplands, covering almost all the terrain, though broken where there was open water and mire accumulation in water-receiving sites. The upper edge of woody vegetation was marked by a hazel scrub. The forest underwent normal processes of death and renewal, which included gaps of various sizes caused by windthrow. The lower edge near streams underwent a sharp transition to an alder wood with other deciduous species, including elm.
2. Openings in the forest with a ground cover of grasses and herbs attracted mammal herbivores differentially on a seasonal basis. This had two consequences:
   (a) the concentration of animals made regeneration of the woodland less likely;
   (b) humans wished to maintain and/or enhance the concentration of animals.
3. Humans thus took over some of the existing openings and maintained them, using fire to keep a grassy sward rather than allow bracken to cover the ground in the early years. Heather grew as well and was burned, though it too attracted grazing animals. Attempts to extend the virtues of a grass-herb sward to the alder woodland were also made.
4. During this time, climatic change and soil maturation brought about the accumulation of *mor* humus over podsolic and gleyed soils. This pre-disposed scrub to turn to a heathy vegetation.
5. Increasing human populations (or more resource-hungry groups) wanted to try to create extra openings in the woodlands or at their ecotones. They did this by killing trees using ring-barking and by opening the canopy by breaking off leafy branches, which were also useful in attracting animals to feed, especially in winter; they could be left behind as a general encouragement rather than be immediate bait.
6. Some of the openings underwent rises in the water-table and became invaded by rushes; thereafter peat accumulation began to get under way even on ground which was water-shedding rather than water-collecting in a micro-topographic sense.
7. None of the human-induced processes totally replaced natural events: natural openings continued to be formed by natural processes and manipulation of the vegetation may have enhanced the rates of the formation of natural openings. None of the relationships necessarily remained the same throughout the period.

The later Mesolithic may thus be something of a turning point in the environmental history of the uplands (*see* Figure 1.5) since it marks the definite end of the human role simply as consumer of usufruct. Instead, there is the precursor condition of the productive and manipulative systems which first came to fruition in the neolithic and which persist to the present.

FIGURE 1.5 *This northern English landscape's essentially open character was initiated by hunter–gatherers in later Mesolithic times. The treelessness and the peat growth are connected with the use of fire and wild animal management in the period 65 000–4500* BC
Source: I.G. Simmons

CASE STUDY: NORTH AMERICAN WOODLANDS AT CONTACT TIME

A number of accounts of Indian practices in the Northeastern woodlands have survived. To these can be added the increasing volume of palaeoecological work which acknowledges the aboriginal population's influence on the ecology, and they can be further illuminated by some near-recent accounts from more remote parts of the continent.

The collections of descriptions of game hunting on land are diffuse in their conclusions, to the extent that no one technique seems to have been favoured over others. Driving into water was very widespread and was found in Great Lakes Indians; fire drives were used in the western third of the continent and in the eastern woodlands. Anell (1969) suggests that the eastern woodlands were virtually free of running down on foot as a practice but Cronon (1983) somewhat contradicts this in his account of the northern New England communities. He suggests that there was great variation: lone hunters with bow and arrow might be part of a spectrum with 200–300 men together at the other end. Traps might be set for a single species; equally, specially planted hedges were set as drive lanes perhaps 1.5 km long to steer animals towards waiting hunters. Hunting was especially the work of the men and they might detach themselves from any other settlement unit for periods of up to 10 days to go hunting or fishing, especially in the autumn when food had to be gathered and stored against the winter. Moving camp and varying group size to maximize the chances of resource acquisition were very important, but camp might also be moved because of war, death or fleas.

Fire management is well detailed in the summary accounts of New England at contact time and soon after, and in statements given relatively recently of the practices of Cree Indians in the Boreal forests of Alberta around 1910 (for New England: Thompson and Smith 1977; Cronon 1983; Russell 1983; Patterson and Sassman 1983; for Alberta: Lewis 1977, 1982). It is summed up for the first region in a nineteenth-century account by a visitor: 'the object of these conflagrations [Indian-set fires] was to produce fresh and sweet pasture for the purpose of alluring the deer to the spots on which they had been kindled'. This was the common practice, it seems, in southern New England. The firing brought about a forest of large, well-spaced trees, few shrubs and much grass and other herbage and this in turn increased the populations of elk, deer, beaver, hare, porcupine, turkey, quail, and ruffed grouse as well as their predators which were often fur-bearers. Mostly ground fires were employed and these might be used to drive game, especially by encircling the animals. Otherwise, the fire cleared away underbrush and also consumed twigs and other crackly litter which gave away hunters. The negative effects were upon the provision of firewood and on the mast of oak and beech; in the long term the size of oaks was decreased and their mortality increased. Fire at 20-year intervals apparently improved oak regeneration but at one- to five-year intervals over an eight- to nine-year period was negative in effect from the viewpoint of a twentieth-century ecologist; the Indians might have thought differently.

In northern New England, the situation was different. Though coastal forests were burned, there is less evidence for it inland, though a recurrence of data allowed Patterson and Sassman (in their 1983 contribution which tries to combine critically all the types of evidence, both palaeoecological and documentary) to wonder if it had been practised by aboriginal groups without access to the coast who were trying to improve their habitat. They note, however, that contact with the fur trade was a form of resource specialization which brought about intensification of land use, which itself might have meant the use of fire in hitherto fire-free areas.

In Alberta, Lewis distinguishes two desirable products of burning the vegetation of aspen woodland. These he calls fire yards and fire corridors. The first produced areas of grass- and sedge-dominated vegetation among the trees which could be called hay meadows, smaller open areas (*swales*) and grassy lakeshores. The second produced grassy fringes to streams, and mires, open areas along ridges, and by trails. Both kinds were collecting and traversing places for game animals. For instance, the burning of brush near water encouraged moose in the autumn, the firing of areas near water produced good trapline conditions and around water helped to provide food for ducks. Early grass in the meadows was an attraction to several species of game. Mires (*sloughs*) were also burned over at the edges to improve the quantities of aspen and birch, the principal food for beavers. In the woods there was the added advantage that natural fires were seen as dangerous and uncontrollable; these latter might start also in deadfall areas and so these openings were converted to swales by being burned every spring or burned for two years to get rid of the high fuel loads and then allowed to return to forest.

The methods of fire management are not well documented for New England. One account suggests that burning took place twice per year, in spring and fall but a counter-argument points to the difficulty of burning deciduous forests (especially oak) twice per year since there is not enough litter production. This might apply with less force if it was abundantly growing grasses that were being burned. In Alberta, spring and fall were the seasons for burning, with some emphasis on the spring since the remaining moisture made fire easier to control. Summer was avoided. Indeed, the question of control of fire was very important: both natural and human-made fire breaks were used and fire was only set in the meadows when surrounding conditions were moist enough to ensure that there was no uncontrolled spread. Fire was, though, never used to run animals.

The bulk of evidence from these woodlands is convincing: when described by Europeans they were cultural products just as much as the forests the visitors had left behind at home. Indeed, when the natives were extirpated in some parts of North America, settlers were taken aback at the speed and density with which the more natural forest reclaimed the ground.

CONCLUSION: THE FIRST GREAT FORCES

The implications of these studies are twofold. First, that fire has been a very important force in shaping ecosystems and that human control of it is long-standing: at least 15 000 years and possibly 0.5 million years. Many ecological systems have therefore spent most of their post-Pleistocene existence in the presence of human-directed fire at the landscape scale. The second finding is that hunter–gatherers were not necessarily passive consumers in the landscape: they manipulated it in order to gain access to resources, notably food. Very low population densities may have meant that in some places, no transformation of the landscape and its biota was attempted. It is, however, possible that minor alterations are not visible in the types of palaeoecological evidence available. But any notion of hunters as submissive children of nature is certainly wrong.

FURTHER READING AND REFERENCES

For a general collection of papers on hunters see Lee and DeVore. The other references are drawn from literature relating to the case studies.

Anell, B. 1969: Running down and driving of game in North America. *Studia Ethnographica Upsaliensia* **30,** 129.

Clark, D.L. 1976: Mesolithic Europe: the economic basis. In Sieveking, G. de, Longworth, I.H. and Wilson, K.W. (eds) *Problems in economic and social archaeology*. London: Duckworth, 449–81.

Cronon, W. 1983: *Changes in the land: Indians, colonists and the ecology of New England.* New York: Hill & Wang.

Lee, R.B. and DeVore, I. (eds) 1968: *Man the hunter.* Chicago: Aldine.

Lewis, H.T. 1977: Maskuta: the ecology of Indian fire in northern Alberta. *Western Canadian Journal of Anthropology* **1,** 15–52.

Lewis, H.T. 1982: Fire technology and resource management in aboriginal North America and Australia. In Williams, N.M. and Hunn, E.S. (eds) *Resource managers: North American and Australian hunter–gatherers.* Boulder, CO: Westview, 45–67.

Mellars, P.A. 1976: Fire ecology, animal populations and man: a study of some ecological relationships in prehistory. *Proceedings of the Prehistoric Society* **42,** 15–45.

Myers, A. 1987: All shot to pieces? Inter-assemblage variability, lithic analysis and Mesolithic assemblage types: some preliminary observations. In Brown, A.G. and Edmunds, M.R. (eds) Lithic analysis and later British prehistory: some problems and approaches. *British Archaeological Reports, British Series* **162.**

Myers, A. 1989: Reliable and maintainable technological strategies in the Mesolithic of mainland Britain. In Torrence, R. (ed.) *Time, energy and stone tools.* Cambridge: Cambridge University Press, 78–91.

Patterson, W.A. and Sassman, K.E. 1983: Indian fires in the prehistory of New England. In Nicholas, G.P. (ed.) *Holocene human ecology in north eastern North America.* London: Plenum, 107–35.

Russell, E.W.S. 1983: Indian-set fires in the forests of the north eastern United States. *Ecology* **64,** 78–88.

Thompson, D.Q. and Smith, R.H. 1977: The forest primaeval in the north east: a great myth? *Proceedings of the Annual Tall Timbers Ecology Conference* **10,** 255–65.

2

THE ORIGINS AND SPREAD OF AGRICULTURE

> Within a few thousand years of the sowing of the first crops, the threshold of literacy had been crossed in several distinct territories and some of the great traditions of mankind [sic] had been launched . . . That is why prehistorians have come to view the attainment of farming as a phenomenon comparable in importance with the industrial and scientific revolutions.
>
> (G. Clark, 1969, 7: *World prehistory: a new outline*. Cambridge: Cambridge University Press.)

INTRODUCTION

No other single change in human history can have had a greater effect upon the landscape than the domestication and global spread of plants and animals. The discovery of agriculture was an important developmental threshold since its associated economies provided the basis for the early stages of civilization. In this chapter we will review current knowledge of the origins of agriculture and its dispersion, in order to understand the variety of manipulated agroecosystems (see also Chapter 13) that have changed the face of the earth.

WHERE?

The question of where useful plants and animals were domesticated and exploited in an agricultural economy has generated heated debate. Nineteenth-century writers from the Christian tradition assumed that the lands of the Bible, commonly known as the Fertile Crescent (*see* Figure 2.1), were the fount of ancient innovation. Such notions have been very powerful, not least because there is much evidence that the Middle East was indeed very important in the domestication of the cereal species, especially wheat and barley, and animals such as sheep and cattle, all of which later were to become key staple items in the diets of western, developed nations.

In 1926 Nikolai Vavilov, a Russian botanical geneticist, proposed that a number of different centres could be identified where groups of plants originated in conditions of ecological diversity, in niches which suited their particular requirements. After some refinement he located these foci in South-west Asia, China, India and South-east Asia, Ethiopia, Mexico, the Andes, and the Mediterranean. His disciples later added other centres, culminating in J.G. Hawkes' codification (Table 2.1) of four nuclear centres, where agriculture emerged at an early date; ten regions of diversity, where genetic variety is substantial but cultivation came later; and eight minor centres, where a few crops were domesticated.

Carl Sauer writing in 1952 identified two primary *hearths*, in tropical South-east Asia and northern South America, with widespread diffusion to the rest of the world. This is a complex question: human beings have always had the singular capacity for picking up ideas from each other and from other groups. This is well-attested,

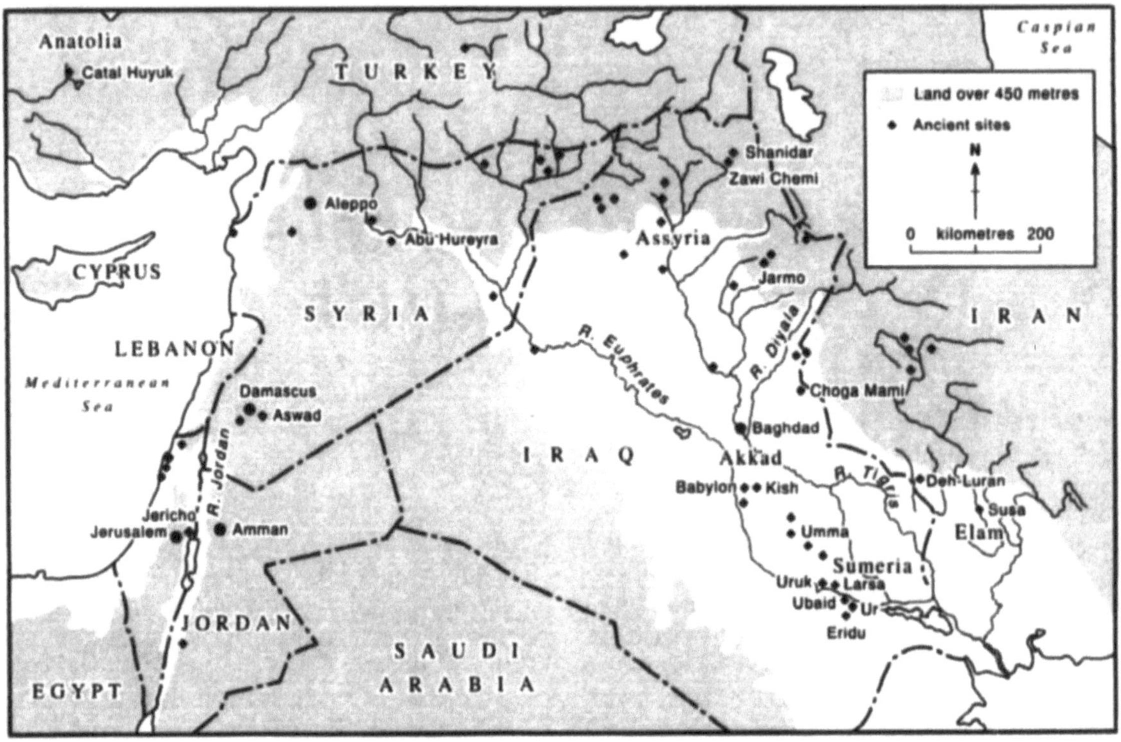

FIGURE 2.1 *The fertile crescent is a tract of land running from the Jordan Valley in the south west, through Syria and southern Turkey, to the valleys of the Tigris and Euphrates in what is now Iraq*

indeed documented, with regard to the alphabet, a system of writing in which sounds are represented by symbols. In a similar manner the idea of plant domestication has spread, involving both the actual spread of domesticates and the application of the concepts to local varieties.

TABLE 2.1 *Hawkes' nuclear centres and regions of diversity*

Nuclear centres	Regions of diversity	Outlying minor centres
Northern China	China India South-east Asia	Japan New Guinea Solomon Islands, Fiji, South Pacific
Middle East	Central Asia Middle East Mediterranean Ethiopia West Africa	North-west Europe
Southern Mexico	Meso-America	North America Caribbean
Central and south Peru	Northern Andes	Southern Chile Brazil

Source: Hawkes, J.G. 1983. *The diversity of crop plants*. Cambridge, Mass.: Harvard University Press

Furthermore, we must not imagine that any of these processes, the use of potential domesticates, their husbandry and domestication or their diffusion throughout their natural range nor their introduction into other environments, were wholly pragmatic. The domestication of plant and animals are both intricately linked with religion and ritual and underlie many myths, a somewhat derogatory word applied to other people's religions.

The implication of the Vavilovian hypothesis was that peoples in different parts of the world were quite capable of domesticating their own local species, independent of imports of genetic materials or ideas from outside. As a compromise, we now think that primary domestications were widespread, but that there was a significant element of secondary diffusion of ideas about domestication and of the plants and animal domesticates themselves.

Geographical diversity is well illustrated by Table 2.2 showing the origins of the major cultivated crops and animals. These can be divided into founder species, such as wheat, barley, maize, sheep and goats, and secondary species which were developed later when the idea of domestication for settled agriculture was well established. Subsequent diffusions of domesticates have broadened their distribution and led to further developments of varieties suited to local conditions. About 1900 crop species have been domesticated all told.

Not all of the qualities of domesticates were necessarily improvements. The body size of domesticated animals, for instance, is always smaller than their wild ancestors. This is due to selection for docility and the inability of early farmers to provide sufficient fodder for bulky animals. Until the advent of biotechnology, breeding has always involved a trade-off of characteristics according to which were most desirable.

The Middle East and domestication

Why did Neolithic peoples bother to domesticate when we know that a wide variety of foods was available to them in the wild? A number of hypotheses have been advanced but it should be noted that none can yet be said to be conclusive. Demographic expansion with consequent pressure upon available resources is one of the most quoted explanations, however, the evidence is not favourable. Archaeological work in both the Old and New Worlds has found that numbers of people remained low and stable until after the establishment of full agriculture. So population increase was an effect rather than a cause. Low fertility among hunter–gatherers continues to be common today.

Environmental factors are more persuasive. In the 1920s Gordon Childe proposed a *stress* or desiccation theory, where people were supposedly forced into close proximity with plants and animals by a change to greater dryness at the end of the Ice Age. Modern environmental historians give some credence to this, for instance in the Middle East where the transition (about 10 000 BC) from the last Ice Age (Pleistocene) to the postglacial (Holocene) saw a change to climatic conditions which encouraged the spread and multiplication of the wild ancestors of domesticates. Winters became warmer but there was also an extreme seasonal variability of moisture which favoured annual plants, such as the cereal grasses, rather than perennials. A subsequent episode of cooling, 9000–8000 BC, caused ecological stress and may have encouraged a shift to more labour-intensive resource extraction and the development of new technologies (Byrd 1994).

Global climatic change after the Ice Age might help to explain why agriculture appeared independently within a relatively short time in several different parts of the globe, but it is only one of several factors. The others according to McCorriston and Hole (1991) were:

- the availability of the technology to use plants effectively, especially harvesting, processing and storage technologies amongst hunting, gathering and fishing communities;
- the social organization (in villages) able to cope with *delayed return* economies, where there were gaps of months between sowing and harvesting;
- an incentive to shift from hunting and gathering.

The balance of resources available to these early peoples was certainly changing. In the southern

TABLE 2.2 *The probable geographical origins of the major crops and animals*

Place	Crops and animals
South-west Asia	Wheat (8000), barley (8000), fig (7800), pea (7800), lentil (7800), chickpea (7500), flax (7500), broad bean (6500), date (4000), olive (3700), grape vine (3200), rye (3000), also cauliflower, kale, garlic, turnip, mustard, rape, bitter vetch, pear, citron, pomegranate, almond, pistachio. Dog (12 000), sheep (9000), goat (7500), dromedary (2000).
Central Asia	Buckwheat, carrot, radish, apple, cherry, walnut, alfalfa, hemp. Horse (4000), bactrian camel (1500), yak.
Central America	Squash (8000), common bean (8000), avocado (7200), maize (5700), amaranth (5200), gourd (5000), runner bean (4000), tomato, sieva bean, papaya, agave, cacao, vanilla. Muscovy duck.
South America	Common bean (8500), pepper (8500), potato (8–6000), Lima bean (6500), oca (6000), quinoa, cañihua, guava, peanut, pineapple, tobacco, rubber. Llama (4000), alpaca (4000), guinea pig (1500).
North America	Sunflower, cranberry, blueberry, pecan. Turkey.
Africa	African rice, sorghum, pearl millet, teff, okra, lettuce, watermelon, cowpea, castor bean, coffee, sesame, kola tree, oil palm. Donkey, guinea fowl.
India	Pigeon pea, moth bean, rice bean, horse gram, asparagus bean, aubergine, cucumber, pepper, lemon, jute, indigo.
China	Rice (6000), foxtail millet (5500), peach (4000), soya bean (1000), mung bean, adzuki bean, leaf mustard, horseradish, Chinese cabbage, spinach, ginger, orange, mulberry, lotus, water chestnut.
South-east Asia	Mango (7200), taro (7000), banana, lime, satsuma/tangerine, tea, clove, nutmeg.
South Pacific/Australasia	Coconut (3000), arrowroot, breadfruit, macadamia nut.
Europe	Oats, beet, mangel, plum, cherry, raspberry, strawberry. Reindeer.
More than one centre	Chili pepper (C. America and Amazonia, 4100), cabbage and other brassicas (China and Europe, 3750), manioc/cassava/yuca (C. America and Amazonia, 7–5000), sweet potato (C. America and Amazonia, 2500), cotton (C. America and Amazonia), yam (Amazonia and Africa), onion (China and S.W. Asia), sugarcane (India, S.E. Asia and S. Pacific), apricot (China, S.W. Asia), muskmelon (Africa, S.W. Asia). Cattle (S.W. Asia, Europe, India, 6500), pig (S.W. Asia, Europe, S.E. Asia, China, 7000), buffalo (India, S.E. Asia, 2500), chicken (S.E. Asia, China, India, 6000), duck (Africa, Europe, China), goose (Africa, Europe).

Sources: various
Note: All dates given are BC.

part of the Fertile Crescent the sea level was rising and several large inland lakes were drying out. Game was probably becoming harder work to hunt and of course it remained problematic how to store the proceeds. Wild plant foods were already providing a bridge for the seasonal hungry gaps but at some stage it may have become essential to turn to cultivation as a means of continued subsistence. Cereal porridge cannot have been much of an alternative to rump steak and it seems to have provided only a meagre existence. The early neolithic farmers were less well nourished than their ancestors and it is unlikely that they would have volunteered to become specialist agriculturalists if realistically they could have continued with their previous way of life.

In this region, wild grasses, including the relatives of modern wheat and barley, thrive in a wide arc from west to east. Figure 2.2 depicts the slightly different distributions of wild einkorn (*Triticum boeoticum*), and wild two-row barley (*Hordeum spontaneum*), which each have their own tolerances of climatic and other limitations. Wild emmer wheat (*Triticum dicoccoides*) has approximately the same distribution as barley. In the years of abundant moisture, stands of these primitive cereals are dense and may yield as much as 500 kg per hectare, more than enough to attract the interest of people with a gathering and storage economy.

Interestingly, the size of wild and cultivated emmer grains is not dissimilar. Domestication's principal contribution was to eliminate the inconvenient characteristic of a brittle rachis, the segmented axis attaching the seed spikelets to the stem, which in the wild ensured that the grains fell from the plant when ripe and thus guaranteed another annual cycle of growth. Grain which was held slightly longer on the plant would have had a higher probability of being selected and, by planting that as seed, early farmers would have perpetuated and eventually concentrated a natural genetic characteristic, enabling them to extend the harvesting season. This process could have been completed within a few decades. Accidental crossing with goatface grass later produced the genetically more complex bread wheat (*Triticum aestivum*), a plant with greater hardiness and a grain which can be more easily threshed.

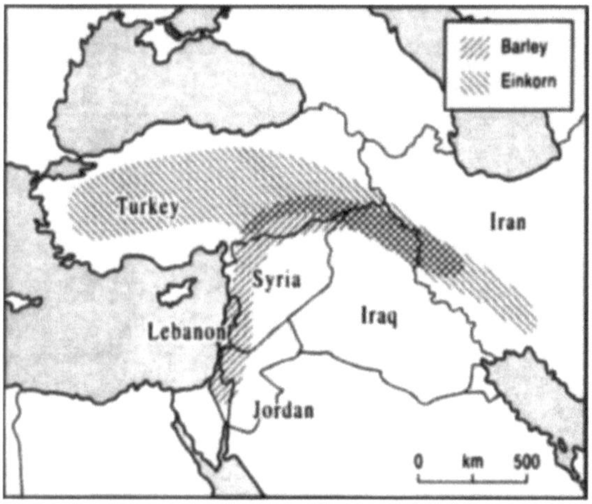

Figure 2.2 *The distribution in the wild of einkorn wheat and barley*
Source: Adapted from Zohary, D. and Hopf, M. 1993: *Domestication of plants in the Old World: the origin and spread of cultivated plants in West Asia, Europe, and the Nile Valley*. 2nd edn, Oxford: Clarendon Press

Two-row barley and emmer wheat appeared first in domesticated form at Jericho (Palestine) and Tell Abu Hureyra (Syria) sometime around 8000 BC (in the cultural phase called Pre Pottery Neolithic A) and emmer, einkorn, peas and lentils at Tell Aswad (Syria) in 7800 BC. Cultivated grasses and pulses became widespread in the Middle East during the following millennium, in the Pre Pottery Neolithic B.

The precise mechanism of how the transition was made from gathering wild seeds to cultivating them and selecting desired traits is not at all clear. The growth of healthy and perhaps high-yielding plants from discarded seeds on nitrogen-rich household waste dumps must have been observed, but not all species have this weedy habit. Another consideration may have been that wild stands were diminishing under pressure from gathering and needed supplementation from additional sowings.

The case of animals is at least as complex: there is a fundamental distance between the hearth animals – the dog and the pig – and herd animals, cattle and sheep. The former were brought to the hearth when their parents had been killed, for curiosity, status and perhaps even the 'Aaah' factor, perhaps suckled by the women (as does happen in simple traditional societies), raised with the family and then either playing an economic role in hunting or scavenging, finally being killed and eaten. Here animals pose more of a problem: the young must have been taken, or perhaps, in the case of cattle, small herds were corralled for ritual purposes, the stock being reduced in size and ferocity by the sacrifice of the king bull. The four key characteristics according to Smith (1995, 26–7) that an animal species needs to become a candidate for domestication are:

- non-specialized feeding habits;
- must reproduce readily;
- non-territorial so they can be kept in groups;
- submissive social animals with a dominance hierarchy.

As villages (in what is now Palestine, Jordan and Syria) supplemented temporary hunting camps during the Natufian era (10 500–8500 BC), a more sedentary lifestyle would have reduced the geographical territory which could be exploited and therefore a more intensive resource utilization was required. Cultivation offered one such option, pastoralism another. Surely, social developments were important here, providing a basis on which people could live together cooperatively.

Natufian villages were five to ten times larger than previously and their component houses had a permanent architecture with provision for both storage and religious ceremony. Their walls were made of stone and mud, with timber posts holding up a roof structured with joists. Cemeteries appeared for the first time, suggesting an attachment to place and a collective thought process.

The inhabitants had sharpened stone sickle blades for harvesting, and even mortars, pestles and grinding stones for processing grain. The heavy tooth wear observable on excavated skulls shows the consumption of a gritty flour. There may even have been some deliberate cultivation (Unger-Hamilton 1989), although at this stage it would have been of *unimproved* plants.

Despite a higher strontium content in Natufian skeletons than earlier periods, evidence of an increased contribution of plants to the diet, the people in these villages continued to derive much of their nutrition from hunting wild game. There is no evidence yet of animal domestication, although it is possible that wild gazelle and fallow deer were managed by selective hunting and perhaps corralling.

By 6000 BC animals had joined the new economy in a morphological form quite distinct from the wild (shorter bones, altered horn shape). They were domesticated first in Turkey/Syria: sheep about 9000 BC, pigs 7000 BC, and cattle 6000–5000 BC. Goats were exploited by 7500 BC in the hilly flanks of the Zagros Mountains where a different economic path had been followed. The opportunities for gathering cereals were less here than in the Levant and instead an initial phase of intensive hunting seems eventually to have led to a form of nomadic animal husbandry, in which sheep and goats were domesticated from locally common feral stocks (*see* Figure 2.3).

David Harris has sketched out an evolutionary continuum (*see* Figure 2.4) which helps us to understand the likely stages of transition to agriculture in its fullest sense. His model draws its motive power from the greater inputs of human energy. Sedentarism, settlement size, population density, and social complexity are all correlated with a greater intensification of the cultivation process, although there are no simple cause and effect linkages, nor any guarantee that the sequence was linear.

CASE STUDY: MESOAMERICA

To move away from a classic hearth in the Middle East, with all that this implies in light of the region's long cultural dominance, the mountainous environment of central Mexico may not sound a promising birth place of agriculture but in fact it presented prehistoric peoples with a wide range of juxtaposed plant resources in a region of ecological diversity. The Basin of Mexico and the valleys of Tehuacan and Oaxaca (*see*

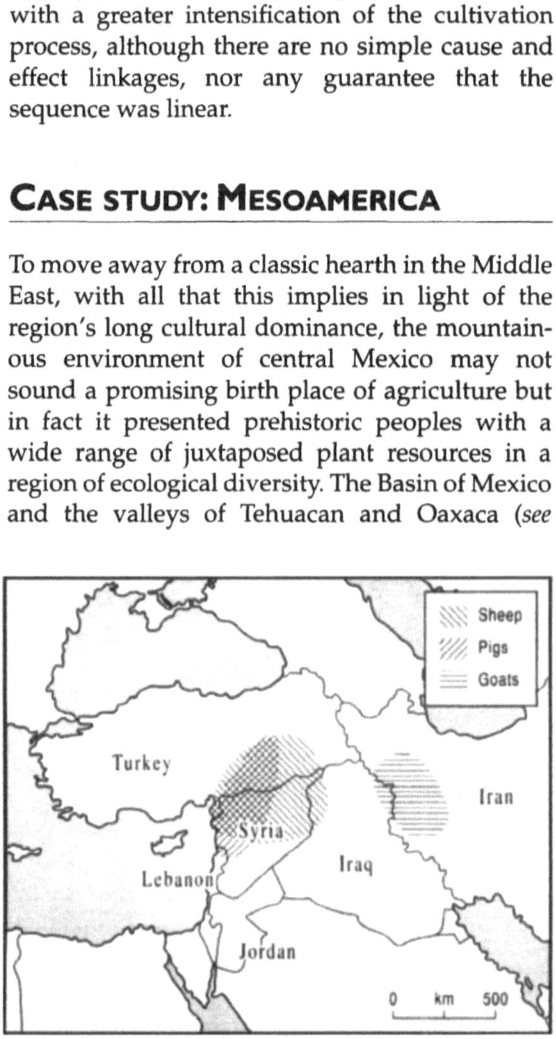

FIGURE 2.3 *The likely distribution of the wild ancestors of domesticated animals*
Source: Adapted from Smith, B.D. 1995: *The emergence of agriculture*. New York: W.H. Freeman

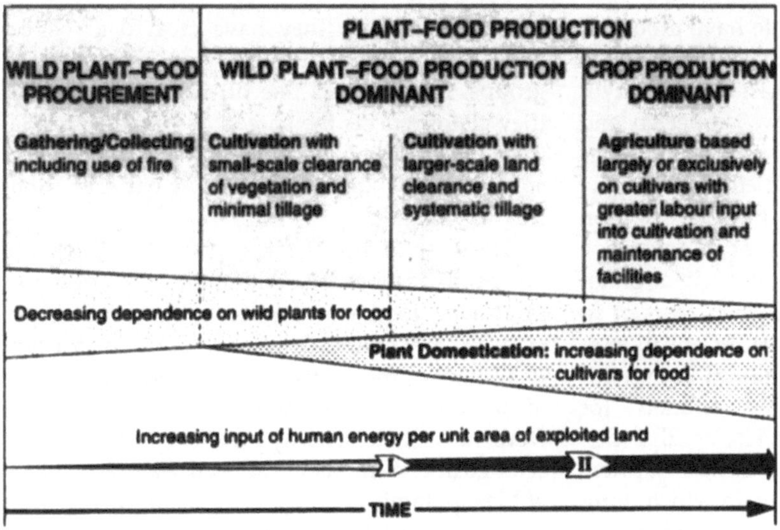

FIGURE 2.4 *A classification and evolutionary model of plant exploitation systems. The Roman numerals show population thresholds in the impact of human energy*
Source: Harris, D.R. 1996: Domesticatory relationships of people, plants and animals. In Ellen, R. and Fukui, K. (eds) *Redefining nature: ecology, culture and domestication*. Oxford: Berg, 437–63

Figure 2.5) in particular offered niches ranging from lake shores to semi-arid vegetated slopes. At first their inhabitants were broad-spectrum hunter–gatherers living in nomadic microbands, who only gradually developed a dependence upon plant foods. We should remember that a generation was probably little more than 20 years, and lifetimes were mainly under 30 years, so that

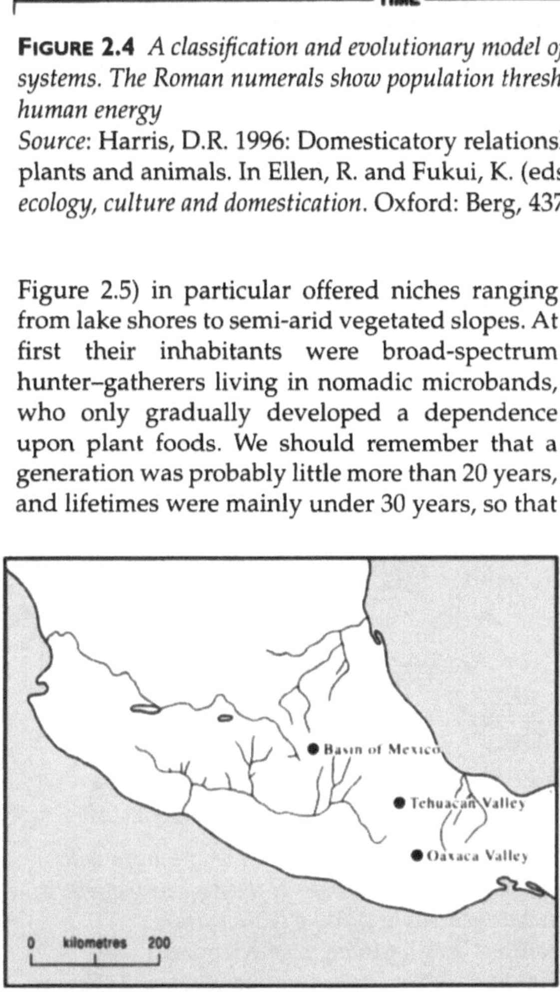

FIGURE 2.5 *Mesoamerica*

the rate of change during an individual's lifespan was infinitesimally slow. Eventually food supplies centred on a trinity of domesticates: maize, beans and squashes (relatives of pumpkins) which, when fully developed, have provided a very successful basis for the agricultural economy in much of Central and South America for the last 3500 years and are still important elements in our own diet. Maize is deficient in the essential amino-acids lysine and methionine, and beans lack tryptophan and cystine, but together they provide a balanced diet. The three became part of a system in which the use of farm land was maximized, with the maize growing tall and straight on the planting mound, a bean planted close so as to climb up its stalk, and a squash plant at ground level with its broad leaves shading out competitive weed growth.

Squash appears to have been domesticated first, but it was the cultivation of maize from about 5200 BC which was of the most significance. The provenance of maize is still the subject of debate, but Doebley (1990) has discovered a close genetic similarity to the wild grass *teosinte*. The main problem is that *teosinte* is not appetizing, nor is it ever likely to have produced as heavy a wild harvest as the ancestors of wheat and barley did in the Middle East. Most likely it was a starvation food which was collected and then cultivated only as a fall-back if all other resources failed in dry years. It either gradually responded to cultivation by human selection each year of the largest cobs or threw up a chance mutation with a much better yield. Either way it is thought to have taken 4000 years or so before yields per hectare were sufficient to support specialized, settled agricultural communities. There were no suitable animals for domestication in highland Mexico, another disadvantage *vis à vis* the Middle Eastern farmers who employed a mixed economy in which dung and crop residues were swapped to restore soil fertility and produce captive sources of animal protein.

SPREAD AND IMPACT

Although the cultivation of founder crops such as wheat and maize was discovered independently in different parts of the world, the geographical gaps were filled by diffusion. This would have been either by the migration of people, taking their farming with them, or by the passing on of ideas, perhaps through cultural contact or trade. The number of crop species was multiplied when enterprising farmers experimented with further cultivars such as pulses, legumes, root crops, leaf vegetables and fruits. The process was very slow however, with Finland about 6000 years behind the Middle East in the adoption of arable farming. In fact, the colder, moister northern latitudes of Europe were so different that the crops themselves had to be modified by the slow process of breeding in order to produce spring-sown varieties that could cope with conditions much colder and wetter than would have suited their wild ancestors.

But did the *idea* of agriculture spread or was there a migration of agriculturalists taking their technology with them? Cavalli-Sforza and Cavalli-Sforza (1995) argue for the latter on the basis of their new and exciting work on human genetics. By taking sample measurements of the frequencies of 95 human genes in different European regions, they have created statistical maps of the similarities between neighbouring groups. Figure 2.6 is one of these maps, which they claim reflects the spread of genetic material from the Middle East. Maps of the evolution and spread of languages show similar patterns (Renfrew 1996).

The slow nature of this expansion diffusion is not at all surprising. Although the notion of producing food was attractive, no doubt especially for any politically minded élite who wished to dominate the increasingly sedentary pockets of population. In summary, by 2000 BC there were, nevertheless, many practical problems. The ecological conditions in the well-watered valleys of Mesopotamia and Egypt, for instance, required more sophisticated techniques of moisture control than the hilly, semi-arid regions of primary origin, and agriculture there seems to have spread only gradually along the belt of *hydromorphic* soils (formed in the presence of water). Wheat and barley made an appearance along the Tigris–Euphrates in the seventh millennium BC, and in the Nile Valley by 6000 BC. It was these fertile, irrigable lands which were to provide a breeding ground for the first civilizations (Chapter 3).

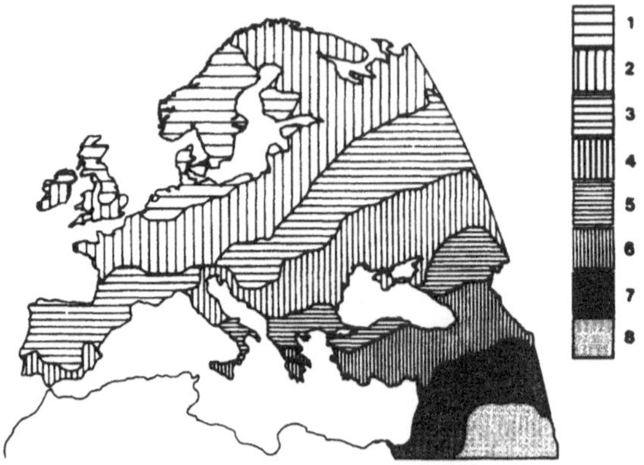

FIGURE 2.6 *The genetic landscape of Europe with regions drawn on the basis of the frequency of 95 human genes. The shading is arbitrary*
Source: Cavalli-Sforza, L.L., Menozzi P. and Piazza A. 1994: *The history and geography of human genes*. Princeton, New Jersey: Princeton University Press

An early diffusion eastwards took barley, sheep, goats and cattle to the Kachhi Plains of Pakistan by 7000 BC, but not to the Indus until 3500 BC. Meanwhile in northern China an entirely separate but parallel history of domestication saw the use of rice, millet, cabbage, pigs, chickens and fruits. In Africa a number of indigenous species were domesticated at an early date and agriculture was widespread north of the Equator by 2000 BC.

In South America the evidence is debatable but earliest estimates have squash, beans and peppers under human control by 8500 BC, with maize entering the diet in the seventh millennium. Manioc was first grown in the upper Orinoco valley sometime between 7000 and 5000 BC, and about 3000 BC agriculture high up in the Andes was based on grains such as *quinoa, tarwi* and *cañihua*, and roots including *ullucu, mashua* and *oca*. From 7000 BC meat came in the form of domesticated *llama, guanaco, vicuña* (all camelids) and the guinea pig. But only the potato from this remarkable Andean complex has been received with enthusiasm in the wider world.

In summary, by 2000 BC most of the major crops used for human food today had been domesticated and some had been taken beyond their natural range by migrating peoples or by the processes of cultural diffusion and commercial trade. The latter is likely to have been increasingly important as well-organized societies sent traders and emissaries further afield to seek resources complementary to their own. In the historic era, the Islamic empire traded widely and collected plants with the intention of their exploitation in different environments. Thus rice and citrus fruits came to the Mediterranean from east Asia. There were further important developments in the age of exploration which was initiated in the fifteenth century by naval powers such as Portugal and Spain (Chapter 21). The real purpose of European outreach, apart from seeking gold and converts to Christianity, was the discovery of routes to the sources of spices and other goods in South, Southeast and East Asia. There was a dissatisfaction with the mix of temperate and Mediterranean products available and a solution was to trade more widely and to dominate the sources of exotic resources. Thus, imperial expansion in the Americas in the sixteenth century yielded crops such as maize and potatoes, which could be grown in European conditions and there were many species exported (consciously or accidentally) in the other direction, including weeds and pests (Clark 1949; 1959).

In recent centuries it has been the global empires, first of colonialism and later of transnational capitalism, which have been primarily responsible for plant dispersals. Cultigens have been carried to new regions both for profit and for ornament. Many of the species in European gardens, for instance, have been gathered from around the world by plant hunters. Botanical gardens played a role, acting as coordinators of expeditions and offering propagation facilities. Their interest was not always altruistic, however, as in the case of the theft of seeds of the rubber tree by the British in 1876, who wished to establish plantations in Malaya, thus undermining an existing Brazilian monopoly.

The exchange of genetic material continues to this day, especially in the form of seeds of high-yielding varieties of crops and pedigree animals for breeding purposes, but the global network of trade, by allowing quick and efficient transport from an area of origin with some comparative advantage such as climate, has reduced the need to carry plants beyond their normal range. The forcing of tropical fruits and flowers in carefully tended and heated greenhouses, so popular among the wealthy in eighteenth- and nineteenth-century Britain, is now a thing of the past. The globalization of trade in food products may have introduced a greater variety of exotic species (known in the trade as *queer gear*) to our supermarket shelves but our diets remain dominated by the narrow range of crops whose origins lie in the Middle East and the Americas (*see* Table 2.3).

In a similar manner, livestock have also been widely spread around the world. Cattle have become almost universal, bifurcating as they spread into *taurus* (short-horn and long-horn) and *zebu* (humped) sub-types. Descendants of the short-horns are popular especially in temperate latitudes and the zebu is better adapted to tropical climates. The South Asian buffalo, another milk/meat/draught species, has been less popular beyond the sub-continent, although one of its products, mozzarella cheese, garnishes pizzas all over the world.

TABLE 2.3 *Percentage of world food crop area devoted to crops from 11 source regions (excluding fruits and nuts)*

Region	Percentage
South-west Asia	31.5
Central Asia	0.1
Central America	13.2
South America	16.4
North America	1.5
Africa	5.3
India	0.3
China	8.0
South-east Asia	4.5
South Pacific	0.0
Europe	0.7
More than one centre	18.5

Source: data from FAO, *Production Yearbook, 1993*. Rome: FAO

THE HISTORICAL ORIGINS OF AGRICULTURAL LANDSCAPES

One of the great puzzles of agriculture is why two areas of very similar environmental characteristics often develop very different crop and livestock systems. We will investigate this further by comparing the rice-growing region of Tamilnadu in South-east India with the rain-fed grain cultivation of eastern Tanzania, which is in East Africa.

These regions are of similar size and lie at the roughly the same latitude north and south of the equator. Both have extensive plains cutting across crystalline rocks, with shallow, seasonally wet valleys and better drained watersheds between (*see* Figure 2.7). There are residual geological features rising from the plains in the form of isolated inselbergs or larger ranges of hills rising to over 2000 metres. Tamilnadu has a higher mean annual temperature, but only by 2°C. Mean annual rainfall varies between 800–1300 mm for eastern Tanzania and 550–1300 mm for Tamilnadu, decreasing inland in both cases. Both have similar natural vegetation, at least, that is, before human interference.

But the agriculture of these two regions is very different (*see* Figure 2.8). It seems that history has had a major role to play. The population density in Tamilnadu is twenty times greater, as a result of the development of a highly sophisticated civilization from an early date. There were cities and trade was extensive, to the west with the Romans and later the Arabs, and to the east with South-east Asia. There were major irrigation works on the coast by AD 900, tanks, wells and canals and these spread to the interior by AD 1350.

Tamilnadu has long had a stable culture, which is strongly religious and based around its magnificent temples (*see* Figure 2.9). Warfare between historic rival kingdoms was less destructive than the inter-tribal conflict experienced in Tanzania. Eastern Tanzania was also on trade routes (for ivory, cloves and slaves), but it was more of a cul-de-sac. Traders did not penetrate regularly inland until the mid-nineteenth century and political integration did not come until German sovereignty was declared in 1885. In the twentieth century cash crop exports have developed, including coffee, cotton and sisal.

There is no reason why rice could not have been grown in eastern Tanzania. Wild rice grows there and rice is cultivated commercially in Madagascar further south. Its present rain-fed agriculture could therefore be greatly intensified. There is also potential for the transfer of simple intermediate agricultural technologies from Tamilnadu to Tanzania. Tank irrigation is one example of this (*see* Figure 2.10). Water resources in Tamilnadu are used to their limit but much water is unused in eastern Tanzania. A limiting factor for such improvements is the lack of population in Tanzania, *not* an excess.

In conclusion, we might say that environmental constraints are important in limiting the potential of agriculture, but in similar environments it is historical development and human organization that matter most.

CONCLUSION: THE DEVELOPMENT OF PRE-INDUSTRIAL AGROECOSYSTEMS

The gradual spread of agricultural plants, animals and techniques to all the inhabited continents replaced what had been successful hunting and gathering economies. Small, scattered bands were no match for the technical superiority, social

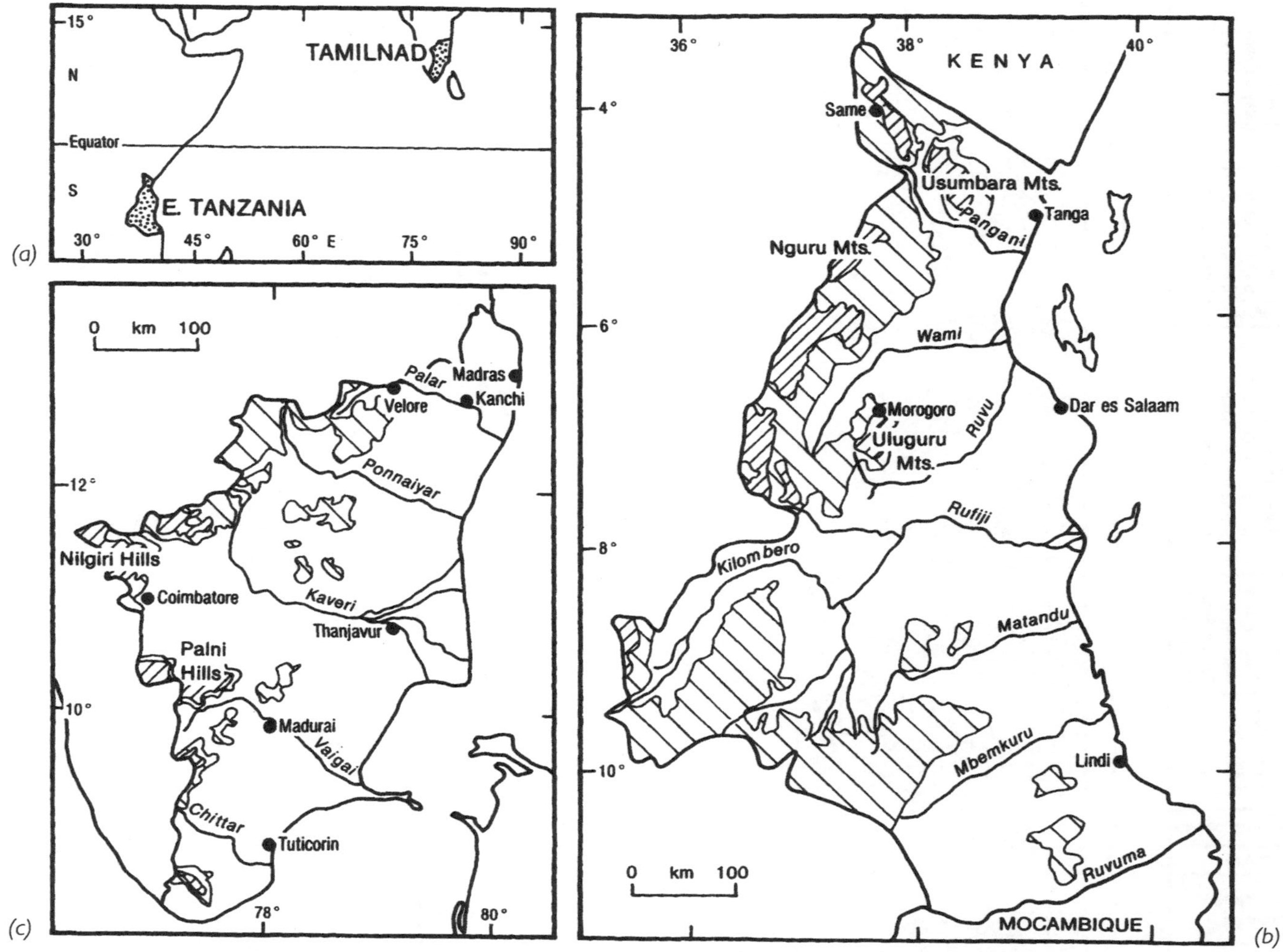

FIGURE 2.7 *(a) Tanzania and Tamilnadu; (b) An idealized diagram of land use in relation to physiography in eastern Tanzania; (c) An idealized diagram of land use in relation to physiography in Tamilnadu*

Source: Morgan, W.T.W. 1988: Tamilnadu and eastern Tanzania: comparative regional geography and the historical development process. *Geographical Journal* **154,** 69–86

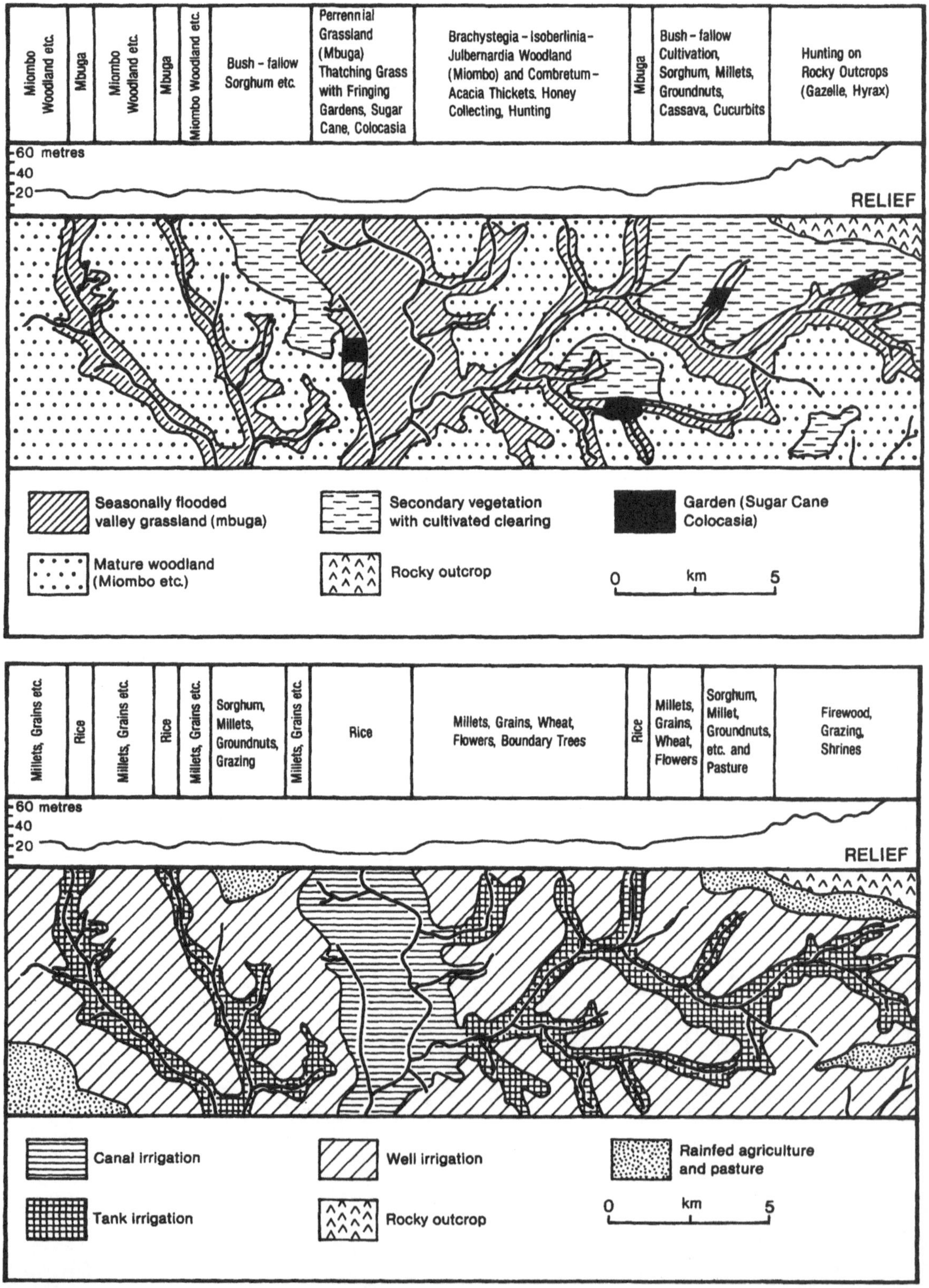

Miombo Woodland etc.
Mbuga
Miombo Woodland etc.
Mbuga
Miombo Woodland etc.
Bush - fallow Sorghum etc.
Perrennial Grassland (Mbuga) Thatching Grass with Fringing Gardens, Sugar Cane, Colocasia
Brachystegia - Isoberlinia - Julbernardia Woodland (Miombo) and Combretum - Acacia Thickets. Honey Collecting, Hunting
Mbuga
Bush - fallow Cultivation, Sorghum, Millets, Groundnuts, Cassava, Cucurbits
Hunting on Rocky Outcrops (Gazelle, Hyrax)
60 metres
40
20
RELIEF
Seasonally flooded valley grassland (mbuga)
Secondary vegetation with cultivated clearing
Garden (Sugar Cane Colocasia)
Mature woodland (Miombo etc.)
Rocky outcrop
0 km 5
Millets, Grains etc.
Rice
Millets, Grains etc.
Rice
Millets, Grains etc.
Sorghum, Millets, Groundnuts, Grazing
Millets, Grains etc.
Rice
Millets, Grains, Wheat, Flowers, Boundary Trees
Rice
Millets, Grains, Wheat, Flowers
Sorghum, Millet, Groundnuts, etc. and Pasture
Firewood, Grazing, Shrines
60 metres
40
20
RELIEF
Canal irrigation
Well irrigation
Rainfed agriculture and pasture
Tank irrigation
Rocky outcrop
0 km 5

(a)

(b)

complexity and aggression which eventually characterized established, sedentary farming. As Chapter 1 has shown, hunter–gatherers remain only in the least accessible parts of the globe, indeed, the very existence of many tribes is under threat, their only chance of survival being to make an accommodation with the agents of so-called civilization, perhaps by agreeing to retreat to regulated reservations.

The agroecosystems which replaced mesolithic economies were slow to mature. Their development depended upon a number of factors, the first among which is the natural environment. The relationship between people and the ecosystem they wish to modify to their advantage is crucial. The difference between success and failure has proved to be a narrow one, especially in marginal areas where over-exploitation can reduce sustainability (see Chapter 6). Finding the right balance was often a matter of trial and error.

Second, technology is crucial, especially in the pre-industrial phase, involving these apparently simple tools used for cultivation, processing, storage and transport. For instance, the plough was a Eurasian innovation of the temperate and subtropical arable lands where draught animals were available: it brought the power of a specialist domesticate, the ox (a castrated bull) to bear in both the varied soils of the Middle East and Mediterranean zones and eventually the rich forest soil of temperate Europe. Universally its use was accompanied by forest clearance, whereas the hoe and digging stick required less disturbance of vegetation and could be used in wooded areas and on steep slopes. Use of the plough does not necessarily imply intensification per unit area, but its use of labour was efficient. At first, the intensification of output was more noticeable in those areas employing water control, especially canal irrigation and water lifting devices. It was these areas where the earliest agricultural-based urban civilization flowered and where population densities were the greatest (see Chapter 4).

FIGURE 2.8 *Although there are similarities in the physical environment, the human use of land in (a) eastern Tanzania and (b) Tamilnadu is very different, as summarized in these idealized models*
Source: Morgan, W.T.W. 1988: Tamilnadu and eastern Tanzania: comparative regional geography and the historical development process. *Geographical Journal* **154,** 69–86

Third, the ownership of land has been important, varying from public control in ancient Egypt, and collective decision-making in many tribal and socialist societies, to the private rights exercised by landlords and owner-occupiers. The power to decide how to farm has been pivotal in history, as has the power to mobilize labour and capital.

Geographical location is the fourth factor. It should be clear by now that spatial diffusion supplemented the independent domestication of plants and animals. The spread of seeds and live animals by trade, along with the migration of peoples, were responsible to a certain extent for the more urgent pace of change in some parts of the globe than others. Even two areas of similar environmental characteristics may thus develop very different crop and livestock systems.

Further reading and references

Much has been published on agricultural origins and spread, but no single volume captures the full range of ideas in non-technical language. Sauer (1952) is still a good read and Harris (1996) is an up-to-date collection of papers on the Middle Eastern hearth. Smith (1995) is a well-written introductory text.

Byrd, B.F. 1994: From early humans to farmers and herders: recent progress on key transformations in southwest Asia. *Journal of Archaeological Research* **2,** 221–53.

Cavalli-Sforza, L.L. and Cavalli-Sforza, F. 1995: *The great human diasporas: the history of diversity and evolution*. Reading, Mass.: Addison-Wesley.

Clark, A.H. 1949: *The invasion of New Zealand by people, plants and animals: the South Island*. New Jersey: Rutgers University Press.

Clark, A.H. 1959: *Three centuries and the island: a historical geography of settlement and agriculture in Prince Edward Island, Canada*. Toronto: University of Toronto Press.

Doebley, J. 1990: Molecular evidence and the evolution of maize. In Bretting, P.K. (ed.) New perspectives on the origin and evolution of New World domesticated plants. *Economic Botany* **44,** Supplement, 6–28.

FIGURE 2.9 *A temple in Tamilnadu. This region of India is of great significance for Hindus*
Source: P.J. Atkins

FIGURE 2.10 *The irrigation of rice paddies in Tamilnadu from a tank (reservoir).* See also Figure 3.8
Source: P.J. Atkins

Harris, D.R. (ed.) 1996: *The origins and spread of agriculture and pastoralism in Eurasia*. London: UCL Press.

McCorriston, J. and Hole, F. 1991: The ecology of seasonal stress and the origins of agriculture in the Near East. *American Anthropologist* **93**, 46–69.

Renfrew, C. 1996: Language families and the spread of farming. In Harris, D.R. (ed.) *The origins and spread of agriculture and pastoralism in Eurasia*. London: UCL Press, 70–92.

Sauer, C.O. 1952: *Agricultural origins and dispersals*. New York: American Geographical Society.

Smith, B.D. 1995: *The emergence of agriculture*. New York: W.H. Freeman.

Unger-Hamilton, R. 1989: The epi-palaeolithic southern Levant and the origins of agriculture, *Current Anthropology* **30**, 88–103.

3

EARLY URBANIZATION AND THE HYDRAULIC ENVIRONMENT

Land goes with water
(Khuzestan proverb quoted in R. McC. Adams 1974: The Mesopotamian social landscape: a view from the frontier. In Moore, C.B. (ed.) Reconstructing complex societies. *Bulletin of the American Schools of Oriental Research, Supplement* **20,** 1–20)

INTRODUCTION

In Chapter 2 we discovered that the hilly and mountainous parts of the Fertile Crescent formed one of the very earliest hearths of domestication. Here we will see that archaeological evidence shows that the growth of villages followed, and that the later development of cities and urban-based civilizations was a feature of the nearby fertile, riverine lowlands of Mesopotamia (Greek for 'between two rivers') washed by the rivers Tigris and Euphrates.

EARLY EVIDENCE

The hunter–gatherers and early farmers in the Middle East were probably not the leisurely and economically secure experimenters envisaged by some writers (see Chapter 2). The analysis of skeletal remains suggests some nutritional stress and domestication may well have been a necessity for them to feed their families. We struggle to find evidence of a surplus of food in the sense of an abundance above and beyond the basic needs of 2400 kcal of energy per day.

An alternative explanation of *surplus* might well be that certain powerful people in a community extracted farm products and perhaps labour service from their neighbours under duress. There is every reason to think that mobilizing ordinary people to devote more time to food production has, from early times, required an institutional setting and maybe a spatial reference point for the concentration of the collective will (Adams 1966). Many writers see the ceremonial centre with its priestly caste as the initial kernel of urbanism (Wheatley 1971). Any accumulation of surplus was likely used for conspicuous consumption and there would also have been investment in physical infrastructure such as shrines and other institutional buildings, military defensive works, residential accommodation, and market places, and also in the means to further enhance agricultural production, especially by the use of irrigation canals.

There has been much academic argument about the definition of *urban*. Size and density of population are popular criteria, as are the non-agricultural occupations of the majority of inhabitants. Economic functions are also important of course, especially marketing, along with public investment in physical infrastructure and record-keeping. The first evidence of an agglomeration looking like a city comes from Jericho in Palestine, at about 7000 BC. There were brick houses, stone defensive walls with massive bastion towers (*see* Figure 3.1), and a 9-metre-wide ditch cut into the

FIGURE 3.1 *The walls of ancient Jericho*
Source: P.J. Atkins

rock. The settlement was capable of holding about 2000 people, although it seems likely that it would have been full only seasonally. The inhabitants were still devoting much of their time to mobile hunting and gathering, supplemented by agriculture. As yet there was little specialization of labour, one of the usual preconditions for ascribing urban status, and Jericho was strictly speaking therefore only a defended village.

More reliable as indicators of primitive urbanism are the *tells* (mounds) of Turkey, Iran and northern Iraq. These are the accumulations of millennia of inhabitation, gradually raised by the repeated rebuilding of mud brick houses. A typical structure would last 70–80 years before collapsing and being rebuilt on the rubble. By 6000 BC about 5000 people lived at Çatal Hüyük in Anatolia, making it the largest neolithic settlement in the Middle East. They grew wheat, barley and peas, made coiled pottery and smelted copper, and traded craft items such as arrowheads, daggers, baked clay figurines, textiles, wood vessels and mirrors made out of a shiny volcanic glass called obsidian. In addition, there seem to have been complex death and fertility cults, expressed in wall paintings and burials.

The major flowering of urban civilization came after 4000 BC in southern Mesopotamia (the core of the Sumerian civilization) with the rise first of Eridu, then Uruk, and from 3000 BC of a number of other city–states such as Ur (*see* Figure 3.2). The extraordinary irony is that this is (and was) an arid and apparently unpromising environment. It lacked both building stone and timber for construction, and the climate made agriculture impossible without irrigation. The towns which emerged shared the following characteristics. First, although they were relatively closely spaced, they had divergent paths of development. The southern Mesopotamian flood plain is far from uniform and the variety of ecological

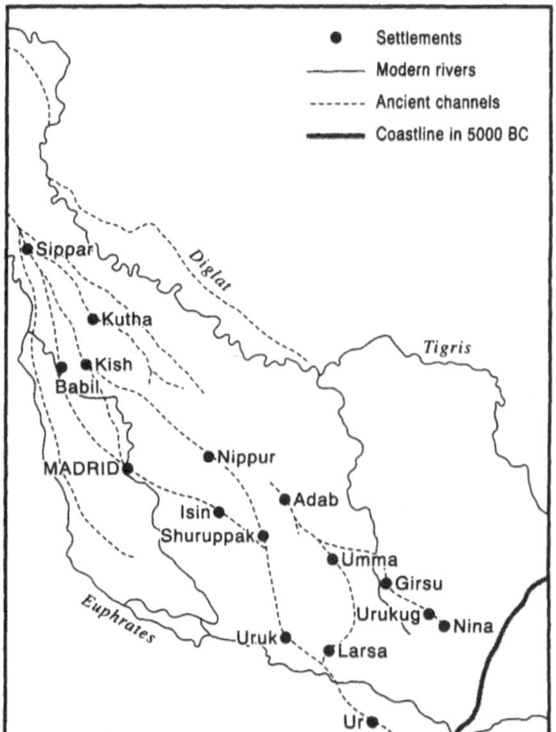

FIGURE 3.2 *Mesopotamia, showing modern rivers and ancient watercourses which are no longer operative*
Source: adapted from Gibson, McG. 1974: Violation of fallow and engineered disaster in Mesopotamian civilization. In Downing, T.E. and Gibson, McG. (eds) *Irrigation's impact on society*. Tuscon: University of Arizona Press, 7–19.

niches (hills, plains, alluvium, blown sand, marsh) demanded different responses. In addition, there is evidence that some of the city dwellers may have migrated from diverse regions such as the Zagros Mountains, the Arabian Gulf and the northern arc of hills from Turkey to Iran, giving rise to a juxtaposition of languages and cultures. Competition for available resources eventually led to conflict and to an inherently unstable regional power structure.

Second, they soon developed complex societies with an advanced division of labour, distinguishing between the farmers and those who did not work in the fields. High quality pottery was produced from the outset and metallurgy was introduced about 3000 BC, especially the working of bronze (at first a copper/arsenic alloy and later copper/tin), later followed by iron. By 2500 BC the techniques of riveting, soldering and casting were applied to a wide range of objects, such as weapons, tools and ornaments.

Third, the Sumerians were quite remarkable in their successful exploitation of innovations, although they probably borrowed the initial ideas from elsewhere. The list is long but among the most significant historically were the use of the wheel (in war chariots and waggons), pictographic writing (later developed as cuneiform and hieroglyphs), the agricultural plough, and the large-scale irrigated cultivation of cereals. Mesopotamia was truly a well-spring of western civilization.

Fourth, there were well-organized élites in each city–state. Originally they were temple-based, drawing their power from their people's religious worship at the ziggurat (stepped pyramid) which formed the symbolic focus of the main settlement. Later there emerged warlords and kings who became the effective rulers. By 2500 BC an elaborate vertical stratification of society seems to have been the norm.

Finally, warfare was a fact of life. Cities were fortified and aggressive behaviour was facilitated by armies of infantry with metal spears and mobile, chariot-borne shock troops. Rivalry between neighbouring cities, due to disputes about tribute territories and the exploitation of scarce resources, was exacerbated by incursions by tribes from the desert margins envious of the wealth created in Sumeria. One of the world's first empires was compiled by Sargon the Great of Akkad (2371–2316 BC) and expanded by his grandson Naramsin (2291–2255 BC). They conquered Assyria (northern Mesopotamia), Elam (south-west Iran), northern Syria and part of Anatolia (Turkey). Later Babylon, north on the Euphrates, rose to prominence (1900 BC) as Sumeria proper declined (Yoffee 1988).

THE PROCESS OF URBANIZATION

> I will proclaim to the world the deeds of Gilgamesh . . . In Uruk he built walls, a great rampart, and the temple of blessed Eanna . . . Look at it still today: the outer wall where the cornice runs, it shines with the brilliance of copper; and the inner wall, it has no equal. Touch the threshold stone: it dates from ancient times. Approach the Eanna Temple, the dwelling of Ishtar, such as no later king or man will ever equal. Go up on the wall and look around. Examine its foundations; inspect its brickwork thoroughly. Is not its masonry of baked brick, did not the Seven Sages themselves lay out its plans?
>
> (*The epic of Gilgamesh*, adapted from N.K. Sandars 1960: *The epic of Gilgamesh*. Harmondsworth: Penguin).

Uruk was a small town, in effect a service centre for the surrounding rural population, with a ceremonial function and craftsmen. In about 3000 BC there was a sudden acceleration of change into a phase of *hyper-urbanization* (Adams 1981). A new defensive wall of 9.5 km was built and the density of the settlement increased tenfold. This was due to an implosion of population, leaving many of the nearby villages deserted and their fields abandoned. Forty thousand people were now crowded into 500 hectares, and there was the very important development of a settlement hierarchy of size and functional range.

Our excavation of luxurious burials suggests that personal wealth had, by the end of the Early Dynastic period (say 2500 BC), become more concentrated in the hands of the few. Such a class-structured society is what we expect in nascent urbanism. There seems little doubt that the rich and powerful were imposing their will upon the majority, most likely through the exploitation of the increasingly technically advanced agriculture. A large temple was already evident at Eridu by 4100 BC, with further elaborations soon after (*see* Figure 3.3).

The radical departure was to apply irrigation water to fertile alluvial soils. Such agriculture was two hundred times more productive per unit area than dry farming but it required technologies and skills that were new. We know that the peoples of northern Mesopotamia had for thousands of years been diverting streams from the mountains on to their crops, but this was on a small scale. The use of mighty rivers such as the Tigris and Euphrates (with significant tributaries such as the Karun, Diyala, Uzaym and Khabur) was an altogether different proposition.

The rivers are fed by snow melt and a Spring rainfall maximum in eastern Turkey. They reach their peak in April/May, which rules out flood irrigation along Egyptian lines because crops planted in May would shrivel in the baking summer temperatures of over 40°C. All irrigation had to be by breaching the river banks (levées) (*see* Figure 3.4) and leading the water to the fields by canal. The civil engineering component of such an agricultural system was high, although the potential rewards were great.

The environmental context is important. The lower Euphrates in particular was a braided river with at least five major channels and a branching network of smaller ones, which distributed water over a wide area. They were not dependable, however, because they shifted their course as silt accumulated. Only a relatively narrow band of the levée backslope could be irrigated because the more distant plains were too flat for effective gravity feed. A 2-year fallow cycle was necessary in order to minimize water-logging and salinization of the soil due to salts dissolved in the river water and those rising from the subsoil by capillary action.

Channel silting and meandering occasionally seem to have deprived an established settlement of its water supply, thus influencing the geopolitics of the region through the rise and fall of individual city–states. By 4000 BC individual canals up to 3–5 km in length had been dug by kinship groups, but all of the large-scale irrigation schemes post-date the crystallization of state control. In roughly 2300 BC river straightening and redirection started and from about 1000 BC large-scale canal networks were under construction. These were apparently efficient enough to maintain a supply of water throughout the summer months and it was possible therefore to introduce thirsty crops such as rice, sesame and cotton.

A water bailiff or *gugallum* was responsible for the organization of water rights at the village level and no doubt arbitrated disputes between neighbours. The Sumerians had a complex system of land tenure and water rights, but the laws (for instance, the Babylonian Code of Hammurapi) were clear: 'If a man has opened his channel for irrigation, and has been negligent and allowed

FIGURE 3.3 *An artist's impression of the temple at Eridu*
Source: Lloyd, S. 1984: *The archaeology of Mesopotamia*. 2nd edn, London: Thames & Hudson

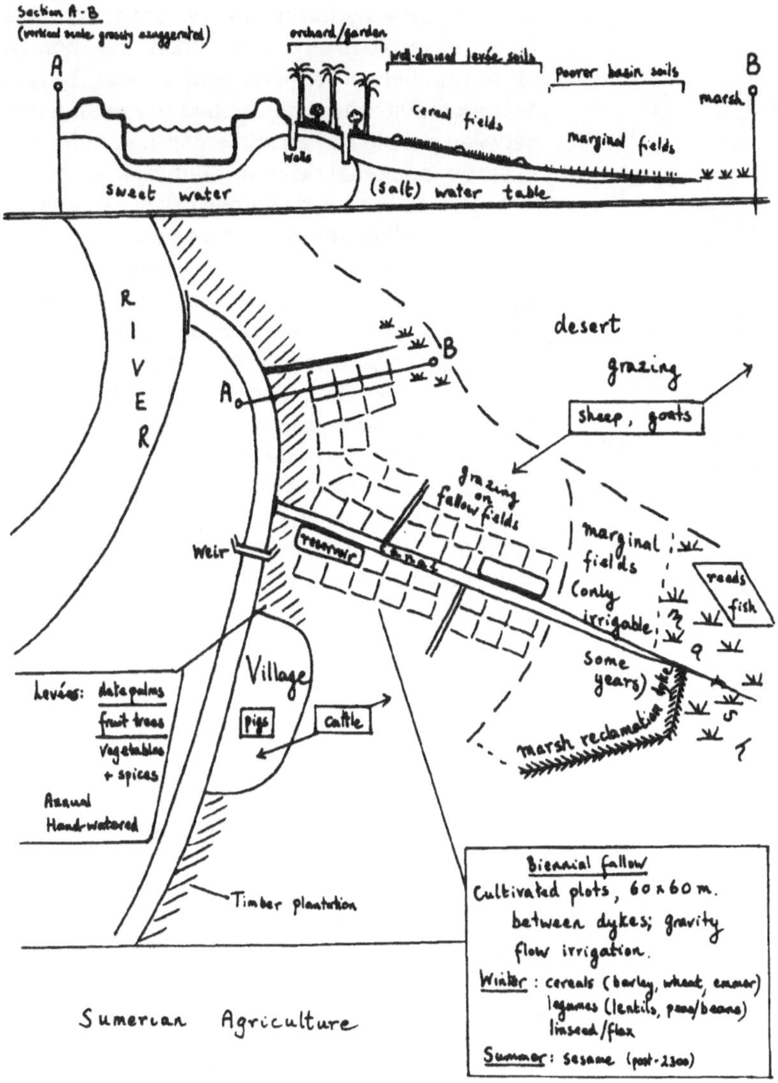

FIGURE 3.4 *The nature of irrigation in southern Mesopotamia*
Source: Postgate, J.N. 1992: *Early Mesopotamia: society and economy at the dawn of history*. London: Routledge

the water to wash away a neighbour's field, he shall pay grain equivalent to [the crops of] his neighbour's' Postgate (1992, 182).

WITTFOGEL AND HYDRAULIC CIVILIZATIONS

The organization of the cooperative effort needed to make irrigated agriculture successful was a complex and difficult task. Levée breaches, canals, drainage ditches and flood control measures all had to be organized and regularly maintained, and scarce water supplies had to be apportioned among the fields. This was practicable only by a large-scale cooperative effort, possibly by coerced corvée labour organized by an urban-based élite with the necessary military muscle and the administrative ability to keep records. In Sumeria a vigorous bureaucracy seems to have evolved, keeping a database written on clay tablets. This was focused on the temple, although the land itself was privately held and could be inherited (Figure 3.5).

One interpretation (Wittfogel 1957) is that the advantages of this *hydraulic civilization* accrued

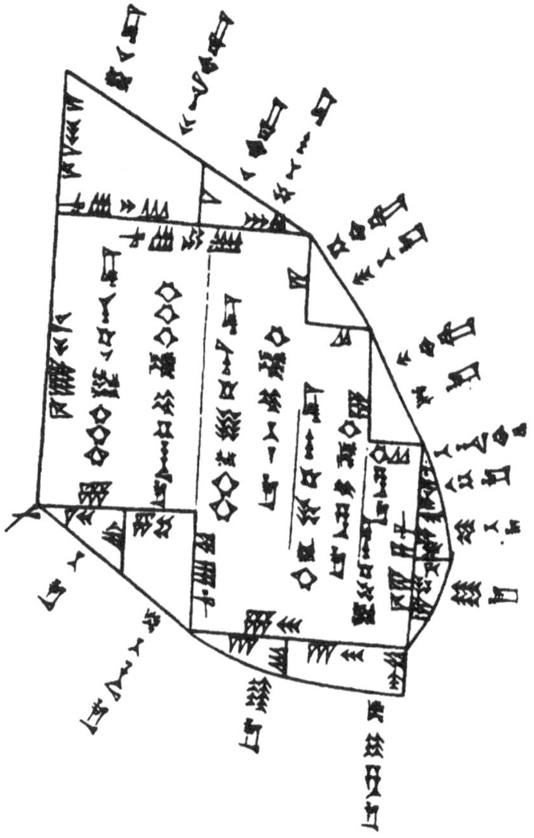

FIGURE 3.5 *A field map of the Ur III era incised on a clay tablet. The hieroglyphs are measurements of length and breadth*
Source: Liverani, M. 1990: The shape of neo-Sumerian fields. *Bulletin on Sumerian Agriculture* 5, 147–86

mainly to the élite. Although the masses did gain some security by living within the city walls, it was their labour and taxes which supported the non-agricultural population. The possibility of an external military threat was convenient for the ruling class because it justified the concentration of population into a space where they could be regulated. Urbanization was therefore an agent of social control. The pattern was thus:

environmental stress → irrigation →
bureaucratic control → despotism

Wittfogel went further and argued that the control of water was the basis of civilization and urbanism was the essential means of organizing the necessary hydraulic works. Thus urbanization was ecologically determined. He pointed to a number of regions within and beyond Mesopotamia where he claimed the relationship between people and an arid environment had led to agro-managerial despotic political systems.

Such a deterministic framework of explanation finds few adherents nowadays because there is plenty of evidence of irrigation systems developing without state control of construction or maintenance. Anyway, in the case of Sumeria it does not appear to be true that the temples controlled everything from irrigation and agriculture to the delivery and storage of crops. We now believe that the temples were important decision-makers but they did not have a monopoly of food supply.

The causalist chicken and egg game, of which came first, the state or the irrigation, has been played for half a century now with little result. It is much more interesting that the employment of irrigation was a crucial step in increasing the productivity of farming systems, and that these evolving systems took a number of forms according to local circumstances of culture, politics, technology, demography and ecology.

In a more sophisticated and altogether more convincing volume than Wittfogel's, Maisels (1990) argues that social organizations *are* the key to understanding the genesis of urbanism, but that the urban history of each region depends upon a unique set of human–environment interactions. To paraphrase his account of Mesopotamia, a dense rural population comes first, supported by settled agriculture on hydromorphic soils, in a stable and richly various environment. Complex, socially stratified societies are predisposed towards the establishment of a central focus, most likely ceremonial, but possibly also connected with a military objective like defence or perhaps an economic one such as trade. This centre is dependent upon its hinterland and vice versa in a symbiosis. True urbanism relies upon the presence of a functional specialism which is not observable in villages, involving craftsmen, soldiers and priests. The rupture of kinship bonds was an important stage, allowing the rulers power and prerogatives without reciprocal obligations to the ruled. Horizontal political relations were replaced by vertical hierarchies and the first vestiges of a state were established when an élite felt strong enough to act autonomously and take the initiative. Finally, the city becomes what Maisels terms a *transformational engine* through which society at large is restructured and continually recreated through an internally generated momentum for change. Thus irrigation is a facilitating rather than a monocausal factor.

Environmental consequences

The economic advances in Mesopotamia had not been won cheaply, for the Sumerians and their neighbours felt that they were part of a struggle against a capricious and vindictive nature. Their dream was to have greater control and in the Epic of Gilgamesh, for instance, the hero subjugates the wilderness by slaying Humbaba, the wild protector of the western cedar forests (Hughes 1975).

As population grew in southern Mesopotamia, the area dominated by city–states such as Ur, Uruk, Kish and Lagash, it became necessary to reduce the extent of fallow land in order to maximize food production. Unfortunately this led to a build-up of salt in the soil (salinization), the progress of which it is possible to trace from the archaeological record, from the study of grain impressions on pottery and from written temple accounts.

It appears that (emmer) wheat was gradually phased out because it is only half as tolerant of salt as barley. As well as being a human staple food, barley was valued as a fodder for sheep. In 3500 BC wheat and barley were grown in equal measure, but by 2500 BC wheat had fallen to 15 per cent of grain supply and in 2100 BC it had virtually disappeared (Jacobsen and Adams 1958). At the same time there was a serious decline in fertility as fields were over-worked. Until about 2400 BC this does not seem to have been a problem, with yields on a par with US yields today. However, they had fallen 40 per cent by 2100 BC and 65 per cent by 1700 BC.

Eventually it became impossible to maintain the vast superstructure of non-producers: administrators, soldiers, and urban-based craftsmen. The weakened Sumerian city–states declined in influence and in 1800 BC were conquered by Babylon, a more northerly power. The number of settlements fell by 40 per cent and the settled area by three-quarters.

In southern Mesopotamia we have probably the world's first example of large-scale environmental degradation as the direct result of human actions, although some caution is necessary because of the uncertain role of climatic change. A quarter of Iraq is still affected by salt, so the impact has been long term.

Other hydraulic civilizations

Sherratt (1980) claims that everywhere early agriculture (and by implication urbanization) occupied only narrow zones of maximum productivity where the water table was high. Soil moisture was as crucial if not more so than rainfall for the growth of crops. This was certainly true in Mesopotamia, and seems also to have been the case in Egypt, China, the Indus Valley, Mesoamerica and Peru. The later spread into temperate latitudes shifted the balance to rain-fed agriculture, which required a different technological environment.

Egypt developed along lines parallel to Mesopotamia but a little later in time. It is difficult to know whether there was any inspiration from Sumeria but anyway the circumstances in Egypt were different for three reasons. First, the geopolitical position of Egypt is more marginal. Being cut off by the Sahara Desert and the Red Sea, it is less threatened by envious neighbours than Sumeria. Second, the Nile has a more convenient hydraulic regime. The water rises in late summer and spreads out through overflow channels and levée breaches into a series of flood basins about 2 metres deep. It recedes by the end of November, leaving a fertile silt deposit on the fields and a moist soil ready for planting crops. The harvest is in the Spring, before the very hot weather. There was canal irrigation in the Nile Valley from the third millennium BC, but it was not as essential and the engineering works involved were less straightforward because the river does not flow above the level of its flood plain, as does the lower Euphrates.

Third, ancient Egypt was never a society *urbanized* in the manner of Sumeria. It was unified at an early date (3100 BC) and the state had control over most aspects of life. In the early phase cities were few, the first walled towns having appeared about 3300 BC in Upper Egypt, and the urban population was scattered in small towns throughout the Nile valley. A local peculiarity was the mausoleum complex which each Pharaoh built during his lifetime near his future tomb, and an extensive 'city of the dead' or *necropolis* for state officials. The ancient Egyptians took the after-life very seriously.

In the Hwang Ho Valley of northern China and the Indus Valley in South Asia (Harappa/Mohenjo Daro civilizations) social control seems to have been looser and urbanism less highly developed. Hydraulic works were important in the advance of agriculture but less crucial in societal evolution. In Peru a succession of civilizations manipulated water to irrigate their crops. The Incas were a small urban-based élite governing a vast Andean empire through an autocracy of astonishing ferocity. Their economy was minutely regulated and labour service was extracted from subject peoples to build roads and irrigation canals and dig mines. The sparse resources of the mountains were carefully harnessed through labour-intensive improvements such as the terracing of hillsides to expand the amount of arable land. Such achievements were without the benefit of writing, the plough or wheeled vehicles. The modest technological base was matched by a vulnerable social system in which the god-king held the reigns of centralized power. This made the Spanish conquest easier because their control of the élite was sufficient to disable opposition. The hydraulic base of the civilizations of the Americas will be dealt with in Chapter 9.

PRE-INDUSTRIAL HYDRAULIC TECHNOLOGIES

The development of methods of applying water to crops was crucial to the origins and spread of intensive agriculture. We have plenty of evidence of the manipulation of moisture conditions from the earliest times, and there are modern parallels, for instance among the indigenous peoples of the south-west USA, suggesting that even before agriculture proper streams may have been diverted to encourage the growth of wild plants. This seems to have begun usually where a stream or spring debouched from a mountain range on to an alluvial fan.

Four types of water use are worth a mention here. First, groundwater flow was exploited from about 700 BC in the form of underground tunnels, variously known as *qanats, aflaj, kariz* and *foggaras*, which tapped the water table. These were excavated in lines leading to centres of population and agriculture over distances from a few tens of metres up to 50 km (*see* Figure 3.6). The visible surface evidence is a line of well-like shafts at regular intervals, for ventilation during construction and maintenance, and piles of excavated spoil (*see* Figure 3.7).

Qanat construction requires great skill and is very labour-intensive. If the gradient of the conduit is too steep, the speed of flow will lead to erosion and collapse. The expense could only be borne by a wealthy ruler, or by some collective system of labour service. The effort is worthwhile because of access to groundwater which would not otherwise be available. Qanats were first developed in western Iran, northern Iraq and eastern Turkey and were spread by the Persians and Arabs to other arid parts of the Middle East, North Africa and Spain. There are also extensive qanat systems in western China. In 1968 English estimated that between one third and one half of Iran's irrigated area was watered by 37 500 qanats, but the pumping of groundwater through tube wells has reduced their dominance since then.

Second, another early technology was water harvesting and retention. We know that the Iron Age people of the Negev Desert built check dams to retain wadi flash floods and were therefore able to cultivate small areas which are now barren (see Chapter 4). In Libya and Tunisia the Romans successfully occupied a dry landscape, building small reservoirs, *impluvia*, which collected surface runoff from micro-scale drainage basins. Their moisture-control technologies were advanced but their deforestation and depletion of soil fertility in the same areas showed less sophistication and population densities eventually declined. The Sassanians (AD 226–800) in Iran used check dams and hillside contour bunding to retain moisture and minimize soil erosion.

After millennia of evolution, this art of water

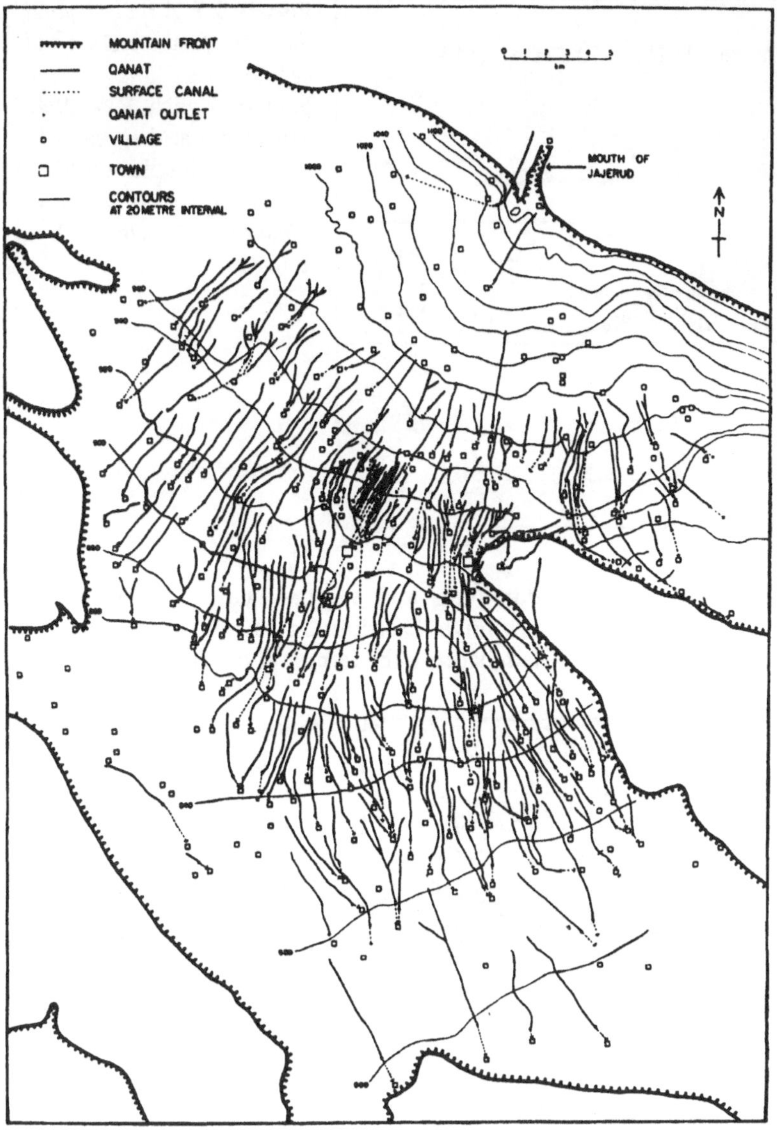

FIGURE 3.6 *Qanats on the Varamin Plain, Iran*
Source: Beaumont, P. 1968: Qanats in the Varamin Plain. *Transactions of the Institute of British Geographers* **45,** 169–80

harvesting and terracing in such arid and semi-arid environments has declined in the last 500 years but the recent problems of drought and famine in the Sahel have revived an interest in low-tech methods of moisture manipulation, and water harvesting is therefore back at the top of the agenda for rural development.

In the wetter, tropical environment of southern India, water is no less precious. In the state of Tamilnadu in particular, a fascinating system has evolved to even out the seasonal deficit of water which occurs before the monsoon replenishes soil moisture and water courses. This is the use of tanks or reservoirs of all sizes, formed by blocking every available watercourse. Tanks serve the multiple purposes of providing irrigation, drinking water, and fish. They pepper the map (*see* Figure 3.8) and make a profound difference to both landscape and life (see Chapter 2). Without readily available moisture, the poor lateritic and gneissic soils of the area could have supported little. Some tanks are very large, such as the Minneriya tank built in Sri Lanka in AD 900, which is 31 km^2, but evaporation then becomes a problem.

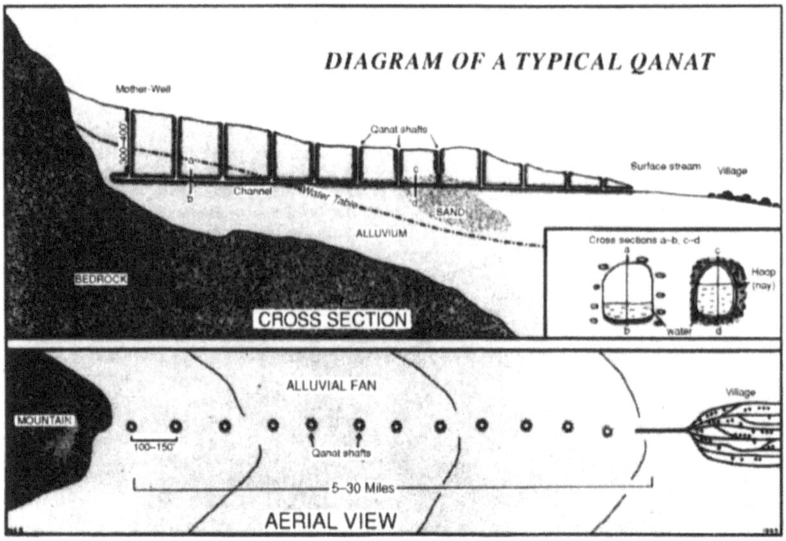

FIGURE 3.7 *A typical qanat, diagram of cross-section and aerial view*
Source: English, P.W. 1968: The origin and spread of qanats in the old world. *Proceedings of the American Philosophical Society* **112,** 170–81

Third, there is surface irrigation from canals. In the pre-industrial era these were usually fed directly from a river or stream, the technology of damming and storage being as yet relatively primitive. Distribution has always been a problem because the farmer at the end of the network

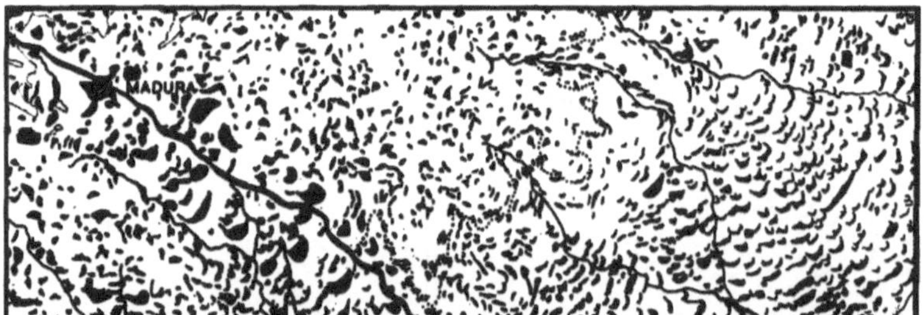

FIGURE 3.8 *The extraordinary number of tanks in this part of Tamilnadu, southern India, makes it one of the most heavily modified environments in the pre-industrial world. See Figure 2.10*
Source: Spate, O.H.K. 1967: *India, Pakistan and Ceylon: the regions.* 3rd edn, London: Methuen

will suffer unless some means of water-sharing is agreed. Fields were irrigated by flooding or, where the labour was available, by leading water along furrows between parallel ridges. Historically, the latter has been widely used because it is adaptable to a variety of slope angles and soil textures.

Last among the ancient technologies of water supply are those which involve a lifting device. The shaduf, a counter-balanced pole with an attached bucket or skin (*see* Figure 3.9), is perhaps the simplest, and there are also human or animal-powered water wheels (saqiya or Persian Wheel), the Archimedean screw, and many other variations on the theme. They allow water extraction from a canal or river, but are labour-intensive and inefficient in terms of the amount delivered.

FIGURE 3.9 *A water-lifting device from China on the shaduf principle*
Source: Sung Ying-Hsing 1637: *T'ien-kung k'ai-wu*

CASE STUDY: THE *HUERTAS* OF EASTERN SPAIN

In Roman times eastern Spain was drawn into a tightly integrated Mediterranean-wide food marketing system which also included Tunisia, Egypt and southern France. Its products were part of an advanced agroecosystem that had gradually evolved, comprising mainly the cultivation of wheat, barley, olives, grapes, and the tending of sheep, goats and pigs. Agronomic practice by the time of Columella (mid-first century AD) was geared to environmental potential, with key elements such as careful selection of plant varieties, intercropping of cereals among tree crops, elaborate seedbed preparation, rules for spacing plants, an agricultural calendar of planting dates, the classification of soil types, and care of soil fertility by manuring and fallowing (Butzer *et al.* 1985).

This system had been intensified in the first millennium BC by the introduction of canal networks, large and small, to deliver water to fields, gardens, orchards and pastures. Roman landlords invested heavily and were rewarded at first by commercial success, although this was later tempered by the agrarian crisis which accompanied the political stagnation and decline of the Empire. Rural depopulation and heavy taxation destroyed the profitability of export-orientated production and the irrigation system in parts fell into disuse. This decline continued under Visigothic rule.

The Arab and Berber invasions in AD 711 and the subsequent make-over of the agricultural landscape in the succeeding 500 years was a tribute to the organizational and technical ability of the new masters. They revived and strengthened the Roman legal framework of water use, extended the canal network and introduced new crops such as rice, sugar cane, citrus, and sorghum. The medieval Christian reconquest merely confirmed the existing practices of the Romans and Arabs and the continuity of irrigated agriculture down to the present is remarkable.

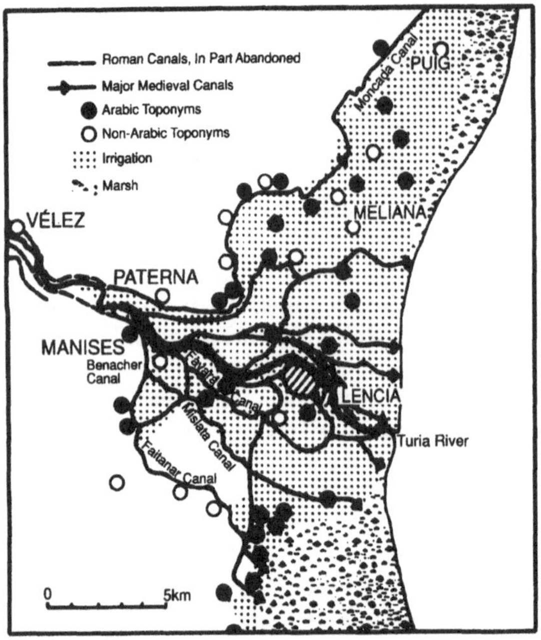

FIGURE 3.10 *The huerta of Valencia*
Source: Reprinted from Butzer, K.W., Mateu, J.F., Butzer, E.K. and Kraus, P. 1985: Irrigation agrosystems in eastern Spain: Roman or Islamic origin? *Annals of the Association of American Geographers* **75**, 479–509 by permission of Blackwell Publishers

The medieval *huerta* (irrigation community) of Valencia at the mouth of the River Turia (*see* Figure 3.10) is a fine example of a region of intensive irrigated agriculture based upon a long-lived institutional framework. Every person owning land was entitled to free use of water in rotation and in proportion to the size of their property, but of course this was limited in volume and disputes regularly arose about how it should be apportioned. In order to deal with such problems, the irrigators from each canal formed a self-governing commune, with assemblies and elected officials (*cequiers*). There were fines for misdemeanours, such as those listed in Table 3.1, and collective works were organized. Each April, before the spring planting, the water was shut off and the canals cleared of vegetation and silt dredged. Regional disputes were settled by a water tribunal in Valencia, the city which dominated its 105 km² huerta.

TABLE 3.1 *Percentage of misdemeanours in each category in the huerta of Valencia, 1443–86*

Misdemeanour	Percentage
Forbidden water	29.0
Wasting water	17.3
Flooding the road	12.5
Flooding fallow field	9.0
Stealing water	8.5
Washing in the canal	5.0
Flooding a crop	3.9
Draining water in wrong place	2.6
Installing or undoing canal check illegally	2.3
Taking water by force	1.6
Irrigating without right	1.0
Miscellaneous	7.3

Source: Glick (1970, 54)

CONCLUSION: IRRIGATION AND INTENSIFICATION

We may reject the extremes of Wittfogel's logic but recent scholarship has certainly found links, direct and indirect, between the deployment of irrigation ecotechnology, agricultural intensification and population growth. Irrigation allows year-round cropping and, if there is sufficient water, multi-cropping. It may permit the expansion of the cultivated area, guard against seasonal drought, and increase the yield of certain key crops.

The world's irrigated area over the last 6000 years has grown, and now covers 17 per cent of cultivated lands. It accounts for half of global crop production. This surge has come particularly in the twentieth century with the advent of energy-intensive pumping technologies and the construction of big dams throughout the developing world. The use of water in some countries will reach capacity in the near future and there may be little further scope for further expansion of the irrigated area without trade offs with other users of water such as industry and domestic consumers. So precious has water become as a commodity, that some commentators think that it will be a source of friction between those neighbouring countries which share rivers. This is already the case in south Asia, where India's Farakka

Barrage has diverted water from the Ganges, to the dismay of Bangladesh which sits downstream.

FURTHER READING AND REFERENCES

The emergence of civilization is well covered in Maisels (1990) and Postgate (1992). For hydraulic organization and technologies, see Glick (1970) and Butzer *et al.* (1985).

Adams, R. McC. 1966: *The evolution of urban society: early Mesopotamia and prehispanic Mexico.* London: Weidenfeld & Nicolson.

Adams, R. McC. 1981: *Heartland of cities: surveys of ancient settlement and land-use on the central flood plain of the Euphrates.* Chicago: Chicago University Press.

Butzer, K.W., Mateu, J.F., Butzer, E.K. and Kraus, P. 1985: Irrigation agrosystems in eastern Spain: Roman or Islamic origin? *Annals of the Association of American Geographers* **75,** 479–509.

English, P.W. 1968: The origin and spread of qanats in the old world. *Proceedings of the American Philosophical Society* **112,** 170–81.

Glick, T.F. 1970: *Irrigation and society in medieval Valencia.* Cambridge, Mass.: Belknap Press.

Hughes, J.D. 1975: *Ecology in ancient civilizations.* Albuquerque: University of New Mexico Press.

Jacobsen, T. and Adams, R.M. 1958: Salt and silt in ancient Mesopotamian agriculture. *Science* **128,** 1251–8.

Maisels, C.K. 1990: *The emergence of civilization: from hunting and gathering to agriculture, cities and the state in the near east.* London: Routledge.

Postgate, J.N. 1992: *Early Mesopotamia: society and economy at the dawn of history.* London: Routledge.

Sherratt, A. 1980: Water, soil and seasonality in early cereal cultivation. *World Archaeology* **11,** 313–29.

Wheatley, P. 1971: *The pivot of the four quarters: a preliminary enquiry into the origins and character of the ancient Chinese city.* Edinburgh: Edinburgh University Press.

Wittfogel, K.A. 1957: *Oriental despotism: a comparative study of total power.* New Haven: Yale University Press.

Yoffee, N. 1988: The collapse of ancient Mesopotamian states and civilization. In Yoffee, N. and Cowgill, G.L. (eds) *The collapse of ancient states and civilizations.* Tucson: The University of Arizona Press, 44–68.

4

RESOURCES, POPULATION AND SUSTAINABILITY

We are just statistics, born to consume resources.
(Horace, *Epistles* ii, 27)

Introduction: extensive, shifting agriculture

The irrigated agriculture discussed in Chapter 3 must have become locationally fixed in order to justify the investment of labour in hydraulic works, but archaeological and historical evidence suggests that many early farmers did not occupy permanent sites and preferred to move every few years to new land. Some practised *swidden* or shifting cultivation, others nomadic pastoralism. Both economies represent an attempt by the farmer to reduce the risk inherent in all low technology husbandry, which relies so much upon the gift of nature. Settled farmers passively wait for the rains and pray to be spared a plague of locusts; nomads and *slash and burn* cultivators actively seek out fresh resources in anticipation of a declining living in their former location.

In this chapter we will develop a discussion of the relationship between resource use and population density by following the spectrum of farming through from shifting and mobile strategies to intensive rice farming. Then the implications will be examined of population growth for land degradation.

Nomadic pastoralism

Nomadic animal herding has ancient origins in the Zagros Mountains of Iran, more or less contemporaneous with the cultivation of cereals in the Levant (see Chapter 2). It was not a precursor to settled agriculture as historians used to think. From the tremendous variations in types of nomadism we may infer that it is not today the remnant of a homogeneous economic form, but rather a disparate collection of human responses to environmental and other local pressures. Although nomads were militarily more than a match for farmers in the past, being mobile and always having the advantage of surprise, in recent times the balance has been reversed and they have been increasingly marginalized in modern urban-centred societies. Nomads survive in a pure form today only on land which is not attractive to anyone else, because of climate, terrain or access. Their way of life is discouraged by governments who wish to vaccinate them, educate their children and collect taxes. In response, many have settled in the last fifty years.

Vertical migration, including seasonal transhumance, is a means of fully using ecological zones at different altitudes in mountainous country. Horizontal nomadism is appropriate in dry environments where localized convectional rainfall brings out a flush of grass which can be grazed for perhaps a few days or weeks, and by its very nature involves less forward planning. Camels are ideal for the latter type of nomad because they can go for up to 3 weeks without water in cool weather (more like 3 days in summer) and can graze vegetation up to 80 km from camp (usually at a source of water) without straying (Ruthenberg 1980). They can carry loads of up to 300 kg and the females lactate for 12–18 months, providing an additional food source. Figure 4.1 models the time–space diary of a hypothetical nomadic group herding sheep, cattle, goats and camels.

Nomadic pastoralists make use of parts of the earth which would otherwise be uninhabited (unless oil or mineral deposits are present). But these tend to be fragile ecosystems, and the nomads are often blamed for over-grazing and its consequences. In many pastoralist cultures concepts of status and wealth are vested in the number of animals owned by an individual and this may lead to stock numbers exceeding range capacity. A large herd is a personal hedge against risk but it may be imprudent in the collective interest. When vegetation is depleted there may be problems of soil erosion and other forms of land degradation which are known collectively by the term *desertification* (Chapter 13).

In reality the evidence for a correlation between nomadic (as opposed to settled) pastoralism and desertification is thin and controversial. The nomadic way of life has been in tune with the environment because it has had to be, simply in order to sustain an economy under difficult conditions. To destroy nomadism by heavy persuasion to settle is both crass and ill-informed.

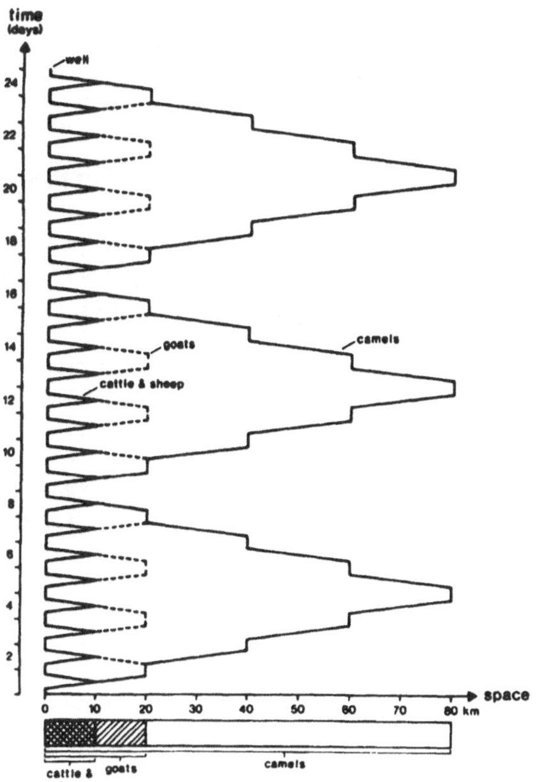

FIGURE 4.1 *The grazing ranges of different species around a water source such as a well, as a function of the speed of movement and frequency of watering. Cattle and sheep are confined to a daily radius of 10 km and goats to 20 km in two days, but camels can wander up to 80 km over a period of eight days*
Source: Carlstein, T. 1982: *Time resources, society and ecology. On the capacity for human interaction in space and time. Vol. 1: preindustrial societies. Lund: Gleerup, 147-86*

SWIDDEN

Another mobile use of resources survives in the form of shifting cultivation, also known variously as *swidden, milpa, jhum*, or *shamba*. Here farmers and their families move their settlement to a new site after a number of years, when crop yields begin to fall. Some scholars believe that from the outset this would have been a standard type of farming, even in areas of relatively dense natural vegetation. Primitive flint axes were sufficient to clear light woodland and so swiddeners would have made a significant impact. Sherratt (1980), however, argues that swidden came later, when population growth pushed cultivation on to marginal, less fertile land.

Shifting cultivation comes in a number of guises (Table 4.1), from a pure form of subsistence-based mobility, shading into semi-sedentary and semi-commercial farming. It remains widespread (*see* Figure 4.2) in the Amazon, Africa south of the Sahara, and parts of South-east Asia, in regions which have relatively low populations, where it is practised by 300–500 million people, tribals and also poor peasant farmers who have no access to land in settled areas. In order to investigate swidden systems and the type of societies they support, we rely less upon archaeology than upon the

TABLE 4.1 *Classifications of swidden*

System	Classification
Vegetation systems	Shifting cultivation is common in tropical forests but is also practised in savanna bush (thicket) and grassland.
Migration systems	Some swiddeners move their houses frequently, others journey to their fields from a fixed village. Frequency of movement and distances covered seem to be a function of rainfall. Migration may be linear or cyclical.
Rotation systems	The length of rotation will depend upon population density, soil fertility, and agro-climate.
Clearance systems	The methods of clearing vegetation and preparing the plot for cultivation vary greatly.
Cropping systems	Some farmers monocrop with a staple, such as rice, maize or manioc, supplemented by a cash crop Others prefer multicropping to cover the risk of crop failure.
Tool systems	The main differences are between digging stick, hoe and plough cultivators.

Source: Ruthenberg (1980)

ethnographic study of present-day shifting cultivators, who still practise this type of agriculture.

Swidden has a number of characteristics. First, it usually involves the clearance of vegetation by slash and burn. Burning trees and undergrowth releases phosphates and calcium in the ashes, making nutrients available to crops and reducing the acidity of the soil. There is, however, a loss of humus and nitrogen. Sometimes trees are merely ring-barked and left *in situ*. This reduces the labour of felling but ensures that the leaves will drop, allowing light to penetrate the canopy, and the tree does not take so much water from the ground.

Second, a minimum of cultivation takes place. The use of a plough is rare because of the difficulty of grubbing out tree roots and stumps. The farmers either sow their seeds straight into the ashes or dibble them into the soil using some form of digging stick.

Cattle are comparatively rare, so the maintenance of soil fertility by manuring is not possible. Plots are abandoned after three to five years because of declining soil fertility or choking with weeds such as the grass *Imperata cylindrica*, and the farmer moves on to a new plot. This amounts to land rotation instead of crop rotation.

Third, the used plot tumbles back into scrub

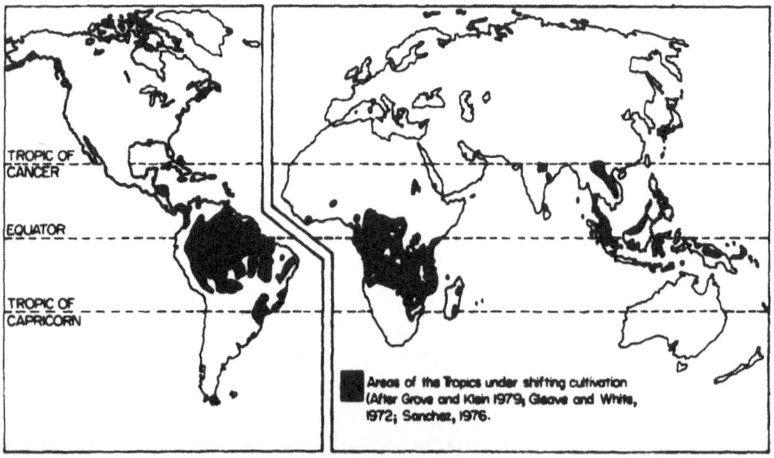

FIGURE 4.2 *The shaded areas show the main concentrations of surviving shifting agriculture*
Source: Okigo, B.N. 1984: Improved permanent production systems as an alternative to shifting intermittent cultivation. In Food and Agriculture Organization, *FAO Soils Bulletin* No. 53

and secondary forest and eventually regains its fertility after a period of years, depending on the soil fertility and other ecological factors. Ten to thirty years is a fairly typical fallow period. Weed infestation, which is a serious limiting factor in the last phase of cultivation, is eliminated with the regrowth of trees in time for the next period of use.

Fourth, shifting agriculturalists are very skilful and display a considerable degree of land cunning. They pick out the best soils to cultivate by recognizing species of trees (although a less fertile plot which is easier to clear may give better returns to labour input). The example is often quoted of the Hanunóo tribe of the Philippines, some of whose members can distinguish 1600 different types of plants (more than Western botanists recognize), including 400 which can be cultivated. It is not uncommon to see 20 crops in a single Hanunóo plot. Polycultural intercropping is the norm in some forms of swidden, in order to reduce the risk of crop failure from attack by pests and diseases. By a system of phased planting throughout the year, tropical shifting farmers can spread their labour to avoid bottlenecks of planting and harvesting, and to make fresh food available at all times.

Shifting agriculture is particularly well suited to the tropics because it simulates the natural conditions of the forest. Farmers plant crops which have different leaf canopies and different nutrient requirements from the soil. The architecture of planting allows the fullest possible use of light, nutrients and water, and prevents the heavy tropical rainfall falling directly on to the soil where it would cause erosion.

Fifth, the spatial organization of cropping is complex. Carlstein (1982) has convincingly shown that a distance-decay of intensity is the outcome of time budgeting by rational farmers. They have a kitchen garden close at hand with vegetable and delicate fruit crops that need attention and which are fertilized by domestic waste. The swidden fields will be further away, at a distance which will change from year to year, with hunting and gathering of wild foods at a further remove.

Typical tropical soils (e.g. latesols) are heavily leached and have a delicate structure. They can easily be ruined if over-exploited. Shifting agriculture gets the best out of the ecosystem because the lush tropical vegetation is the main storehouse of nutrients. These are released on burning, but can be regenerated relatively quickly if the forest is allowed to re-establish itself.

Slash and burn, when properly managed, causes less damage to the environment than many other forms of tropical agriculture. Crop yields under shifting agriculture are quite high per unit of labour input and even per unit of cultivated land, but low per hectare if the fallow land is included. For this reason shifting agriculture is said to be *land extensive* and is usually associated with low densities of population.

In the tropics the forest may grow again quickly after a plot is abandoned, although it is estimated that the full climax may take 500 years to recover. Excessive cutting may prevent regeneration and a savanna scrub vegetation will become established. Heavy tropical downpours may erode fragile soils on even slight slopes and then environmental degradation can be severe. This type of result is most evident, however, in regions where pressures of population or of commercial exploitation have led to clear-felling of the forest.

In temperate Europe trees grow very slowly and burning vegetation is not as productive of nutrients. In the neolithic, the fallow period would have been much longer and shifting agriculture was less well suited, because polyculture throughout the year is not possible for climatic reasons; and because any farmers who kept animals would have grazed them on the tender young saplings and this would have meant that grass was more likely to follow a swidden than forest regeneration.

POPULATION AND INTENSIFICATION

There are two opposing views about the significance of population for agriculture and food in pre-industrial societies (Table 4.2). The first was suggested by Thomas Malthus as long ago as 1798 in his classic text *An essay on the principle of population*. He proposed that population growth in certain sections of society is likely to outstrip food production, with the result that standards of living would fall at the very least, and at worst starvation might ensue.

TABLE 4.2 *Theoretical traditions about the population–agriculture relationship*

		Population growth under different economic conditions	
		Subsistence	**Market**
Expectation of outcome	Optimistic	Boserupian	neo-liberal economics
	Pessimistic	neo-Malthusian	neo-Marxist neo-Malthusian

Source: Modified from Turner, B.L., Hyden, G. and Kates, R.W. (eds) 1993: *Population growth and agricultural change in Africa*. Gainesville: University of Florida

Malthus' ideas were overtaken in the nineteenth century because improved farming technology partially solved the problem of low yields, and sea transport allowed an international trade in foodstuffs. Britain came to rely upon grain from the prairies of North America, beef from Argentina, butter from New Zealand and tea from India. Cheap food was a major policy of successive governments for 130 years from the repeal of the Corn Laws in the 1840s until joining the European Community in 1974. Neo-Malthusian ideas have recently reappeared in late twentieth-century literature concerned about rapid population growth in poor countries.

Ester Boserup took a more optimistic view in her 1965 book *The conditions of agricultural growth*. She recognized that population growth means increased demand for food but argued that the outcome will not automatically be disaster. In short, food scarcity will lead to higher prices and therefore create greater incentives for farmers to intensify their production. There is a wealth of historical and contemporary evidence in subsistence societies which seems to back Boserup (Netting 1993). Even in technologically primitive societies intensification has been possible:

- by the application of more labour, for instance through the reduction of leisure time;
- by adopting technologically more advanced implements and farming methods, which farmers had known about before but hitherto had not thought worth using. The transition of harvesting in nineteenth-century England from the sickle to the scythe may appear trivial but it saved 50 per cent of labour for that task. The use of irrigation or fertilizers would be an equivalent jump to greater productivity in the Third World today;
- by the reduction of fallow land and the introduction of multicropping.

Returning to the issue of swidden, Boserup argues that the variety of practices observable are really evidence of stages of an evolutionary sequence towards intensification. She identifies three broad types which have progressively shorter fallow periods, with increasing numbers of people: forest-fallow, bush-fallow and short fallow. Respectively, they average 23, 7 and 3 years of fallow and support 5, 5–65 and 16–65 people per square kilometre. Definitions of intensification vary according to:

- the frequency with which a particular plot is cultivated;
- the intensity of economic inputs invested in terms of capital, labour and skills;
- the ratio of outputs, such as food harvested, to inputs in the form of seeds.

In bush-fallow the fertility of the soil is not restored before the next cultivation. Yield per hectare therefore can only be maintained by additional labour in weeding and manuring the fields. In the transition to short-fallow, technical innovations become necessary, for example, use of the plough to cultivate the soil thoroughly and the hoe to assist weeding.

Short-fallow may eventually progress to annual cropping or multicropping (Figure 4.3) and support a still higher population with social re-organization, for instance the introduction of a communal system of agriculture in medieval England, where labour, tools, and land were to a certain extent shared or at least centrally organized. This encouraged a permanently settled population.

In effect, Boserup is proposing a model in

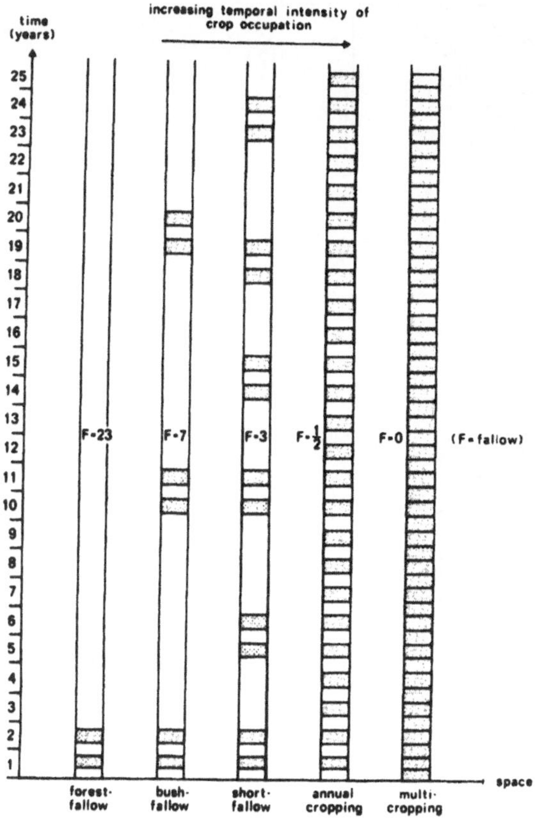

FIGURE 4.3 *As population increases, so the fallow period in years (F) is reduced. In modern settled agriculture usually lies between 0 and 1*
Source: Carlstein, T. 1982: *Time resources, society and ecology. On the capacity for human interaction in space and time. Vol. 1: preindustrial societies.* Lund: Gleerup, 147–86

which population growth is the prime cause and technology a secondary effect. Because this is the reverse of conventional wisdom, her work has stimulated an important debate which has yet to be fully resolved. Implicitly she challenges the view that peasant agriculture, with its so-called primitive technology, is backward and therefore unable to hold its own as an efficient form of food production. On the contrary, small holders are demonstrably rational in their enterprise, capable of levels of intensification superior to the modern sector in every respect except capital investment, and, other things being equal, they are more likely to maintain a sustainable usage of environmental resources (Netting 1993).

Beguiling though it is, there are a number of problems with Boserup's theory (Grigg 1979). Two are worth mentioning here. First, there is the chicken and egg problem. Might not population growth have been a consequence rather than a cause of improved agriculture? It is impossible to be sure, but it seems inconceivable that demography should always be the independent variable. It is equally unreasonable to rule out responses to population pressure other than intensification: expanding on to marginal lands is an obvious alternative, as happened in medieval England (Chapter 11), and another is outmigration, for instance by poor Irish and Scots overseas in the Victorian period. On the other side of the coin, intensification may result from authoritarian dictat or from urban demand, irrespective of total numbers of people.

Second, Turner *et al.* (1977) performed a statistical analysis of data for 29 tropical subsistence farming groups and found that population pressure was indeed important. Yet they criticized Boserup for holding environmental factors constant in her model, when climate, soils, topography and ecology are so clearly vital sources of variation in the real world. After all, the high yielding varieties of crops which were the basis of the Green Revolution were adopted most rapidly in regions that were highly favoured environmentally, such as the Punjab in India and Pakistan.

Turner and his colleagues prefer a modified version of Brookfield's (1972, revised 1984) interpretation of intensification in terms of agricultural feasibility, where inputs are most likely to be enhanced in environmental conditions which are perceived to be advantageous. Boserup's interpretation is most appropriate in unconstrained environments, but in marginal environments population pressure may actually cause productivity decline and degradation (see Chapter 5) rather than innovation and intensification.

CASE STUDY: WET RICE

One of the best possible examples of a correlation between population density and agricultural intensity is that of the wet rice areas of Asia. But which came first: the population pressure or the technical change? Rice (*Oryza sativa L.*) was

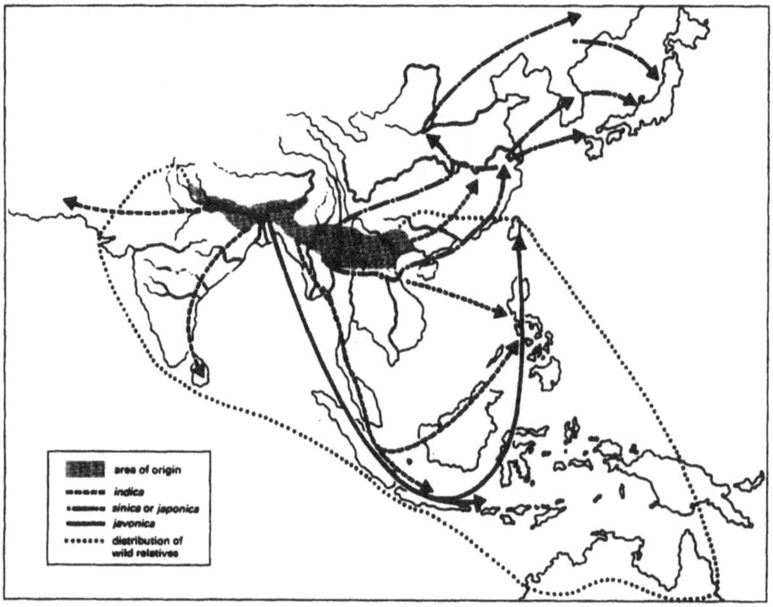

FIGURE 4.4 *The origins and spread of rice* (Oryza sativa) *in Asia. There are three major races of rice:* indica, sinica/japonica *and* javonica
Source: Chang, T.T. 1989: Domestication and spread of the cultivated rices. In Harris, D.R. and Hillman, G.C. (eds) *Foraging and farming: the evolution of plant exploitation*. London: Unwin Hyman, 408–17

originally a swamp plant. Domestication probably took place in poorly drained river valleys of the hilly country in an arc from the fringes of the Himalaya in the west to southern China in the east or in the middle Yangtze Valley in central China. One date for the latter is 6500–5800 BC. By 2000 BC it had spread extensively into South-east Asia and westwards across India and Pakistan (*see* Figure 4.4). There are a great number of strains of rice, including varieties that float in deep water and upland rice which is entirely rainfed. African rice (*Oryza glaberrima*) is a separate domestication with a different history, and will not be discussed here.

Paddy (as rice is called when still part of the plant) requires a temperature of 20°C for 3–4 months, and will stand a wide range of soils. It needs to have its roots in water for at least three-quarters of its growing season. Providing sufficient water and maintaining moisture levels need irrigation, flat fields and an impervious subsoil. These conditions are best found in river valleys and deltas. The dramatic pictures of staircases of hillside terraces are not typical of most rice-growing areas.

Five other features of the wet rice system are worth listing:

1. The farms are often small, perhaps only 0.5–1.0 hectares and often fragmented into parcels.
2. It is a highly labour-intensive system, using mainly family labour. In Japan for instance, where there is no water buffalo to pull a plough, cultivation was by spade until the recent innovation of the mechanical rotovator.
3. Potential output per unit area is high and, where the growing season allows, two or even three crops a year are possible.
4. Wet rice agriculture supports a high population density, with 400–500 per km^2 not unusual and over 1,500 per km^2 in some areas.
5. Rice plants are often grown in a nursery field and then painstakingly transplanted into the main paddy fields, which have been heavily manured/fertilized.

We cannot be sure when people began to create artificial rice swamps by the manipulation of water. Certainly flood control in the Hwang-Ho valley in 2200 BC was associated with rice and irrigation was known in Honan in 563 BC and the Yangtze Valley by 548 BC. It then spread to east and south China. Three hundred years later a mature system of cultivation was widespread in the south, with terraced hillsides, fallowing, and water buffalo for deep ploughing. Other improvements followed, such as manuring with animal and human waste, green manuring, transplanting, and harrowing. The introduction of early maturing varieties in the tenth century AD, enabling two crops per annum, put into place the last piece of the wet rice regime that we know today.

There is historical evidence that early rice cultivation was not especially intensive, but became more so as population increased. Population densities are now greatest where rice cultivation has the longest history (Japan, China, Java, Korea, Vietnam, India) and less so in Burma and Malaysia. According to Geertz (1963) intensification, which he terms *involution*, does not guarantee improved standards of living. In Java, where wet rice cultivation has been perfected for centuries, an increasing population had instead merely forced the adoption of more elaborate, labour-absorbing farming techniques. Productivity per hectare had risen but not productivity to a unit of labour input. More people were supported by the land but per capita real incomes had not risen in the previous 150 years.

The lesson of the history of rice cultivation is that although extraordinary intensification is technologically possible, in reality each region has adopted only as much as is necessary to feed its current population at a bare level of subsistence. Different rice technologies sometimes exist side by side in South-east Asia. Hanks (1972) found a succession at Bang Chan, Thailand: shifting cultivation (1850–90), broadcasting (1890–1935) and shifting cultivation (1935–70). Only in the last 150 years has there been any real sign of population pressure, especially in China and Java, when increased labour input was yielding worryingly diminishing returns.

The impact of a permanent, terraced wet-field system upon the local community is well summarized by Spencer (1974, 65):

> The need for staying with the water control problem day by day and year by year perhaps tends to increase the sedentariness of the families involved. Once the terrace groups are laid out, meaningful and effective cooperation is required ... The lasting stability of a well-maintained assemblage of terrace groups makes for social and economic stability and for local autonomy in the political sense.

Case study: Intensification and environmental degradation in the Mediterranean of classical times

> In the time of the Trojan War the Argos was marshy and able to support a few inhabitants only, while Mycenae was good land ... Now the opposite is the case ... for Mycenae has become ... unproductive and completely dry, while the Argive land that was once marshy and unproductive is now under cultivation.
>
> (Aristotle, *Meteorologica*, Book 1, Chapter 14)

Agricultural intensification as a result of population pressure and technological innovation usually involves the replacement of vegetation by permanent fields and the regular disturbance of the soil which, as time progresses, has a lower and lower organic content. Inevitably, it seems, on the most vulnerable soils this results in the modification of geomorphological processes so that the rate of erosion increases. There may be other factors in changing rates of erosion, such as climatic change, but human impact has undoubtedly been the principal cause of such landscape modification over the last 5000 years in Europe.

As an example let us take the Mediterranean basin, in which context there has been much discussion about soil erosion in ancient history. Vita Finzi's (1969) classic study identified two episodes of erosion, the Older and Younger Fills, both of which he took to be the result of variations of geomorphological process due to climatic change. Such a broad interpretation has in recent years been superseded, however, by more detailed analyses of local stratigraphies, with complex results.

For Greece, archaeological interpretations have indicated, contrary to Vita Finzi's model, that human activity as a result of population pressure has been of greater significance than climatic fluctuations. Over-exploitation of the land in the Argolid, such as pastoralism and tree felling, seems to have exposed vulnerable soil horizons to erosion, especially in five key periods: the early Bronze Age (*c.* 2500 BC); classical Hellenistic times (*c.* 350–50 BC); the late Roman period (*c.* AD 400); the middle ages (AD 950–1450); and in the modern era (*see* Figures 4.5 and 4.6). On each occasion population fell with the destruction of the resource base, but recovery came eventually and re-initiated a cyclical process. The construction of terraces and check-dams across streams during periods of prosperity was effective in reducing soil erosion, but both need regular maintenance and an economic/demographic downturn led to their abandonment. There is debate about the relative significance of gradual processes including soil creep and extreme meteorological events such as torrential downpours, but either way 40 cm of soil depth has been lost in the hard limestone region of the Argolid mountains, and 100 cm in the more easily eroded lowlands. Not all soils were equally affected, the entisols and inceptisols on steep slopes suffering most. For Plato the landscape of neighbouring Attica was equally denuded: 'what now remains is the skeleton of a sick man, all the fat and soft earth having washed away, and only the bare framework of the land being left' (Plato, *Critias*).

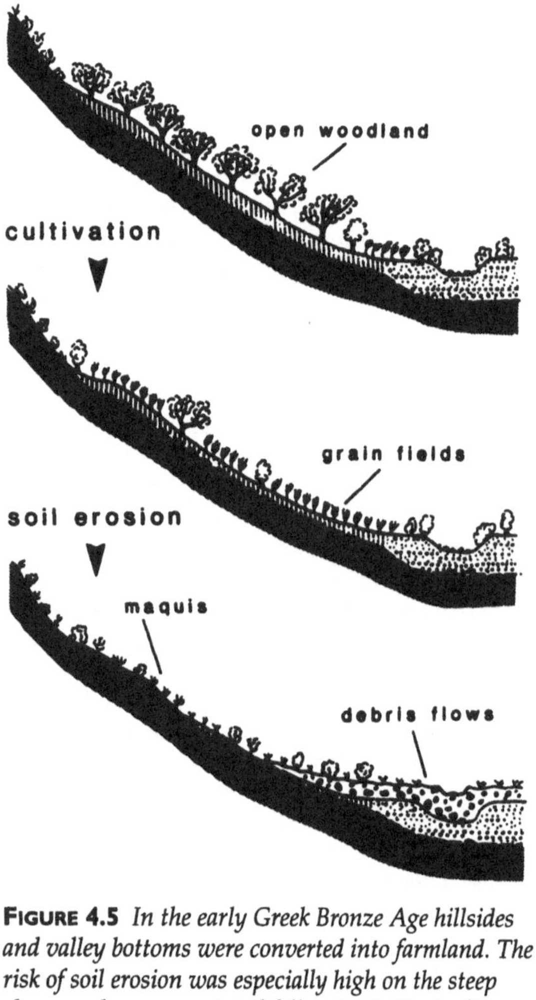

FIGURE 4.5 *In the early Greek Bronze Age hillsides and valley bottoms were converted into farmland. The risk of soil erosion was especially high on the steep slopes and on unvegetated fallow land. Gradually debris was washed downslope, forming layers of sediment on the lower ground*
Source: Andel, T.H. Van and Runnels, C. 1987: *Beyond the Acropolis: a rural Greek past*. Stanford: Stanford University Press

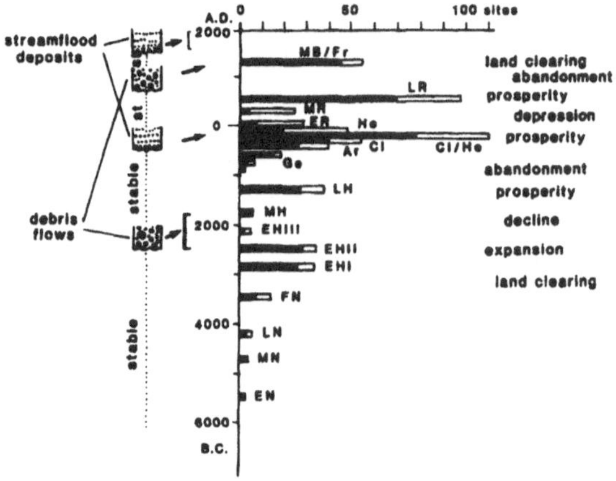

FIGURE 4.6 *Soil erosion and alluviation events in the southern Argolid, Greece, graphed along with the number of sites at each period and an indication of the prosperity of the rural economy*
Source: Andel, T.H. Van, Runnels, C. and Pope, K.O. 1986: 5000 years of land use and abuse in the southern Argolid, Greece. *Hesperia* **55,** 103–28

Frequently the eroded soil was deposited on river deltas, which then expanded. This was inconvenient where it silted up established port facilities and occasionally classical writers complain of the disruption of trade. A well-known example is the flood plain of the River Meander (Turkish Menderes) where Priene, a city founded in the fourth century BC with a harbour, is now 15 km from the sea (Figure 4.7). Pausanias (Book VIII, 24.11) commented (*c.* AD 175): 'The Meander: flowing through the lands of Phrygia and Caria, which are ploughed every year, it has in a short time turned the sea between Priene and Miletus into dry land.'

The Greeks did not consciously mine the soil without regard for the sustainability of their agriculture. They were aware of the need for contour ploughing and terracing to limit the problem but their hillsides were overgrazed and they were engaged in an ever-accelerating race between their growing population and the resources available for sustenance. Overseas colonies were a solution because they yielded some tradable commodities such as timber but most of all they attracted colonists away from Greece.

Beyond Greece in other parts of the Mediterranean the history of erosion and deposition has been similar in process but very varied in timing. This is the strongest indication of all that human impact (such as deforestation) has over-ridden any natural process, such as climatic change, which would have been widespread. But this view is far from universal. Some have argued that climatic change in the Holocene *has* had a significant impact upon erosion, deposition and landscape change; and others believe that degradation (human or natural) has had much less effect than often thought.

One issue that remains to be resolved is the temporal form taken by environmental change in the Mediterranean and, for that matter, in other parts of the world. Most commonly, human impact is described as progressive (*see* Figure 4.8), implying linear and unidirectional change and, presumably, uniformity of process. More recently, catastrophism has become popular, with morphological modifications, again in one direction, ascribed to extreme events of a long return interval. Thirdly, in some circumstances it seems that cyclical interpretations are possible because natural soil-forming processes might replace eroded horizons and vegetation might regrow.

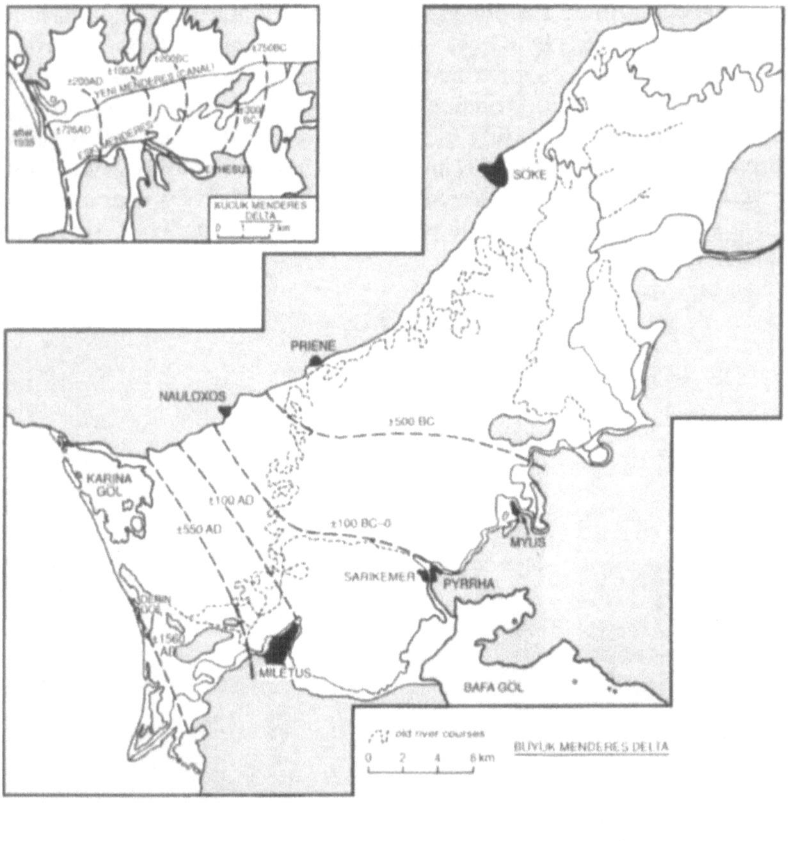

FIGURE 4.7 *The silting of the River Meander (Turkish – Büyük Menderes) in classical Greek times*
Source: reprinted from Eisma, D. 1978: Stream deposition and erosion by the eastern shore of the Aegean, in Brice, W.C. (ed.) *The environmental history of the Near and Middle East since the last Ice Age*, by permission of Academic Press Limited, London

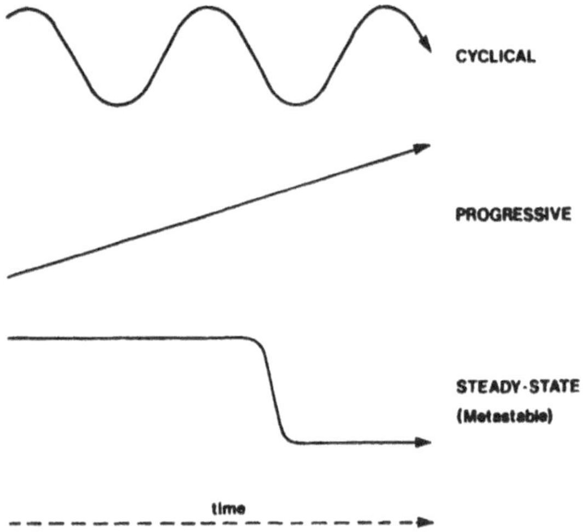

FIGURE 4.8 *Temporal models of human-induced environmental change*
Source: reprinted from Bottema, S., Entjes-Nieborg, G. and W. van Zeist (eds) 1990: *Man's role in shaping the eastern Mediterranean landscape*

Malta is another example of resource depletion in the Mediterranean. It had the world's first stone buildings, a series of massive temples (*see* Figure 4.9), dated *c*. 3600–2500 BC and therefore built well before the pyramids of Egypt. They were the product of a remarkable megalithic culture of the Copper Age. In the Bronze Age which followed, the Maltese farmers were responsible for an indelible imprint on the landscape. In the period around 1500 BC a number of so-called cart-ruts were made in the rock by their slide-cars or by loosely fitted wheeled carts carrying heavy loads (Trump 1990). One hypothesis is that this heavy traffic was moving soil for the construction of new fields, others suggest the carriage of building blocks from quarries or salt from coastal salterns. The rock in Figure 4.10 is Upper Coralline limestone which is relatively resistant to wear when exposed to the air but under a cover of soil it remains soft, and we can reasonably infer that these deep incisions were formed before the layer of top soil was eroded or carried away. Widespread erosion seems likely because Malta was cleared of trees early on by a growing population. Nowadays it is the world's third most densely populated country and for a long time has been unable to feed itself from its own meagre agricultural resources.

The Romans had an even more profound impact upon the landscape of their far-flung empire. Their self-confident exploitation of nature was counter-productive and their parasitic imperialism was especially destructive of natural resources in the periphery. Brückner (1986) describes what he calls an ecological catastrophe in southern Italy where over-exploitation by Greek and later, by Roman farmers, from approximately 700 BC to AD 200, caused the destruction of the protective vegetational cover leading to such extensive and severe soil erosion that a 12-metre-deep horizon of alluvial fill may be detected in the valleys of rivers running into the Gulf of Taranto. This experience was not

FIGURE 4.9 *Maltese megalithic temple, Hagar Qim*
Source: P.J. Atkins

FIGURE 4.10 *The cart tracks at Clapham Junction, Malta*
Source: P.J. Atkins

repeated until the nineteenth century as rural population densities again increased.

Natural rates of erosion in west central Italy of 2–3 cm per thousand years were accelerated to 10–100 cm from the second century BC. And all around the Mediterranean deposition of eroded material caused modifications in shorelines. The deltas of the Ebro, Rhône, Tiber, Arno, Po, Medjerda have all grown substantially during the historic period. Ostia, the former port of Rome, was partially silted up as a result. Deforestation for ship-building, charcoal, construction timber and the creation of agricultural land was mainly responsible, with grazing by sheep and goats preventing regeneration. Some writers argue that climatic fluctuations have also been important.

As a counterpoint to the somewhat negative message of this chapter, it is worth remembering that there is plenty of evidence of the sophisticated and sensitive use of marginal lands in the past (see also Chapter 6). One example is the Nabataean/Byzantine civilization in the Negev Desert of southern Palestine, which exploited a hyper-arid environment successfully for approximately 800 years up to about AD 700. The economy of this state flourished, based on runoff farming of three types: terraced wadis, hillside conduits, and diversion systems. Water was diverted from the occasional flows of desert streams and collected on fields where sufficient cereals, vines and olives were grown, along with supplementation from a caravan trade, to support six cities in the desert.

In one sample survey area of 130 km^2 in the Negev 17 000 dams have been found of all shapes and sizes according to the local topography, modifying valleys in order to gather moisture, even in areas where the current annual rainfall is below 80mm (Issar 1995). These dams are often set as close to each other as 40 metres, the object being to slow the wadi flow sufficiently to encourage the deposition of silt and infiltration of moisture. In addition, the surface of the hamada desert has been altered by the collection of stones into hundreds of thousands of mounds, probably in order to accelerate the rate of surface erosion. Otherwise the natural rate of 8mm per annum would have taken centuries to provide sufficient material to create fields behind the check dams.

It seems likely that the present desertified aspect of the Negev was created after this civilization collapsed, when incoming pastoralists overgrazed and removed the vegetation cover. Conventionally, the explanation for the collapse of agriculture in the Negev was the Arab invasion of AD 636, but recent work suggests that later climatic change may have been responsible. Research on oxygen isotopes in carbonate deposits in lakes and caves indicates a significant rise in temperature in the region between AD 700 and 900, correlated with a fall in the level of the Dead Sea due to reduced rainfall. This warm, dry phase was sufficient to tip the environmental balance against a finely tuned marginal agricultural system.

Conclusion

The relationships between population, the intensification of agricultural production, and environmental degradation are complex. We have seen that the available evidence has been interpreted differently by scholars and we are probably at too early a stage in our research yet to be able to create satisfactory causal models. Suffice to say at this stage that population density is an important variable in the evolution of landscapes and environmental change. In the next chapter we will see how civilizations may be constrained in their ultimate success by environmental factors.

Further reading and references

Ruthenberg (1980) is good on swidden agriculture. Grigg (1979) dissects Ester Boserup's in terms that are easy to grasp and the population/intensification debate is well covered in the various publications of Turner. Glover and Higham (1996) have the latest information on rice origins and Grigg (1974) has a scholarly chapter on rice agriculture.

Boserup, E. 1965: *The conditions of agricultural growth*. Chicago: Aldine.

Brookfield, H.C. 1972: Intensification and disintensification in Pacific agriculture: a theoretical approach. *Pacific Viewpoint* **13**, 30–48.

Brookfield, H.C. 1984: Intensification revisited. *Pacific Viewpoint* **25,** 15–44.

Brückner, H. 1986: Man's impact on the evolution of the physical environment in the Mediterranean region in historical times. *Geojournal* **13,** 7–17.

Carlstein, T. 1982: *Time resources, society and ecology. On the capacity for human interaction in space and time. Vol. 1: preindustrial societies.* Lund: Gleerup.

Geertz, C. 1963: *Agricultural involution: the processes of ecological change in Indonesia.* Berkeley: University of California Press.

Glover, I.C. and Higham, F.W. 1996: New evidence for early rice cultivation in south east and east Asia. In Harris, D.R. (ed.) *The origins and spread of agriculture and pastoralism in Eurasia.* London: UCL Press, 413–41.

Grigg, D.B. 1974: *The agricultural systems of the world: an evolutionary approach.* Cambridge: Cambridge University Press.

Grigg, D.B. 1979: Ester Boserup's theory of agrarian change: a critical review. *Progress in Human Geography* **3,** 64–84.

Hanks, L.M. 1972: *Rice and man: agricultural ecology in southeast Asia.* Chicago: Aldine-Atherton.

Issar, A.S. 1995: Climatic change and the history of the Middle East. *American Scientist* **83,** 350–55.

Netting, R.M. 1993: *Smallholders, householders: farm families and the ecology of intensive, sustainable agriculture.* Stanford: Stanford University Press [esp Ch. 9].

Ruthenberg, H. 1980: *Farming systems in the tropics.* 3rd edn, Oxford: Clarendon Press.

Sherratt, A. 1980: Water, soil and seasonality in early cereal cultivation. *World Archaeology* **11,** 313–29.

Spencer, J.E. 1974: Water control in terraced rice-field agriculture in southeastern Asia. In Downing, T.E. and Gibson, McG. (eds) *Irrigation's impact on society.* Tucson: University of Arizona Press, 59–65.

Trump, D.H. 1990: *Malta: an archaeological guide.* 2nd edn, Valetta: Progress Press.

Turner, B.L., Hanham, R.Q. and Portararo, A.V. 1977: Population pressure and agricultural intensity. *Annals of the Association of American Geographers* **67,** 384–96.

Vita Finzi, C. 1969: *The Mediterranean valleys.* Cambridge: Cambridge University Press.

5

ENVIRONMENTAL DEGRADATION AND THE COLLAPSE OF CIVILIZATIONS

> [Environmental] devastation, with all its consequences, becomes particularly intense among civilized peoples.
>
> (J. Brunhes 1952: *Human geography* London: Harrap)

INTRODUCTION

This chapter follows on from the discussion and the end of the previous chapter. There the environmental degradation of early Mediterranean civilizations was reviewed. Here we will argue that the human abuse of nature can lead to severe economic imbalance and possibly even to social and political collapse.

COLLAPSE

Archaeologists, historians and historical geographers are like the general public; they prefer to dwell upon success rather than failure. Studies are made of the rise and florescence of the great ancient civilizations, such as Egypt, Rome and Greece, with scholars devoting careers and research resources to the minute investigation of art, material culture, polity and economy. This work certainly has been important for understanding the foundations of our own modern civilizations which have been built upon the achievements of our distant forebears. It has also assisted in developing general principles of societal evolution, although these have been powerfully unidirectional in their explanatory structure. As a result, until relatively recently much Anglo-Saxon writing has been of the liberal and whiggish tendency, steeped in an optimistic, onward and upward outlook. Notably there has been less enthusiasm for research into failure, whether it be into the medium-scale societies which did not quite make it to international significance, or the fall of the major powers. Perhaps we do not like to be reminded of the fragility of advanced societies because there are implications for potential disintegration in our own age.

By collapse we mean a rapid loss of a major part of the common elements of civilization: social differentiation, central political control, and economic specialization. One theory is that civilizations are organisms which follow a natural life cycle of birth, maturity, old age and death. This is most clearly articulated in the writings of Toynbee and Spengler, but such organic analogies have been popular since the age of the Greek philosophers.

There is also a school of romantic writers on the subject of lost civilizations who feed our fascination with the mysterious and the exotic. We are all interested to know how civilizations could have prospered in the unpromising settings of the Iraqi desert or the jungles of central America.

Thus an industry has grown to satisfy the demand for tourism among ruined cities, and for books decoding the *true meaning* of Stonehenge or the Pyramids.

A third intellectual drift has been the delight taken by popular writers and some scholars of collapse in shopping at the supermarket of causalism. Simple, penetrating explanations are obviously preferable to the inconvenience of grasping complex reality, and therefore we are tempted by the list of causes which have been advanced to account for the historical puzzles such as the fall of the Roman empire. This week's special offer may be the lead poisoning induced by a piped water supply. Next week it may be an equally bizarre theory which has its own band of advocates. But explanations based upon single causal factors are both inadequate and misleading. The systems view, in which change is seen as the result of the interacting elements of a whole, is far more convincing because it can cope better with understanding the highly complex relationships which characterize all advanced civilizations.

According to Tainter (1988) it is this very complexity which may lie at the heart of the explanation of collapse. After reviewing a number of civilizations widely separated in time and space he concludes that, for pre-industrial economies at least, complexity has disadvantages beyond a certain point (*see* Figure 5.1). In fact the likelihood of a collapse is increased when to add further investments of capital and labour to bolster an existing power structure would bring a poor return. Any political or economic problems may then become more difficult to solve because cheap and accessible resources are not available by way of a solution.

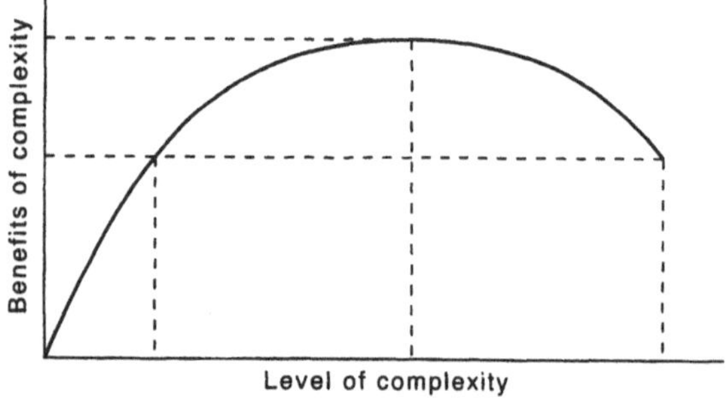

FIGURE 5.1 *The marginal return on increasing complexity*
Source: Tainter, J.A. 1988: *The collapse of complex societies*. Cambridge: Cambridge University Press

The classic example of this is the Roman empire. It expanded by a series of conquests, each of which brought a windfall for the treasury in terms of confiscations, slaves and taxes. Once all of the fattest milch cows were within the empire, however, the costs of administration became a burden which could be maintained only by the extraction of an excessive surplus from the peasantry. A debased currency and weakened agricultural base eventually took their toll and barbarian incursions became difficult to repulse.

COLLAPSE AND THE ENVIRONMENT

Having pointed to the danger of relying upon limited causalist reasoning, we are happy to acknowledge that in pre-industrial societies environmental considerations were exceptionally important because their agricultural economies were reliant upon the vagaries of the elements for stable food supplies. Particularly where population pressed upon resources, the gearing of agricultural technology and human muscle power to climatic, soil and topographical conditions was finely balanced and any disruption could provoke a crisis of subsistence, with political consequences. For many among the oppressed working population the eventual collapse may have actually seen an improvement in their standard of living. It is therefore not unreasonable to ask 'a collapse for whom?'.

Some economies survived for hundreds or thousands of years in balance with their environment. This was due to a number of factors, the most important of which is a keen appreciation, on the part of apparently primitive people, of the dangers of over-exploiting their resources.

The examples of societies which have damaged their environments are unfortunately more numerous. Making a direct connexion between such environmental degradation and the collapse of civilizations is dangerous because we may stray on to the ground of the environmental determinists who were convinced that human affairs were subordinate to grosser physical forces such as climate. Although the environment was crucial to the lives of pre-industrial peoples, we must be careful not to make it the principal god in the pantheon of explanation without very careful consideration.

Case study: the Maya

The Maya formed a long-lived and highly developed civilization in central America, from about 3000 BC to fourteenth century AD. They were the only native American civilization to have a fully literate culture and they knew mathematics and made astronomical observations. They calculated the year to be 365.2420 days long, which differs from the true figure of 365.2422 days only in the fourth decimal place. Their art and architecture were rich and elaborate, expressed characteristically through a monument cult. They carved their history and calendar in hieroglyphs on stelae and built monuments of a size and complexity rivalling any other New World group. Their mausoleums were of stepped pyramid shape (*see* Figure 5.2), made of limestone blocks, and were supplemented by temples, ritual ball courts, plazas and palaces.

There is a certain mystery about the Maya. Until relatively recently many aspects of their society were unclear. Even their material culture was little known because their abandoned cities were overgrown and inaccessible. Most mysterious of all was the sudden collapse of the civilization from the ninth century AD which, within 100–150 years saw a reduction in population by at least 70 per cent in the southern lowlands and the desertion and decay of many of their cities.

Figure 5.2 *Tikal's temple I, the funerary pyramid of Ruler A*
Source: The University Museum, Philadelphia

Greater clarity has come in the last few decades as archaeologists have investigated a wide range of sites, by digging, field walking, and using technologies such as remote sensing to map features on a regional scale.

Figure 5.3 shows the region occupied by the Maya. It covers western Honduras, Belize, Guatemala and four states (Yucatán, Quintana Roo, Campeche and Chiapas) in Mexico's Yucatán peninsula. We can distinguish between the mountainous areas in the south and the lower relief, mostly under 400 metres, to the north. These lowlands are comprised of two distinct zones: the flatter areas of the northern Yucatán and the rolling topography of the Petén. Our discussion will focus on the southern lowlands in Guatemala. It was here that population density was greatest.

This area has (and as far as we know also had in prehistory) a tropical climate with a lengthy dry season, ranging from luxurious forest vegetation in the south to xerophytic scrub in the north. Mean annual rainfall varies from 500 mm in the north west to 3000 mm in the south, but there are significant annual variations (Turner 1983). The northern part of the peninsula exhibits typical karst limestone scenery, with little surface water and occasional solutional depressions such as large poljes (locally called *bajos* when they contain seasonal wetlands) and smaller dolinas (here known as *aguadas*).

This environmental context is interesting. The lowland region is not especially attractive, with a lesser diversity of ecological niches than surrounding mountainous and coastal regions. This would have reduced the probability of local trade. The large majority (88 per cent) of the soils are of moderate or high fertility, particularly the brown/black mollisols (rendzinas), a much higher proportion than in most of the tropics, but a third (37 per cent) are classified as easily erodible. In Fedick and Ford's (1990) sample areas most (89 per cent) of Maya structures were found in association with these soils, avoiding the acidic and more difficult-to-work vertisols.

The most important period from our point of view is the Classic period, from roughly AD 250–930 (Table 5.1). The civilization had evolved into a number of independent city states, each with massive temples, palaces, paved roads, ball courts and public plazas, though often with regional variations on the cultural theme. Until recently these were thought to be merely ceremonial centres at which the surrounding population would occasionally assemble, but we have now found signs of relatively mature urban settlements

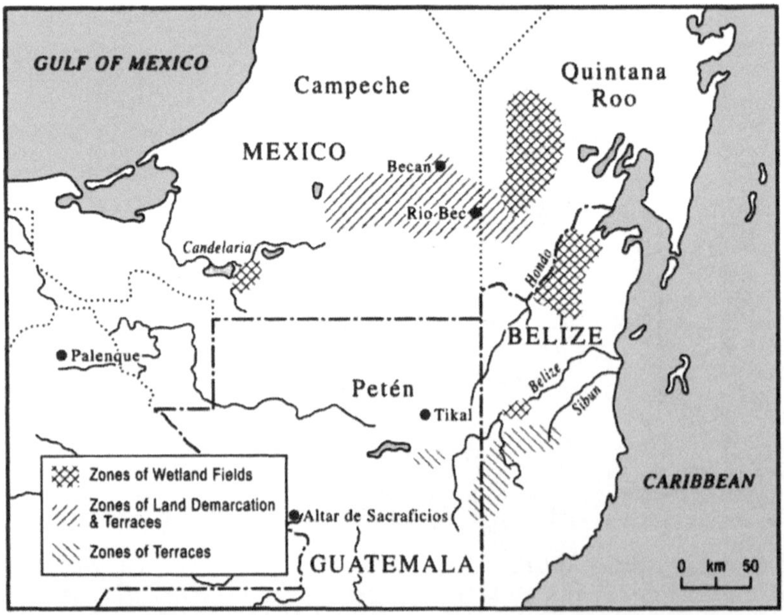

FIGURE 5.3 *The Maya region, showing selected sites and the main areas of wetland fields and terracing. The most densely populated area was probably the Petén region of northern Guatemala but there were significant settlements also in the Mexican states of Chiapas, Campeche, Quintana Roo, and in Belize, El Salvador and western Honduras*

TABLE 5.1 *Periods of Maya development*

Period	Date	Stage of development
Early Preclassic	1800–1000 BC	Early agriculture.
Middle Preclassic	1000–300 BC	Population growing, uniform pottery.
Late Preclassic	300 BC–AD 250	Writing and calendar, chiefdoms.
Early Classic	AD 250–600	Monuments, towns.
Hiatus	AD 534–593	Construction ceases.
Late Classic	AD 600–800	High peak of Maya civilization.
Age of expansion	AD 600–700	More sites erect monuments.
Age of interconnexion	AD 700–750	Greater contact between sites, spread of lunar cult. Acceleration of monument construction.
Incipient collapse	AD 750–790	Breakdown of alliances and decentralization of power. Fewer sites erect monuments.
Decay	AD 790–830	Cultural decline. Invasion of Seibal.
Terminal Classic	AD 830–925	Population collapse in south, construction ceases, calendar abandoned. Puuc sites flourish.
Postclassic	AD 930–1200	Maya civilization continues in north. Toltec invasion.

Source: after Coe (1993)

with permanent residents and extensive residential suburbs. The economy was sufficiently advanced to support a politico-religious hierarchy and a small army of non-agricultural specialist workers. There were flint and obsidian workers, potters, woodworkers, dentists, stoneworkers, monument carvers, textile weavers, human carriers, feather workers, leather workers, musicians, manuscript painters, merchants, basket makers, bark cloth makers, and many others.

The largest city, Tikal, at its peak, probably had an urban population of over 60 000, with a further 30 000 rural fringe dwellers. It covered 16 square kilometres and contained 3000 structures, from temple pyramids to thatched huts. The majority of the other people were rurally based, living in hamlets of 5–12 houses, with a minor centre for every 50–100 houses. The total lowland Maya population is difficult to estimate, but 8–10 million people at the apogee seems possible (Coe 1993).

The Maya food supply had originally been based on shifting agriculture. The bulk of their carbohydrates came from maize, and they also grew or collected beans, squash, avocado, cacao, and root crops such as malanga (*Xanthosoma violaceum*), and tree crops like sapodilla (*Achras zapota*), nance (*Byrsonima crassifolia*) and guava (*Psidium guajava*). Ramón (breadnut) trees (*Brosimum alicastrum*) were also plentiful and potentially an important source of protein. There is no evidence that they were cultivated, but they were a useful famine food, producing fruit in the driest of seasons. Altogether, 150 economic plants were either grown or collected, many of them encouraged by the management of vegetation in what amounted to an *artificial rainforest* (*see* Figure 5.4).

The analysis of carbon and nitrogen isotopes and mineral content in bones of a known age has allowed the reconstruction of the Maya diet (White and Schwarcz, 1989). Maize accounted for about 50 per cent of nutrition in the pre-Classic, falling to about one third in the Terminal Classic due to a diversification of the diet with other plant foods. This may have been due to success in broadening the agricultural base or the desperation of a malnourished population eating famine foods. After the collapse the dietary role of maize rose again to 70 per cent, suggesting a reversion to a narrowly based subsistence agricultural economy, possibly due to a decline in trade.

As population grew, an intensification of agriculture was necessary, as early as the Late Preclassic period in the central and northern Petén. Several possible adaptations have been suggested, all of which would have had landscape implications:

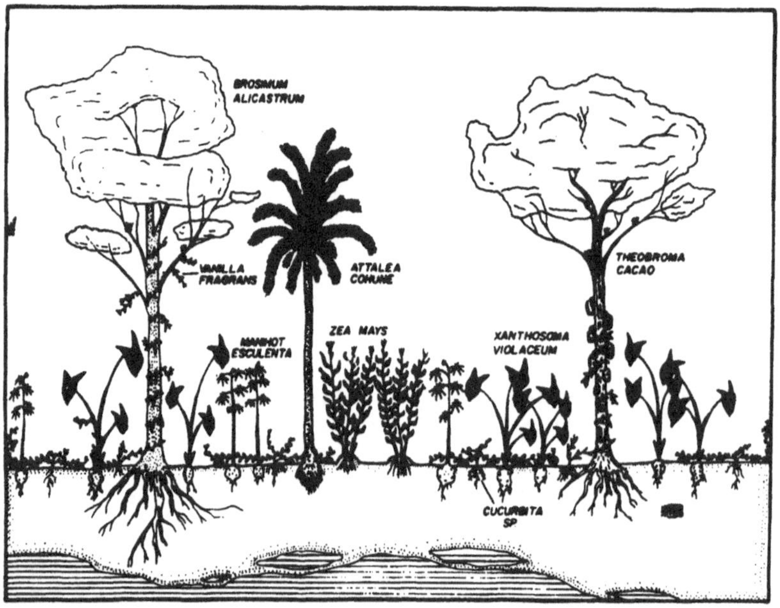

FIGURE 5.4 *The artificial rainforest created by the Maya Indians*
Source: Wiseman, F.M. 1978: Agricultural and historical ecology of the Maya lowlands. In Harrison, P.D. and Turner, B.L. (eds) *Pre-hispanic Maya agriculture*. Albuquerque: University of New Mexico Press, 63–116

1. home or kitchen gardens
2. wetland agriculture
3. hillside terracing.

Home or kitchen gardens (*solares*) could be cultivated intensively close to settlements because labour was at hand for weeding and mulching. Night soil and household waste would no doubt have been used as fertilizers. Maya urban structure left sufficient room for this sort of activity, with 0.16–0.20 ha available in central Tikal, for instance.

These *infields* produced fruits, nuts and vegetables, with the staple maize crop being cultivated in *outfields*, perhaps by less intensive swidden (milpa) cultivation. Ethnohistorical research shows that gardening has been common among the Maya in recent times although it is very difficult to know if this is the result of continuity from the prehistoric.

Wetland agriculture was an intensive adaptation, in seasonally inundated *bajos*, on lake margins, and in lowland river valleys. It was attractive because it offered the possibility of long-term cultivation and removed the tree-felling effort involved in slash and burn (Pohl *et al.* 1990). Flood plain soils would have been fertile and relatively weed-free in comparison the uplands. First, raised fields (*see* Figure 5.5) were constructed between a network of canals or ditches. Their surface was raised by the addition of material excavated from the water courses. Second, channelized fields are formed between long trenches sunk into the margins of wetlands to enhance drainage.

Turner and Harrison (1981) argue that wetland agriculture is likely to have started after upland milpa cultivation had proved insufficient as a food source. The effort of digging canals and moving earth would simply not have been worthwhile unless population pressure elsewhere encouraged a search for intensive alternatives. They also propose an evolutionary model in which the margins of wetlands would have first been used with (*marceño*) cultivation of rapidly maturing crops as the water receded during the

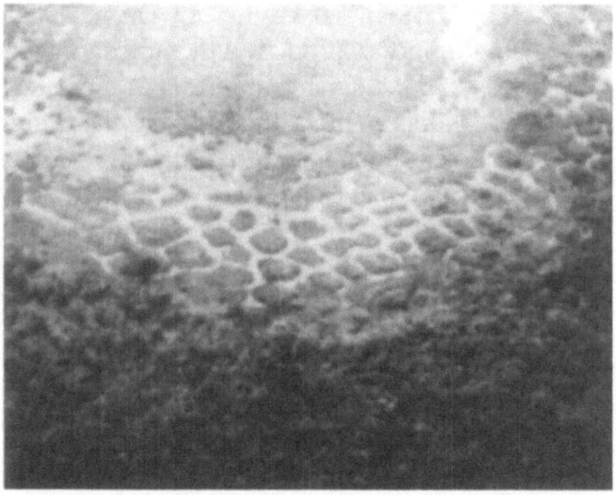

FIGURE 5.5 *Raised fields, Pulltrouser Swamp, Belize*
Source: Turner, B.L. and Harrison, P.D. 1981: Prehistoric raised field agriculture in the Maya lowlands. *Science* **213**, 399–405

dry season. Only subsequently would intensification have seemed justified. It is thought that channelized fields probably pre-dated raised fields.

Wetland agriculture is thought to have become intensive in the Late Pre-Classic. Farmers were then unwilling to write off the significant labour input and came to depend upon élite groups for the protection of that investment (Pohl 1990). As little as 0.5 ha was sufficient to support a family. The canals were used for fish farming and their silt for fertilizer. The high water table kept the root zone of the plants moist.

Satellite radar remote sensing has shown these areas of raised fields to have been extensive and there is a correlation between swamps and the large centres of population (Tainter 1988). Water availability must have been an issue because of the annual season of drought and due to a karstic landscape lacking surface water in about half of the lowlands. Recent work has indicated that the Maya were practising sophisticated water harvesting and storage methodologies. In the immediate vicinity of Tikal, for instance, 75 reservoirs have been found, including six central precinct tanks with a total capacity of 100 000–250 000 cubic metres.

The expansion of the cultivated area into the dry interior of the Petén was a risk and much of the Yucatán peninsula is limestone, with little surface water. The Maya built canals, dams, underground cisterns waterproofed with a plaster lining (*chultúnes*), wells, and they modified naturally occurring limestone swallow holes (*cenotes*) to provide a perennial supply. The water was used for irrigation, drinking, and for water transport.

Hillside terracing was constructed to prevent the erosion of thin soils, to minimize the leaching of soil nutrients and to control soil moisture (*see* Chapter 10). Level planting platforms for ease of cultivation and irrigation are not common, however, the terraced fields being typically dry and uneven.

Two types of terraces have been identified from detailed field work in the Río Bec region: linear sloping, dry field terraces and channel bottom check dams or silt traps (Turner 1983). The terrace walls are either broadbase or stone-slab embankments. These terraces are found on slopes varying from 4° to 47°.

Turner (1983) has calculated for the Río Bec area that a total of up to 16.5 million days of construction work would have been necessary to establish the terraces in that region. This means 400 people more or less full-time over a 400-year period. Clearly, a high level of commitment and planning was required, perhaps in response to population pressure and the dangers of environmental degradation. Towards the end of the Classic period it is conceivable that demographic instability and reduced work efficiency due to malnutrition may have made it difficult to maintain such a labour-intensive system of terraces.

Social differentiation and social control

In the early and mid-Classic period increased socio-political complexity was a solution to shortages for some, with the emergence of élite groups who had privileged access to resources. The outward sign of such differentiation was public architecture on a scale so massive that slave or corvée labour must have been involved. The transition from egalitarian to stratified societies may have been facilitated in areas of settled, wetland agriculture. Swidden farmers would have been more difficult to control and exploit.

Before about AD 400 urban centres seem to have been evenly spaced, with an approximate equivalence of status. Tikal seems to have emerged as the dominant central place with a subservient hierarchy of settlements on perhaps two levels. In the hiatus period of AD 534–593 there was a decentralization of the erection of stelae to the peripheries of the Petén, perhaps indicating a lessening of central power, and in the subsequent Late Classic a number of major centres were visible, each with their own secondary centres. Tikal was clearly no longer dominant.

The collapse

The degree of intensification was remarkable. At the peak of its powers, in about AD 800, the Mayan civilization supported the densest population of any pre-industrial society in history. In the Rosario Valley, De Montmollin (1989) estimates that there were 400 people per square kilometre, and a conservative calculation puts the average for the Petén at 200 per square kilometre (Culbert 1988). Compare this with the population densities of countries today: China 120, India 264, United Kingdom 231, United States 27. Actually, the Maya population had probably ceased to rise sometime earlier on, perhaps about AD 650, although the date would have varied from area to area.

There were limits. Short-fallow milpa could have supported at most 30–60 people per square kilometre, so the bulk of the population relied upon intensive, settled agriculture and interregional trade. By the Late Preclassic much of the central Petén had been deforested and population densities were such that competition for resources must have occurred. It cannot be coincidental that evidence of conflict (fortifications, carvings showing prisoners) is common at this time. Warfare meant the ebb and flow of territories and alliances, with the capture of high status individuals for sacrificial purposes as one objective.

There was some trade with the hill regions: pottery, cacao, honey and salt from the lowlands, in exchange for jade, obsidian, quetzal feathers, and granite (for grinding stones). Scope for short-distance trade in the lowlands was limited, however, because of the lack of topographic and ecological diversity. In times of subsistence crisis, such as during a prolonged drought, to raid neighbouring groups may have been the most logical short-term solution.

This militarization seems to have been entwined with population growth and concentration. It was in the defensive interests of each state to have its people in nucleated settlements rather than dispersed and therefore vulnerable. The archaeological evidence reveals several such periods of aggregation, which presumably must have been times when productive food resources were not fully exploited. In addition, we think that population increase may have been encouraged to guarantee sufficient numbers of able-bodied soldiers for deterrence and enough labourers for the construction of public works such as monuments. In turn, the architectural display was probably a means of indicating power and wealth to friend and foe alike in a seemingly endless competitive spiral.

This polity and economy were essentially non-sustainable. The necessary superstructure of authoritarian control was inherently fragile. The system never reached a stable state but kept on intensifying, with a larger and larger population, until eventually the system collapsed under an unsupportable weight.

The evidence of environmental stress is conclusive. Abrams and Rue (1988) have calculated that pine trees were fully cleared from the Copan area by AD 800. Pollen analysis confirms the absence of deciduous forest and the decline of montane pine forest, with a recovery of trees only after AD 1250. In turn this must have affected the habitats of game species such as the white-tailed deer, a valuable source of protein.

Immediately prior to the collapse it seems that social complexity and monumental architecture grew significantly, yet there were signs that all was not well. Population growth slowed and skeletal remains from this period often show evidence of disease as a result of specific nutritional deficiencies, such as shortages of vitamin C, which caused scurvy, and iron leading to anaemia. The Late Classic males were an average of 7 cm shorter than their Early Classic forebears, and may therefore have been undernourished when children or at least were short of protein. A calculation of the age of death at burial shows a reduced life expectancy in the Late Classic. Parasites were probably common and may have contributed to a reduction in work efficiency.

The collapse took place around AD 800–830. During the Terminal Classic population declined by two-thirds, at Tikal to as little as 1000, and shrank even further in the Postclassic (*see* Figure 5.6). Most of the elements of the civilized society were lost: administrative and residential structures, erection and refurbishment of temples (*see* Figure 5.7), tomb burials, stela construction, the manufacture of luxury items, writing and the maintenance of the unique and elaborate calendar (Tainter 1988). People were living in the remains of the former glory, throwing rubbish in previously restricted areas. They did not abandon their religious ceremonies but lost the specialized knowledge of the rules of stela erection.

The last inscribed date at Copan was AD 800 and at Tikal AD 869. The collapse was therefore not sudden but it was extensive and fundamental. A few cities, such as Altar de Sacrificios and Seibal, continued to grow, with dated monuments as late as AD 889, and in the northern Yucatán region of Puuc, Chichén Itzá, Mayapán and other cities flourished for a few centuries more, as did some sites in Belize with access to marine resources.

The precise reason for the collapse is unknown. Both Culbert (1973) and Tainter (1988) offer various single factor explanations for the collapse:

1. Resource depletion due to soil exhaustion and erosion.

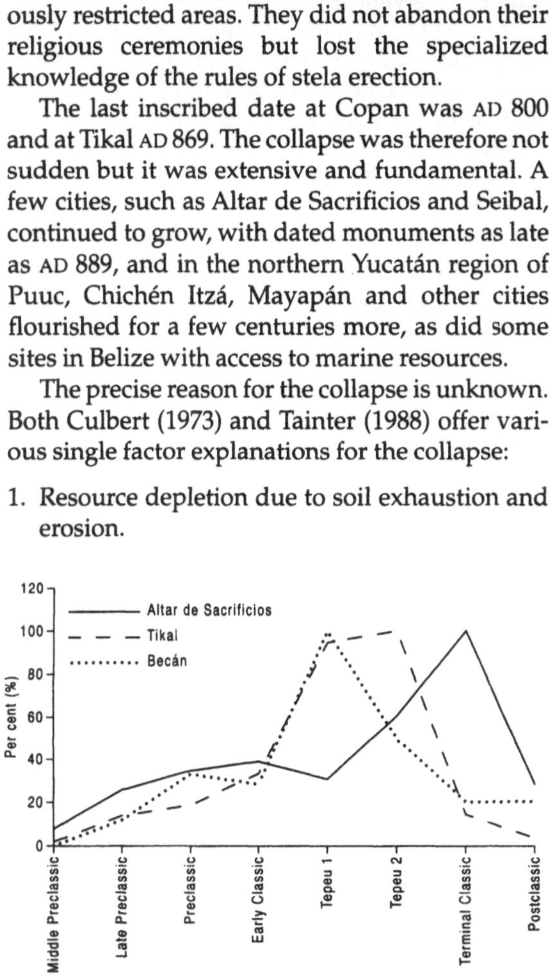

FIGURE 5.6 *Population change at selected Maya sites in relation to maximum population*
Source: Culbert, T.P. 1988: The collapse of Classic Maya civilization, in Yoffee, N. and Cowgill, G.L. (eds) *The collapse of ancient states and civilisations.* Tucson: University of Arizona Press, 69–101

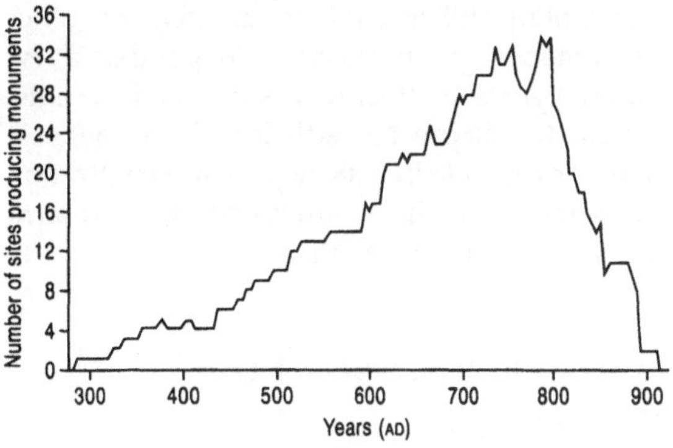

FIGURE 5.7 *The number of sites erecting monuments in the Maya southern lowlands*
Source: Lowe, J.W.G. 1985: *The dynamics of apocalypse: a systems simulation of the Classic Maya collapse.* Albuquerque: University of New Mexico Press

2. Fields choked by weeds.
3. Long-term climatic change.
4. Catastrophic events such as earthquakes or hurricanes.
5. Disease.
6. Social dysfunction leading to revolution against élite.
7. Intersite warfare, invasion from north or west.

Soil exhaustion may have been a contributory factor or perhaps soil erosion. There is evidence of the latter in the silting of lakes in the central Petén. A series of seasons when drought reduced crop yields might have tipped the balance or the introduction of maize mosaic virus. Alternatively the dominant priestly élite may have been overthrown in a revolution by their oppressed subjects or an invasion may have come from the north or the Gulf coast. Whatever the immediate cause, the collapse was undoubtedly exacerbated by the imbalance between people and the land, and the inherently slow natural regeneration.

CONCLUSION

We must take care not to ascribe monocausal significance to the environmental problem facing the Maya. It seems likely that unstable political structures, due to internecine strife within and between élite groups, were contributory, and De

Montmollin (1989) reminds us that no other Mesoamerican polity lasted for longer than a few hundred years. In other words they had a predisposition to collapse. Nevertheless, this is the most dramatic example in history of how a civilization that over-exploits its environment for short-term gain will eventually pay the price.

FURTHER READING AND REFERENCES

The mysterious Maya have attracted many scholars, especially American archaeologists. Their vast outpouring of data and interpretations are best summarized in Coe (1993). The collapse is dealt with by Culbert (1973, 1988) and Tainter (1988).

Abrams, E.M. and Rue, D.J. 1988: The causes and consequences of deforestation among the prehistoric Maya. *Human Ecology* **16,** 377–95.

Coe, M.D. 1993: *The Maya.* 5th edn, London: Thames & Hudson.

Culbert, T.P. 1973: The Maya downfall at Tikal. In Culbert, T.P. (ed.) *The classic Maya collapse.* Albuquerque: University of New Mexico Press, 63–92.

Culbert, T.P. 1988: The collapse of Classic Maya civilization. In Yoffee, N. and Cowgill, G.L. (eds) *The collapse of ancient states and civilizations.* Tucson: University of Arizona Press, 69–101.

De Montmollin, O. 1989: *The archaeology of political structure: settlement analysis in a Classic Maya polity.* Cambridge: Cambridge University Press.

Driever, S.L. and Hoy, D.R. 1984: Vegetation productivity and the potential population of the Classic Maya. *Singapore Journal of Tropical Geography* **5,** 140–53.

Fedick, S.L. and Ford, A. 1990: The prehistoric agricultural landscape of the central Maya lowlands: an examination of local variability in a regional context. *World Archaeology* **22,** 18–33.

Pohl, M.D. (ed.) 1990: *Ancient Maya wetland agriculture: excavations on Albion Island, northern Belize.* Boulder: Westview.

Pohl, M.D., Bloom, P.R. and Pope, K.O. 1990: Interpretation of wetland farming in northern Belize: excavations at San Antonio Rio Hondo. In Pohl, M.D. (ed.) *Ancient Maya wetland agriculture: excavations on Albion Island, northern Belize.* Boulder, CO: Westview, 187–254.

Tainter, J.A. 1988: *The collapse of complex societies.* Cambridge: Cambridge University Press.

Turner, B.L. 1983: *Once beneath the forest: prehistoric terracing in the Río Bec region of the Maya lowlands.* Boulder, CO: Westview.

Turner, B.L. and Harrison, P.D. 1981: Prehistoric raised field agriculture in the Maya lowlands. *Science* **213,** 399–405.

White, C.D. and Schwarcz 1989: Ancient Maya diet: as inferred from isotopic and elemental analysis of human bone, *Journal of Archaeological Science* **16,** 451–74.

6

SUSTAINABLE RESOURCE MANAGEMENT IN PRE-INDUSTRIAL SOCIETIES

It may not be too unfair to suggest that previous models of the development process have tended to assume that the 'future will look after itself', whereas the sustainable development approach acknowledges that the ability of the future to do this can be seriously impaired by actions taken now.
(D. Pearce, E. Barbier and A. Markandya 1990: *Sustainable development*. London: Earthscan)

INTRODUCTION

A major reason for the collapse of the various civilizations discussed in Chapter 5 was that their economies (i.e. their agricultural systems) were not *sustainable* in the long term. That is to say, they were unable to maintain a stable production of food and other necessities, for reasons connected with environmental change and/or social and political organization. Many writers tell us today that even our own high-tech age is susceptible to problems of sustainability, where we are drawing down our non-renewable resource base and degrading our surroundings by pollution, deforestation and desertification. It will therefore pay us in this section to consider the concept of sustainability and to look at some examples of historical economic systems which have important lessons for us today.

SUSTAINABILITY

Sustainability is a fashionable word that has crept into mediaspeak and become somewhat devalued as a result. Let us be clear what we mean by it. According to Repetto (1986, 15):

> Sustainable development is a strategy that manages all assets, natural resources, and human resources, as well as financial and physical assets, for increasing long-term wealth and well-being. Sustainable development as a goal rejects policies and practices that support current living standards by depleting the productive base, including natural resources, and that leave future generations with poorer prospects and greater risks.

Rather than the rapacious and greedy over-use of resources so that they are degraded and destroyed, sustainable development seeks long-term development and is willing to sacrifice short-term profit. This may be illustrated by considering the properties of agro-ecosystems (ecological systems modified by humans to produce food and fibre) (Table 6.1).

Some pre-industrial economic systems have been based on the management and use of resources held in common. If properly regulated, such systems can be sustainable, but the lesson of history generally is not encouraging. The interests

TABLE 6.1 *Agro-ecosystem properties*

Property	Definition
Productivity	Yield per unit of resource input.
Stability	The degree to which productivity is constant in the face of small disturbances caused by normal fluctuations of climate, market prices, etc.
Sustainability	A system's ability to maintain productivity in the face of major disturbances, e.g. soil erosion, desertification, pest attack, farmer indebtedness, earthquake, etc.
Equitability	The evenness of the distribution of the benefits of production among the people.

Source: after Conway (1987)

of the individual and society as a whole may not coincide, and commons may be over-exploited when the goal of private accumulation and consumption overrides notions of the collective good. Thus we continue to pollute the atmosphere (a resource common to us all) with car fumes and emissions from power stations even though we are well aware of the health risks and the dangers of global warming.

The degradation and depletion of common resources have been called the 'tragedy of the commons'. Because appealing to morals or environmental ethics has little impact on individual behaviour, one of four approaches must be considered (Johnston 1994). First, the description and replication of good practice from around the world, with a view to sponsoring low-tech, labour-intensive rather than resource-intensive agricultural and industrial techniques. Second, collective agreements may be entered into by all current resource users in order to establish a set of management rules and to limit the number of newcomers. Third, the common resources may be privatized, because owners are thought to protect their own property assiduously. Fourth, an outside body, such as an organ of the state, may regulate by the use of legal sanctions. The examples we will describe here come from the first two approaches.

CASE STUDY: THE MEDIEVAL MANOR AS A SUSTAINABLE SYSTEM

Historians have been able to recover a surprising amount of detail about the nature of the European medieval economy from documents such as the manorial rolls. These were compiled by the manorial court of each landed estate in the course of regulating the use of resources and monitoring the interests of the Lord of the Manor in relation to those of his feudal tenants. The parchment rolls enable us to follow the nature of land tenure and the land market, cropping decisions, and sometimes facts relating to the marketing of commodities surplus to local subsistence needs. According to Pretty (1990) we can also reconstruct a picture of a farming system which exhibits certain interesting features of sustainability.

The manor was the fundamental building block of power in the medieval countryside. Large manors controlled several villages and drew upon the varied resources of the surrounding countryside to fill the Lord's table. All available resources were valued, from arable, pasture and meadow, to wood and waste. Many were held in common between the villagers and these represented an equitable form of resource management. To give an example, a key aspect of the minimization of the risk of poor crops or harvest failure was the scattering of strips in the open fields (see Chapter 11). There was a greater cost in time travelling between scattered strips but a reduced risk of complete crop failure.

According to Neeson (1993) and Pretty (1990), Common resources from the waste were as follows:

- fuel: turf, furze (gorse), bracken (fern), peat, wood;
- construction timber;
- roofing material: turf;

- ash: used in soap-making, glass-making, bleaching;
- reeds: thatch;
- rushes: bedding, woven goods, rush lights;
- nettles: linen;
- holly, thorns: threshing flails;
- sand: abrasive for scouring pots and pans, floor covering in cottages;
- collected food: nuts, herbs (some medicinal), berries;
- pasture: for geese, cows, sheep, pigs;
- fodder: grass, hay, holly;
- hunting: wild boar, deer, rabbits, hares, birds, fish.

Apart from the carrot of sharing out the available land, there was also a stick of regulations ensuring that no individual took more than was due. These were the bye-laws which were enforced by the manorial court and its officers in order to prevent long-term damage to village resources (Table 6.2).

The feudal system mobilized a convenient source of labour for the Lord through the compulsory labour duties and obligations of each peasant. They were required to work for a certain number of days each year on the demesne, the Lord's own farm. The serf may have had few legal rights, but he was able to farm and was certainly better off than a landless freeman. In times of hardship the Lord helped his tenants. The peasants helped each other through the sharing of plough teams.

The agricultural productivity of the system was poor by modern standards but production was sacrificed in order to encourage stability, sustainability and equity. Yields on the Bishop of Winchester's manors 1283–1349, for instance, were very low (Table 6.3) but lower productivity

TABLE 6.2 *The management of village resources*

Activity	Form of management
Hunting, snaring, gathering, collecting	Licences
Pigs	Nose rings to prevent deep rooting Fines for owners of destructive pigs Pannage season limited to protect saplings Elected swineherd responsible
Cattle, sheep	Stinting (numbers limited)
Trees	Regulation of cutting and sale Villagers to carry own wood only Fines for possession of unlicensed wood cutting tools Lopping of oak, beech, apple forbidden Replacement trees to be planted
Hedges	Regular repairs
Fencing and gates	Around gardens to prevent livestock damage
Reeds and rushes	Cutting controlled Gathering for own use only
Manures	Not for sale off manor No removal from meadows
Fishing	During daylight only
Watercourses	Regularly cleaned Pollution prohibited by human waste, animal offal, and hemp or flax residues

Source: after Pretty (1990, 15)

TABLE 6.3 *Harvest ratios and stability of production on the Bishop of Winchester's manors, 1283–1349*

	Harvest ratio (seeds/seed sown)	Coefficient of variation (%)
Wheat	4.0	37
Oats	2.3	34
Barley	3.5	37

Source: after Pretty (1990)

went with greater stability. The manors with the highest yields also had the greatest variability (as expressed by the coefficient of variation) from year to year.

Even apparently unpromising environments were brought into productive and sustained use by careful management in the Middle Ages. Marshy areas, for instance, which we now value for their ecological richness but little else, were fully exploited (Figures 6.1 and 6.2). There were cattle and sheep pastures; fishing, for instance for eels; salt making; fowl such as ducks and geese; reeds, rushes and sedges for thatching; willows for basket making; and peat for burning. These resources were used in an organized fashion, quite often being shared between neighbouring parishes by a pattern of agreed intercommoning (Figure 6.3). The Fens sustained some of the wealthiest settlements listed in the Domesday Book (1086).

BOX 6.1 *The best aspects of the medieval resource system*

1. The use of a very wide range of resources: cereals and other field crops, livestock, fowl, timber, stone, peat, turf, rabbits, fish.
2. Mixtures of grains sown in the same field to smother weeds and to ensure against the failure of any one species. Examples are maslin (wheat + rye) and dredge (barley + oats).
3. Mixed agriculture, which maintains productivity by swapping animal manure, then applied to the soil to improve its fertility, with waste arable products such as straw for fodder.
4. Attempts to enhance agricultural productivity by the application of chalk/lime marl, burnt turfs, human waste, crushed shells, and seaweed.
5. Oxen were kept because they needed less fodder than horses. Oxen can survive on straw and hay. Horses need 6-20 times more oats, but could work faster and lived longer.

A problem was that, as for all organic agriculture, there was damage from pests (insects, birds, wolf attack on livestock) and diseases (mildew on wheat). Cooter (1978) has also shown that open field arable cultivation would have led to a gradual draining of nitrogen and other nutrient reserves in the soil. Nevertheless this system of

FIGURE 6.1 *Very little of this original fen landscape now survives. The closest approximation which is accessible to the public is Wicken Fen, Cambridgeshire*
Source: Skertchly, S.B.J. 1877: *The geology of the Fenland*. London: Geological Survey

FIGURE 6.2 *Fen slodgers in the early nineteenth century. These fenmen made their living from the sustainable resources which were still available.*
Source: Thompson, P. 1856: *History and antiquities of Boston*. Boston: Author

medieval farming lasted several hundred years, until the fourteenth century when a series of stresses and shocks forced change (Chapter 11). There was population growth before 1300, to the point where further expansion of the arable had to be on marginal land. Average yields were even lower here and the risks correspondingly higher. Between 1315 and 1321 there were a series of severe famines due to extreme weather conditions of wet and dry years, and the Black Death (bubonic plague) in 1347–9 attacked an already weakened population.

CASE STUDY: CONTEMPORARY SUSTAINABLE PEASANT FARMING

For much of the twentieth century the experience and wisdom of peasant farmers have been ignored. They have been regarded negatively, as ignorant and irrational decision-makers, or patronizingly as constrained actors who need help with advice on the latest in hi-tech innovations. It is thought that if only they would listen to extension workers and marketing experts, peasants could become successful capitalist farmers.

What was forgotten is that peasants are not only extraordinarily resilient under economic pressure from the wider world, but that they also have access to a deep fund of agricultural lore, much of it accreted gradually by trial and error over centuries. They have a good idea what will work in their own micro-environmental conditions and they are rightly wary of trying anything new that will involve a risk of crop failure and therefore a loss of livelihood for their families. That is not to say that they are not willing to experiment, as discovered by Richards (1985) among the rice farmers of West Africa. The use of existing indigenous knowledge is vital if we are to build a more productive and sustainable agro-ecosystem in poor countries.

Researchers are just beginning to unearth the varied range of landscape adaptations which have been developed by pre-industrial farmers around the world. By pre-industrial we mean those peasant small-holders who have yet to join a wholly market-orientated agriculture, with high levels of industrial inputs such as chemical fertilizers, pesticides, and heavy machinery. Most do not have tractors for shaping the land, but still work with simple ploughs or even hoes and digging sticks.

Work by Wilken (1972, 1987) and the contributors to Klee's (1980) book are exemplars of this type of research (Box 6.2), and there is good reason to think that many of the low-tech solutions that they describe have scope for duplication in other parts of the world. A south–south transfer of such good ideas could help to boost production and reduce environmental degradation, without the involvement of Western aid donors and the sales(wo)men of multinational seed companies.

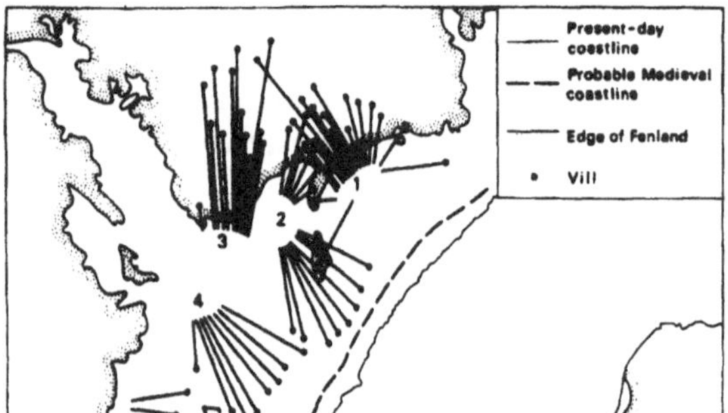

FIGURE 6.3 *The main clusters of intercommoning in the Fens. Parishes shared the available resources in an organized fashion*
Source: Darby, H.C. 1983: *The changing fenland*. Cambridge: Cambridge University Press

A CASE STUDY OF BRITAIN: SUSTAINABLE WOODLAND MANAGEMENT

Although Britain was denuded of much of its wild wood from an early date (Chapter 8), nevertheless scraps survived. This was not an accident because pre-industrial society required wood in considerable quantities for a wide range of purposes, many now all but forgotten to us. Every region had some resources of trees, which were only supplemented by large-scale imports during the industrial era of the last 200 years. The management of these woods was highly skilled and sustained over long periods of exploitation.

Woodland with agriculture

The first type of woodland management was its use in association with agriculture. In the Middle Ages one function was for *pannage*, or the pasture mentioned below (Chapter 8) in the context of the Domesday Book. The Anglo-Saxon place name *denn* meaning woodland pasture, usually for pigs, indicates that this practice has an ancient pedigree. Pannage was a feudal right of pasture in wooded common land or on the stubble of harvested fields common fields, exercised roughly from the autumn equinox in late September, when the acorns and beech nuts were ripe, to Martinmas in mid-November, when the fat animals were killed and salted down for the winter. In return for this right the Lord of the Manor received a tithe (one-tenth) of the produce. Pigs are prolific breeders and are ideal for this system of fattening and slaughter. They were kept only for their meat, unlike other livestock whose hides, milk and wool were also valuable.

Cattle pasture in grassy glades, again with ancient origins, as the Anglo-Saxon place names *steall* (stall) and *stoc* (stock) attest, was also common, although oak forests were not appropriate because of the poisonous nature of acorns for bovines. Such wooded commons are a kind of paradox because, on the one hand, dense tree canopies may shade out the grass which the animals thrived on while, on the other, the cattle may eat enough foliage and even bark to damage young trees. Solutions include *pollarded* trees and *coppices* surrounded by high banks or fences.

The New Forest in Hampshire and Epping Forest in Essex are surviving examples of wooded commons, but *wood-pasture* and *wood-meadow* can still be seen throughout western Europe on private land (Peterken 1993). It now takes the form of permanent pasture edged by hedges and trees in a landscape that was neatly described by William Marshall in 1787 for Norfolk: 'The eye seems ever on the verge of a forest, which is, as it were by enchantment, continually changing into enclosures and hedgerows' Darby (1973, 326).

Box 6.2 *Traditional agricultural resource management techniques*

Soil amendments, organic
animal manure, guano
night soil
ant-nest refuse
seaweed, shells, fish offal
peat
green manure
ash
organic mulch
nitrogen fixation with leguminous plants/trees

Soil amendments, inorganic
silting from uncontrolled/controlled runoff
silting fields (pantles)
muck from ditches, canals
sand
lime

Other soil management
fallowing
crop rotation
livestock stinting
ploughing
harrowing
clod-breaking
puddling
hoeing
drainage
rock removal
soil transport

Vegetation management
fire
shifting cultivation
composting

Slope management
check dams
sloping terraces
flat terraces, including tablones
pit terraces (cepas)

Field-surface management
raised beds (camellones)
lazy beds, ridges
maize mounds, manioc mounds
tie-ridging
asymmetrical ridging for maximum insolation

Water management with intermittent sources
rainwater harvesting
runoff management
flood/basin irrigation

Water management of surface systems
pot irrigation
scoop irrigation
canal/furrow irrigation
tank irrigation

Water management of sub-surface systems
well irrigation
qanats
sub-irrigated/drained fields, including chinampas
sunken fields

Pest and disease management
fences/hedges
bird scaring
weeding
control of insect habitats
biological control of insects

Climate management
shade trees to reduce heat of sun
shelter belts against wind
mulching to preserve moisture and prevent frost
shading arbours
tillage to conserve sub-surface moisture and heat
hotbeds

Space management
field layout
plant spacing
intercropping
vertical architecture
kitchen gardens
crop scheduling, including multiple cropping and phased planting
seed/nursery beds

Supplementary foods
hunting
fishing
collecting

Sources: Klee (1980), Wilken (1987)

Rackham's ancient landscape (Chapter 7) has many examples of this enclosed version of wood-pasture, called by the French *bocage*. It is characterized by small, irregular fields, accessible only by narrow, winding, often sunken lanes (*see* Figure 6.4). The hedge banks have a core of rubble removed from the field surface and are rarely as ecologically simple as the hawthorn boundaries of the Midlands. In fact, it is possible in some cases to demonstrate from the surviving variety of flora that the hedges are in fact linear remnants of the primeval wild woods and therefore represent a precious reserve of biodiversity. There are certain species which mark out ancient woodlands and hedges: lime, service, woodland hawthorn, herb Paris, anemone, and wood sorrel. Recently planted woods have plants rarely found among ancient flora: ivy, cow parsley, hedge garlic, and lords and ladies (Peterken 1993).

Woodland for pleasure

Second among the historic forms of woodland management is its use for private pleasure. In early times this equated mainly with hunting and many royal and baronial estates were preserved for this purpose. William I institutionalized the crown's existing hunting reserves as *Forests* which were legally set aside for his exclusive use. Anyone taking the king's deer or wild boar was subject to severe punishment. These Forests were not necessarily wooded, some being open moorland or heath, and this causes some confusion for modern travellers who assume that *Forest* on the map will mean trees.

FIGURE 6.4 *Sunken lanes and ancient banks are characteristic features of the 'ancient' countryside. This example is from Devon/Dorset*
Source: P.J. Atkins

Fallow deer were imported to stock these hunting Forests, being easier to keep than the native red deer. In AD 1200 there were 143 Forests occupying a fifth of the country, after which they declined due to piecemeal disafforestation. Following the example of the crown, regional magnates created their own chases and deer parks as status symbols. In AD 1300 there were about 3200, averaging 80 ha each, but they gradually waned in popularity.

When revived in the eighteenth century, parks were altogether different in design and intent. There were carefully landscaped as works of art by professionals such as 'Capability' Brown. These will be described in more detail in Chapter 18.

Wood for fuel and timber

The taking of wood was a third, and economically the most important, use, employing timber of various species and thicknesses according to the purpose. Mature oak trees were managed and felled for ships and for the framework of houses and barns, but most trees were cut before they reached their full size in order to avoid the cost of sawing. A typical English sixteenth-century farmhouse, for instance, was made of 330 oak trees, but half of the timbers were less than 23 cm in diameter, against an average today of nearer 50 cm.

The provision of fuel was the most important use of non-pannage woodlands, for domestic uses and increasingly during the medieval and early modern period for industry. Charcoal was widely used in rural industries such as iron and salt-making and was itself produced by skilled woodland craftsmen. Coal, the obvious alternative in those parts of the country fortunate enough to have reserves, was far too bulky and expensive to transport to other areas before the transport revolution ushered in by the railways. The exception to this was the large-scale coastwise trade in household coals from Tyneside to London.

Fortunately most species of native British trees are able to regrow when cut. So long as the stump or stool is left, regeneration is only a matter of time and successive harvests of regrowths can be taken for hundreds of years. Figure 6.5 shows three means of harvesting which rely upon this marvel of nature:

- Suckering. Here the stump may die but the roots send up new shoots. Species: blackthorn, wild cherry, smooth-leaved elm, poplar, service.
- Coppicing. The tree is cut to the ground every 20 years or so and from the woody base or stool grow conveniently sized branches for use as rods and poles. Species: alder, crab apple, ash, beech, birch, wych elm, hawthorn, hazel, holly, hornbeam, lime, maple, oak, rowan, sallow, service, whitebeam, sweet chestnut, sycamore. Commercially coppiced woods are divided into compartments or *fells* and then cropped in rotation, one fell each year while the others recovered.
- Pollarding. The tree is cut at a height of 2–5 metres above ground level to prevent the browsing of young shoots by cattle or deer. The trunk then sends out regrowths as for coppiced trees. Species: ash, beech, wych elm, smooth-leaved elm, English elm, holly, hornbeam, maple, oak, black poplar, willow.

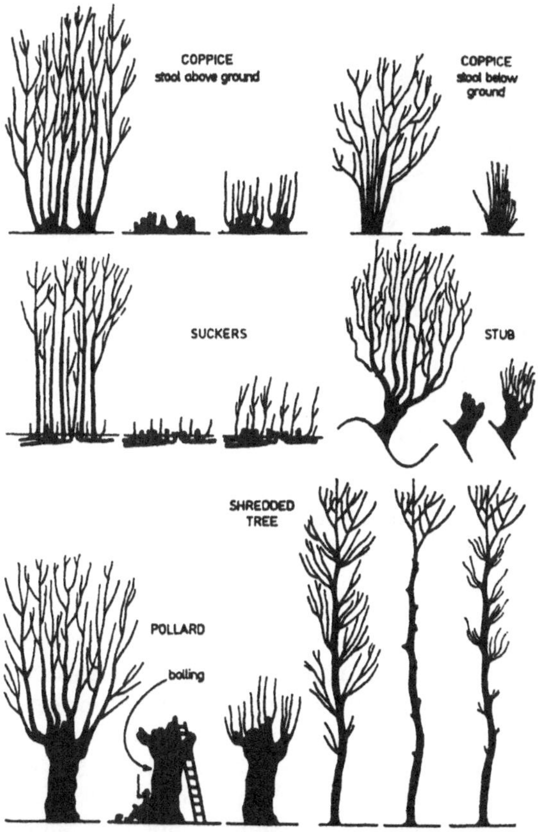

FIGURE 6.5 *Wood production from managed trees*
Source: Rackham, O. 1976: *Trees and woodland in the British landscape*. London: Dent

All too few managed woodlands survive, but a few are very old indeed. According to tree ring evidence, some pollarded trees have yielded crops for over 500 years. Among these are the oaks of Windsor Great Park (Berkshire), and the beeches of Burnham Beeches (Buckinghamshire) and Epping Forest (Essex) (Figure 6.6). Coppice stools can survive for a thousand years or more. Coppicing as a management practice probably originated in the Bronze Age. Excavations of the Sweet Track in the Somerset Levels have shown that a footpath was first laid in the marsh in about 4000 BC and maintained by sophisticated woodsmanship. Planks of oak, ash and lime were used, along with standard-sized rods, poles and stakes, cleverly interwoven to give strength and resistance to floods. There is clear evidence of coppice products in the later Walton Heath and Rowland's Tracks (3100 BC), all cut with stone tools.

The management of woods eventually attracted statutory support. The Laws of Ine passed in seventh-century Wessex, for instance, decreed that 'if anyone destroys a tree in a wood by fire, and it becomes known who did it, he shall pay a full fine. He shall pay 60 shillings, because fire is a thief' (Hooke 1985, 155).

Fears about shortages of timber later became such that in 1543 the Statute of Woods attempted its preservation. Young trees had to be left when woodland was cut and the area fenced against animals browsing the regrowth, and there were fines for grubbing out coppice or underwood for the extension of agriculture. In 1558 another Act sought to prevent the taking of large trees suitable for naval uses within 20km of navigable water and the Acts of 1581 and 1585 controlled the use of timber as fuel in iron-making. Such legislation probably had little direct influence on woodland management but the fears of diminishing resources that they represented did affect the

FIGURE 6.6 *Coppice woodland, Bradfield Woods, Felsham, Suffolk*
Source: Committee for Aerial Photography, University of Cambridge

intellectual climate of the day and encouraged those who wished for the more careful husbanding of those resources and even the planting of trees.

John Evelyn's *Sylva*, published in 1664 at the behest of the newly founded Royal Society and the first major book in English about woodland management, was not an isolated tract but represented a general concern of the day about resource depletion. The main era of demand for naval shipping, however, came in the late eighteenth and first half of the nineteenth centuries when oak was still the main species for ships' timbers, supplemented by imported masts and spars from the Baltic. This was also a time of demand for oak bark, which was used in the tanning of leather. Rackham concludes that domestic supplies were sufficient for the dockyards and tanneries until about 1815, after which oak was imported, and he pours cold water on the idea that there was ever a serious crisis of wood shortage. More likely the economics of woodland exploitation were difficult when accessible woods were under pressure because of the high cost of hauling felled trees to navigable rivers or canals.

The eighteenth century saw a reassessment of woodland exploitation. In the age of the Enlightenment many landowners sought to rationalize their estates and were open to new ideas. In Switzerland, Germany and France notions of regulated and sustained yields, and density control, which had been in operation in Italian states such as Venice and Florence since at least the fourteenth century, were adopted and further developed in response to local crises of timber shortage. Here were the early stages of scientific forestry. Strategic considerations were sufficient for naval timber to become a product as prized as oil is for our age. A single ship of the line required 700 mature oak trees or 4000 cubic metres of timber.

From 1758 the Royal Society of Arts gave prizes for tree planting and the age of afforestation was well under way by the 1790s judging by the county reports commissioned by the Board of Agriculture. Some would have been for aesthetic enhancement of landscaped parks but even here tree management for profit was common. Half a million hectares were planted in Scotland alone between 1750 and 1850. Thus by the mid-nineteenth century timber trees and coppice were gaining ground, often with scientifically planned regimes of silviculture, but this never matured into a national policy as in France. Quite simply, the need for large-scale forest planning was not pressing. The opening up of the British empire yielded an endless source of raw material, supplemented by softwood from Scandinavia and, on the demand side, wood was overtaken by iron for battleships (from the 1860s) and coal as a fuel. The age of the improver saw woodland removal as farmers grubbed them out to enlarge field sizes, and on the urban fringe woods made way for housing.

In 1990 there were 1 179 981 ha of woodland in England and Wales, of which 397 702 ha were ancient woodland (origins pre-1600) (Figures 6.7 and 6.8) and 236 650 ha of semi-natural woodland. Much of the managed ancient woodland that has survived is in the south of England, but even here little coppicing took place after 1930. Less than 50 000 ha remains under traditional management, although there has been a recent revival. About 8 per cent of ancient woodland was cleared between 1930 and 1990, mainly for agricultural land, but the problem is worse in some areas than others (South Yorkshire 15 per cent, 1 per cent in County Durham). Of the ancient, semi-natural woodland about half been lost since the 1930s (7 per cent cleared, 39 per cent replanted) and most of the rest has been neglected. The situation has not been helped by the loss of 20 million trees to Dutch Elm Disease in Britain since 1967. This has thinned the tree content of hedges in the south of England.

Between 1850 and 1913 there were six enquiries into British forestry by Parliament, but with little practical result. It took the shock of the First World War to puncture the prevailing complacency and the Forestry Act of 1919 established the Forestry Commission with a view to acquiring land and establishing fast-growing softwood plantations. Thus arose the 1.5 million hectares of exotic coniferous landscapes of spruce, pine, fir and larch that dominate in the Breckland, Wark Forest in Northumberland, and in parts of Scotland and Wales (*see* Figure 6.9). The ancient reverence of native oaks, dating back to the sacred groves of the druids and reinforced by the status invested in owners of oaks since the sixteenth century, was not easily overturned and a battle royal has been joined between those who profit from softwoods and those who claim that they are both aesthetically unpleasing and ecologically damaging.

FIGURE 6.7 *Coppice woodland in 1905. Each black circle is ten times the area of coppice wood in the county at the scale of the map*
Source: Rackham, O. 1976: *Trees and woodland in the British landscape*. London: Dent

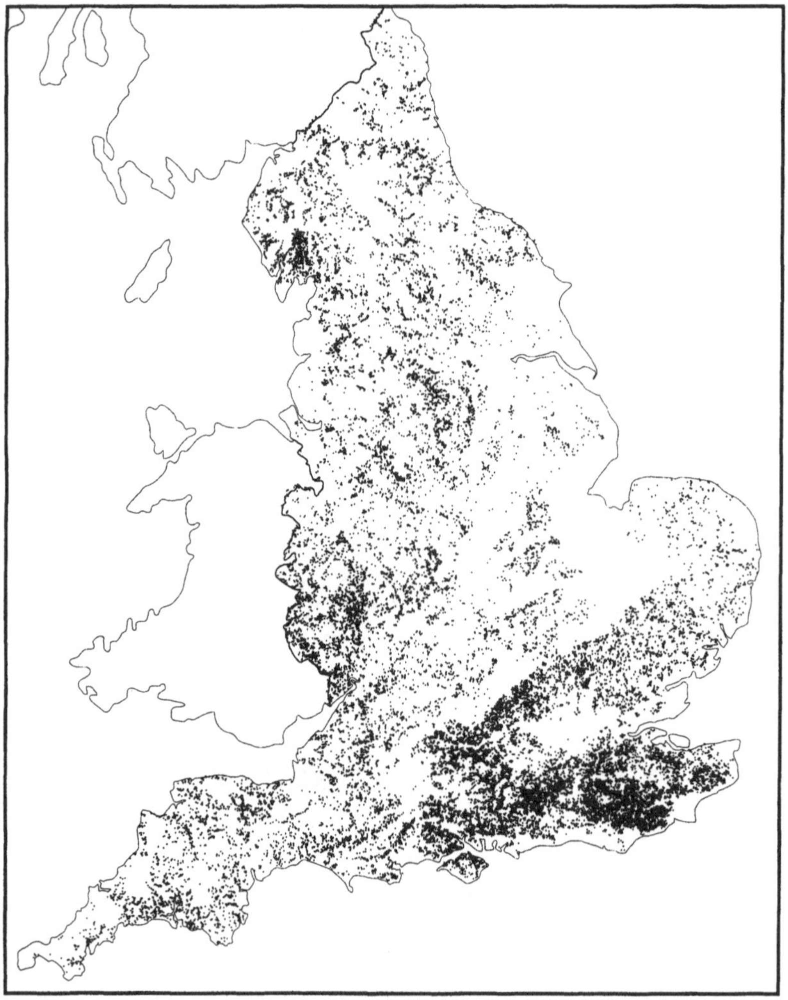

FIGURE 6.8 *Surviving ancient woodland in England*
Source: English Nature

In 1996 plans were announced by the Countryside Commission and Forestry Commission to increase the woodland cover from 10 to 15 per cent, including the establishment of a National Forest in the English Midlands (Figure 6.10). This would help reduce the 85 per cent of timber needs which are currently imported.

Of course trees represent far more than just an economic resource. They are of great significance in many pagan religions and Stephen Daniels (1988) has shown that they have long represented power and status in western societies. The slow-growing and long-lived oak was a symbol of strength and security in Britain, appealing particularly to those wedded to the ancient order of aristocratic domination. Trees have also played a part in the *picturesque* view of landscape which has been important since the eighteenth century (Chapter 18).

CONCLUSION

The definition of sustainable economic systems is far from straightforward and our case studies in this chapter have shown that sustainability rarely lasts for ever. There are internal and external pressures upon even the most skilfully balanced use of resources. We might reasonably conclude that some mechanism of planning is needed at the wider community or state level to ensure the sensible use of resources. It is to the notion of

FIGURE 6.9 *Kielder Forest 1967*
Source: Committee for Aerial Photography, University of Cambridge

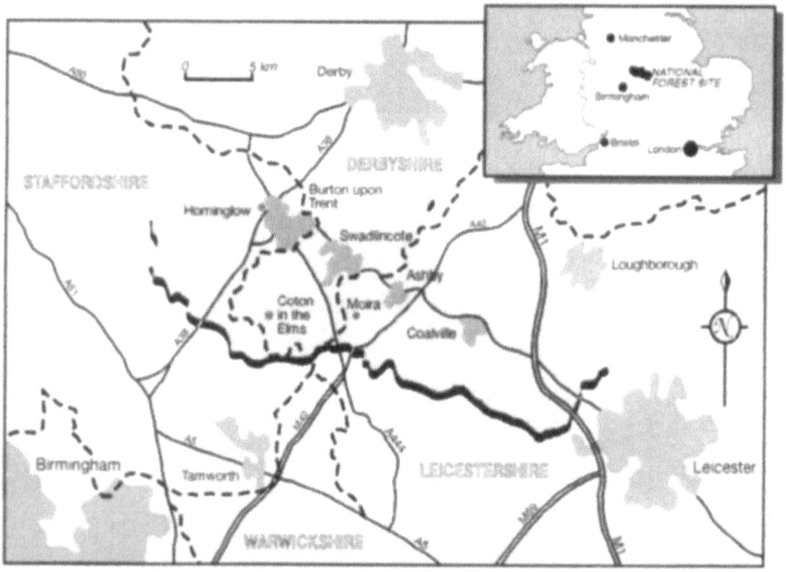

FIGURE 6.10 *The English National Forest*
Source: Cloke, P., Milbourne, P. and Thomas, C. 1996: The English National Forest: local reactions to plans for renegotiated nature–society relations in the countryside. *Transactions of the Institute of British Geographers* N.S. **21**, 552–71

planning that we turn next, especially to the landscape impact of planning in pre-industrial societies.

FURTHER READING AND REFERENCES

Pretty is a well-known exponent of sustainable resource management in the modern world and his account of the medieval manor is therefore especially interesting. Peasant agriculture in less developed countries is well covered in Klee (1980) and Wilken (1987). For woodland see Rackham (1986).

Conway, G.R. 1987: The properties of agroecosystems. *Agricultural Systems* **24,** 95–112.

Cooter, W.S. 1978: Ecological dimensions of medieval agrarian systems. *Agricultural History* **52,** 458–77.

Daniels, S. 1988: The political iconography of woodland in later Georgian England. In Cosgrove, D. and Daniels, S. (eds) *The iconography of landscape: essays on the symbolic representation, design and use of past environments.* Cambridge: Cambridge University Press, 43–82.

Darby, H.C. (ed.) 1973: *A new historical geography of England.* Cambridge: Cambridge University Press.

Hooke, D. 1985: *The Anglo-Saxon landscape: the kingdom of the Hwicce.* Manchester: Manchester University Press.

Johnston, R.J. 1994: Tragedy of the commons. In Johnston, R.J., Gregory, D. and Smith, D.M. (eds) *The dictionary of human geography.* Oxford: Blackwell, 639.

Klee, G.A. (ed.) 1980: *World systems of traditional resource management.* London: Arnold.

Neeson, J.M. 1993: *Commoners: common right, enclosure and social change in England, 1700–1820.* Cambridge: Cambridge University Press.

Peterken, G.F. 1993: *Woodland conservation and management.* 2nd edn, London: Chapman & Hall.

Pretty, J.N. 1990: Sustainable agriculture in the Middle Ages: the English manor. *Agricultural History Review* **38,** 1–19.

Rackham, O. 1986: *The history of the countryside.* London: Dent.

Repetto, R. 1986: *World enough and time.* New Haven: Yale University Press.

Richards, P. 1985: *Indigenous agricultural revolution: ecology and food production in West Africa.* London: Hutchinson.

Wilken, G.C. 1972: Microclimate management by traditional farmers. *Geographical Review* **62,** 544–60.

Wilken, G.C. 1987: *Good farmers: traditional agricultural resource management in Mexico and central America.* Berkeley: University of California Press.

7

LARGE-SCALE LANDSCAPE MODIFICATION

PAYS AND PRE-INDUSTRIAL PLANNING

> Watch out for the fellow who talks about putting things in order! Putting things in order always means getting other people under your control.
> (D. Diderot 1796: *Supplement to Bougainville's 'Voyage'*)

INTRODUCTION

We tend to think of planning in the landscape as a twentieth-century phenomenon. In reality, the visual appearance and functional organization of landscape features have always been evolving, with two scales of action particularly noticeable. First, there is small-scale and localized change. The term *organic* is often used to describe this gradual accretion of features in the cultural landscape, which may concern perhaps the rebuilding of a few cottages, the sub-division of a field, the exploitation of a new resource such as the opening of a quarry, or any of the myriad run-of-the-mill decisions which people make to improve their lives. Doolittle (1984) makes out a case that such incremental change is the norm in agriculture, involving 'gradual upgrading through small units of input over long periods of change'. There may well have been back-breaking work involved in making such minor alterations but no-one would have felt threatened by revolutionary change.

Second, much less frequently but more fundamentally, there have been key points in history when whole landscapes and parts of landscapes have been very substantially changed; one might almost say re-planned *de novo*. Such wholesale reorganization has been by no means uncommon in the past, even the distant past. In southern England they were certainly present in the late Bronze Age, as we know from the *reaves* of Dartmoor, and in Neolithic Ireland when reave-like field walls were created.

On Dartmoor reave field boundaries were laid out (*c*. 1700–1600 BC) in parallel lines approximately 100 metres apart (*see* Figures 7.1 and 7.2), in groups covering up to 3000 hectares, each presumably representing the territory of a clan or community group (Fleming 1988). There are *terminal reaves* at the edge of each group and a stretch of grazing beyond to make up a balanced farming system. This is an astonishing feat of common will for an allegedly primitive people, no doubt directed by an élite with power on a regional scale. Similar, *coaxial*, field systems and swarms of linear earthworks and ditches have been identified in other upland parts of England, and the suspicion is that such boundaries were also present in the lowlands but have been swept away by the tidal waves of history. In the case of Dartmoor the surviving reaves and their associated huts imply farming, at first entirely pastoral but later with some arable, at

FIGURE 7.1 *The reaves of Dartmoor. Note how they are arranged in swarms, suggesting territorial divisions*
Source: Adapted from Fleming, A. 1988: *The Dartmoor reaves: investigating prehistoric land divisions*. London: Batsford

250–400 metres above sea level, well above the current limit of cultivation. They probably show an expansion into a mixture of light woodland and open heath during a phase of warm climatic conditions, and successive generations of farmers would certainly have been put under pressure by the deterioration of temperatures from about 900–500 BC.

THE ORIGINS OF THE GREAT LANDSCAPE PROVINCES OF EUROPE

Western Europe has a fascinating variety of landscapes that cannot be explained by geology and topography alone. The human imprint has been marked from an early date, leading by a process of steady change to today's distinctive cultural regions.

A satisfying approach to describing these regions, or *pays* as he called them, was developed by the French geographer Paul Vidal de la Blache (1845–1918). To him the study of landscape was best achieved locally, by analysing the connexions between people and the resources available to them in their immediate physical environment. France had retained its regional lifestyles and typical products such as local cheeses and wines, culinary dishes, vernacular architecture and traditional

FIGURE 7.2 *Reave system, Mountsland Common, Ilsington, Devon (1973)*
Source: Committee for Aerial Photography, University of Cambridge

dress. These were outward expressions of a popular culture which gave a focus of identification for local inhabitants. French geographers, inspired by Vidal de la Blache and others, wrote a series of regional monographs which represent a very strong tradition in the subject. To this day the country is a patchwork of hundreds of pays which, although increasingly drawn together by the processes of industrialization and the homogenizing influence of modern culture, remain clearly identifiable.

Cultural regions or pays have survived in all the European countries despite the devastation of wars and the restructuring of economy and society which has accelerated in the twentieth century. Their resilience suggests an inertia of both physical capital and cultural investment which is not easy to write off.

Meeus (1990, 1995) and his collaborators have classified European landscapes into 30 types (*see* Figures 7.3 and 7.4) on the basis of vegetation,

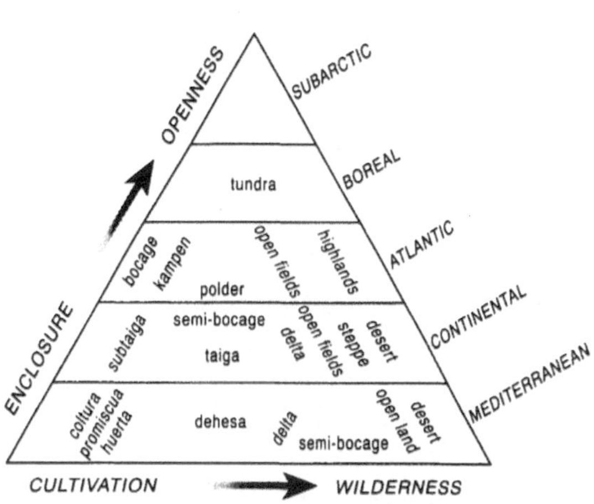

FIGURE 7.3 *A model of European landscape types*
Source: adapted from Meeus, J.H.A. 1995: Pan-European landscapes. *Landscape and Urban Planning* **31,** 57–79

EUROPE: MAJOR LANDSCAPE TYPES

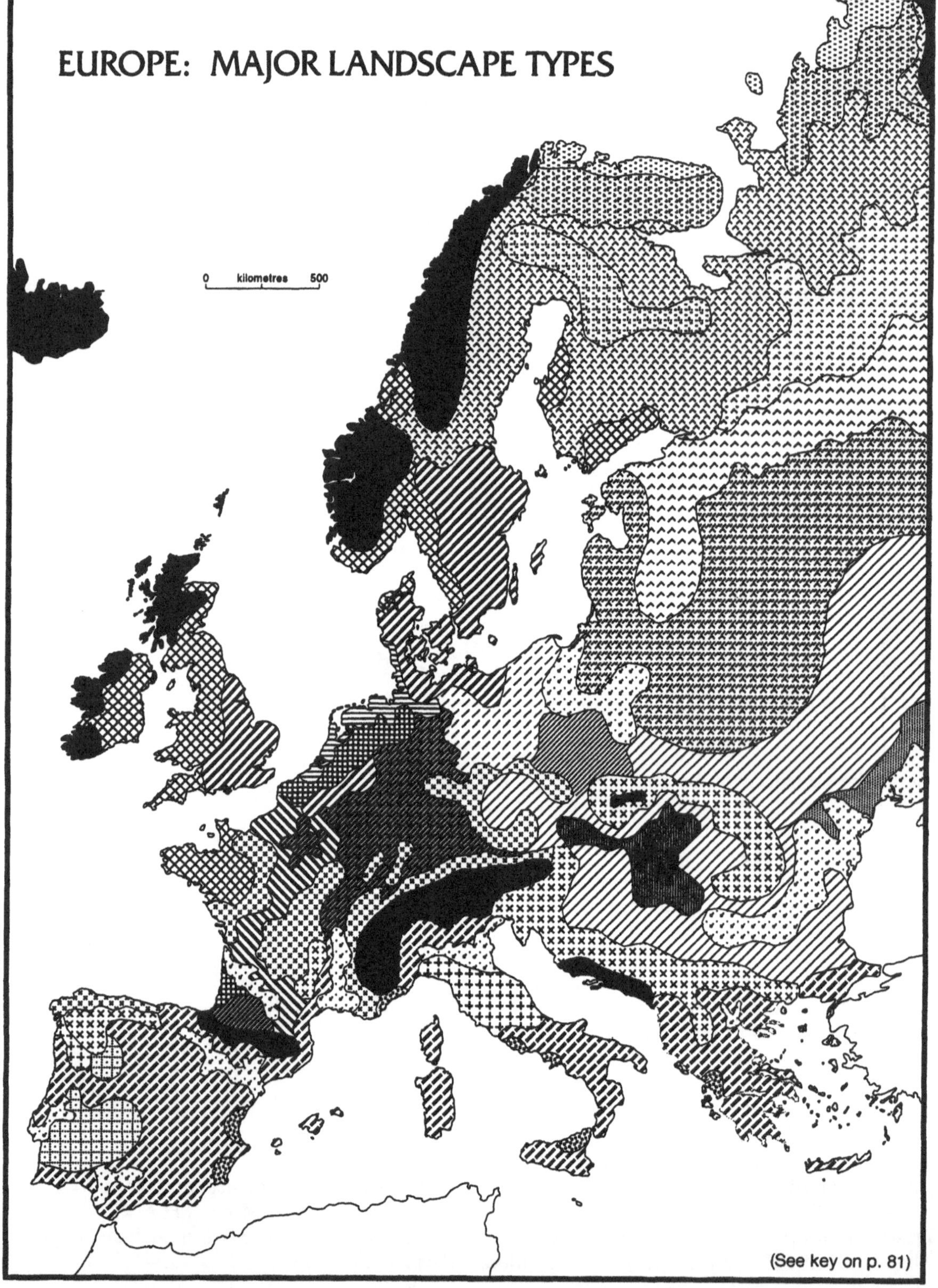

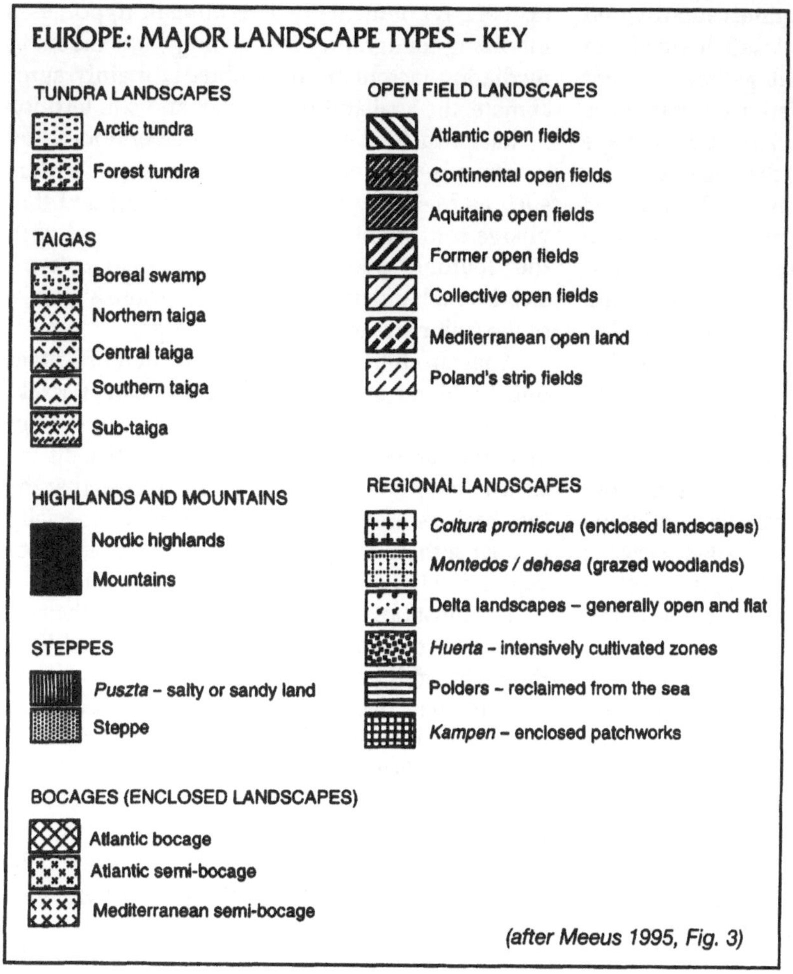

FIGURE 7.4 *Pan-European landscapes types*
Source: adapted from Meeus, J.H.A., Wijermans, M.P.L and Vroom, M.J. 1990: Agricultural landscapes in Europe and their transformation. *Landscape and Urban Planning* **18,** 289–352

degree of enclosure/openness, and the intensity of cultivation. While we might not agree with the precise boundaries that he has delimited, Meeus has performed a considerable service in alerting us to the underlying logic of the landscapes that we tend to take for granted. One of the faults of historical geography has been its tendency to be parochial, perhaps because of the degree of specialization needed to analyse the documentary record of any one country, and this map gives us a first draft of a Europe-wide view, based upon map analysis and the skilled visual impressions of a team of landscape architects, which need to be refined in the light of the local scholarship of historians and geographers. The reality is likely to be a great deal more complex than this simple model.

CASE STUDY: CHAMPION AND ENCLOSED

To the casual observer looking out of a car or train window, there are clear but puzzling differences between the major landscape types in north-western Europe. There are vast tracts of open country with large fields, sometimes sub-divided into strips, juxtaposed with enclosed, hedged and walled areas which have an altogether different ambience. There are many variations within these broad groupings but the fundamental cleavage remains between the open and the enclosed.

The open-field landscape has also been called the *Midland system* after its characteristic region of England, and also *fielden, champion* or *champagne*. On the continent it flourished in northern and eastern France, Denmark, and in parts of Germany and Sweden. Agricultural organization varied: although most of the arable was subdivided into narrow strips, and still is in some areas, land was not always held collectively as implied by the confusing term *common fields*. Contrast this with the texturally dissimilar landscape of wooded or hedged enclosures (called *bocage* in Britanny) of western Britain, East Anglia, western France, the Low Countries, northern Germany, and the Baltic fringes.

Various explanations have been put forward for the evolution of these landscapes. Agricultural specialization, for instance, has certainly been important because a pastoral-based economy requires less cooperative effort for its success and is therefore less likely to need an associated settlement pattern of nucleated villages. The links between arable farming and low-lying fertile land, on the one hand, and between pastoralism and hilly country, on the other, inevitably led to a rash of environmentally deterministic hypotheses involving topography and geology, but the main mediating factors are undoubtedly mainly agroclimate, the availability of water, and soil fertility.

Racial factors have also been mentioned, by nineteenth- and early twentieth-century writers such as Meitzen who 100 years ago argued that village settlement patterns were associated with the Teutonic peoples, whereas isolated farmsteads and hamlets were said to be more a feature of the Celtic fringe of western Europe.

Apart from the well-known division of Britain into highland and lowland zones, Oliver Rackham (1986) has identified within lowland England another boundary between what he calls the *ancient* and *planned* landscapes. He argues that the two regions are significantly different especially in their settlement patterns and the disposition of vegetation in woods and hedges (Table 7.1).

The most obvious differences between the two arise from the widespread enclosure of fields in the eighteenth and nineteenth centuries, but the contrast has deeper historical roots. Rackham postulates that the *planned countryside* (central belt) was in fact the legacy of the Roman occupation. Their fields were rectangular and probably hedged, but these hedges were removed by the Anglo-Saxons when they initiated the open field system with its very different type of open landscape.

The *ancient countryside* (west, north and south

TABLE 7.1 *Today's landscape elements of the 'ancient' and 'planned' countrysides*

Ancient countryside	Planned countryside
Many antiquities	Few antiquities
Heathland common	Heaths rare
Many ponds	Few ponds
Open fields few in number and enclosed early	Strong tradition of open fields, lasting until Parliamentary Enclosures
Pollarded trees away from human habitation	Few pollards
Non-woodland trees: oak, ash, alder, birch	Non-woodland trees: elders, thorns
Winding hedgerows of mixed species	Straight hedges, often hawthorn
Many small woods	Fewer woods, but larger in size
Many footpaths	Few public footpaths
Dense network of narrow, winding lanes, often sunken	Fewer roads, mainly straight
Isolated farms of ancient origin	Isolated farms created upon enclosure
Dispersed settlement	Nucleated villages

Source: after Rackham (1986, 4–5)

east) was less affected by the Romans. It was a landscape of great antiquity which visually remains essentially unchanged today. Here fewer open fields were created and then enclosed. A more detailed view of the landscape provinces of England from the rural settlement work of Brian Roberts is shown in Figure 7.5.

LAND REFORM

Most land reforms, both historical and modern, have been instigated by two types of argument. The first is that greater efficiency of production will be gained by the restructuring of land-holdings, leading to reductions in costs, improved profitability and increased quantities of output. Experience has shown that such arguments stand on shifting ground because fluctuations in economic variables and technological developments may within a few years undermine a neatly planned reorganization. Second, political and social logic has demanded land reform in the interests of greater equitability of land-holding, especially, but not exclusively, in socialist countries such as the former Soviet Union and the People's Republic of China. Land reforms drawing upon this second strand of arguments were rare before the French Revolution.

Land reforms may be classified according to their physical and functional results:

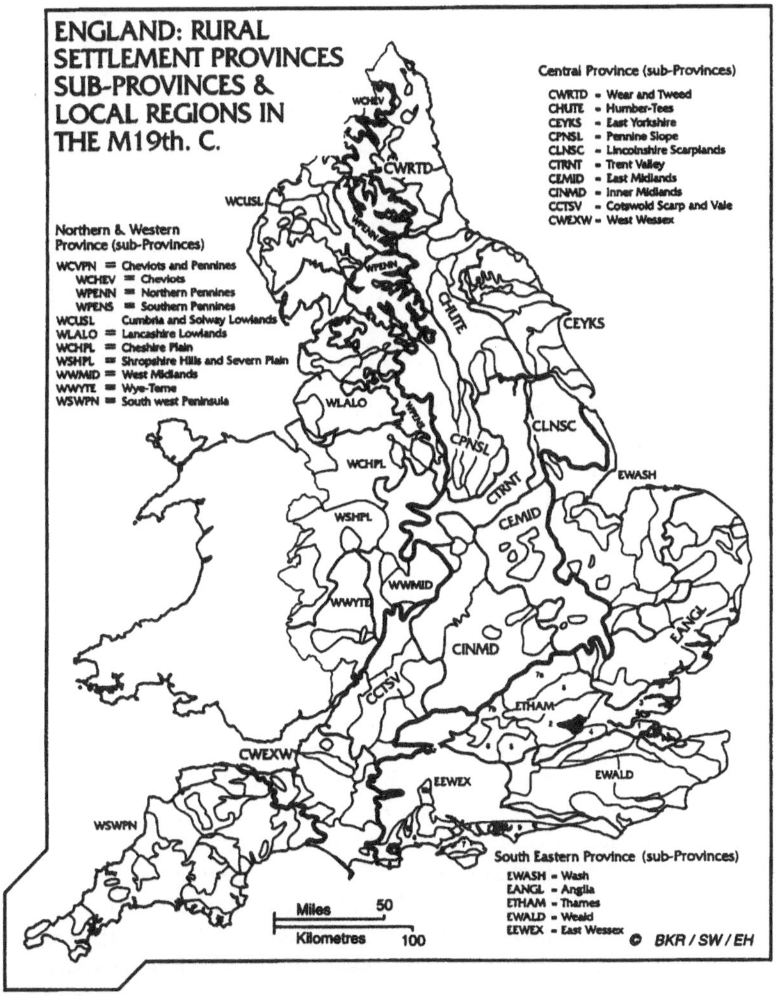

FIGURE 7.5 *Settlement provinces in nineteenth-century England*

- A re-ordering of landownership patterns which, according to the system to be replaced, will usually mean collectivization or privatization.
- A consolidation of scattered parcels of each farmer's land into a compact unit, for instance the French *remembrement* which since 1941 has been responsible for the reorganization of 12 million hectares, about 40 per cent of the country's agricultural area, into consolidated blocks of fields which are large enough for modern machinery. Here the state pays the administrative costs and heavily subsidizes physical activities such as hedge removal and the laying of new roads.
- Engrossment, where once independent farms are merged to make more viable economic units. Again, the French have pioneered this process as a policy to modernize their farming structure.

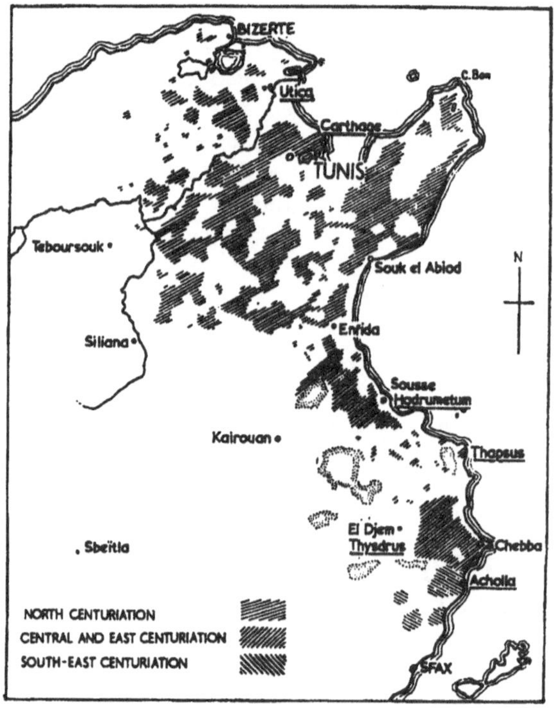

FIGURE 7.6 *Centuriation in Tunisia, showing the three major systems established by the Romans*
Source: Reprinted from Dilke, O.A.W. 1971: *The Roman land surveyors*. Newton Abbot: David & Charles

CASE STUDY: IMPERIALISM AND ROMAN CENTURIATION

The Roman empire was administered with a methodical and practical efficiency that has had few equals in history. This extended to agriculture and the land division that formed the intersection between the farmer and the landscape. The practice of *centuriation* was widespread, from the Po Valley in northern Italy to the semi-desert of southern Tunisia (*see* Figure 7.6), and from the air we can immediately see it as a very important cultural signature. The chessboard appearance of the Roman field system was not repeated on any scale until the land division of North America in the nineteenth century.

The basis of the grid of *centuriae* was a pair of base lines (roads) set at right angles. From these lines a network of minor intersecting boundaries were laid out using the Roman surveying cross-staff, the *groma*. Secondary roads were provided for at regular intervals of 2400 Roman feet (708 m). Each *century* was subdivided into two hundred *iugera* (0.25 ha) which were then grouped into square or rectangular farms.

In Tunisia the whole province was laid out in several groups of centuriae, but land in other parts of the empire was dealt with in smaller packets. Professional land surveyors, the *agrimensores*, were responsible for the technical work on the ground, but decisions about the orientation of the pattern and its extent were presumably made at a high bureaucratic level. The centuriation of the landscape epitomizes the Roman will to dominate the natural world but it also has implications for the control of the economic and political spheres.

Many centuriated landscapes were laid out for colonists. In other words the farmers, often retired soldiers and their families, were used as a means of establishing control, as in the recently conquered North African territories of Carthage in 146 BC. Government policy was 'that all lands that are suitable for olives and vines as well as for grain crops should be brought into cultivation', and to encourage this, tenants on imperial estates were allowed to keep all of their olive harvests for the first ten years, after which one third had to be delivered to the estate manager. According to the third-century North African writer Tertullian: 'Wildernesses have been replaced by most attractive estates; woods have yielded to the plough; the cover of wild beasts has become grazing land; sands are sown, stones are broken up, and marshes are drained' (*De Anima* 30.3, quoted in R. Meiggs 1982: *Trees and timber in the ancient Mediterranean world*. Oxford: Clarendon Press).

Taxes were tied to the land and became ever more extortionate in the later empire as the authorities despaired of alternative means of raising revenue. Taxes paid in grain were stored by the state and some surpluses went to feed the population of Rome. Their need to pay these taxes forced farmers to use extractive techniques that were damaging to the fragile semi-arid environment. Classical mosaics suggest a rich flora and fauna, but this was later greatly depleted and it seems likely that the fall in population since Roman times is due less to climatic desiccation than to the over-exploitation of the land and the cutting down of trees, leading to soil exhaustion and erosion.

CASE STUDY: AGRICULTURAL ENCLOSURE IN BRITAIN, 1500–1900

In the Europe of AD 1500 there was still a widespread distribution of subdivided, open fields, with a tremendous variety of form and function (Chapter 11). Over the next four centuries land reform was responsible for the transformation of the appearance and function of such landscapes out of all recognition in some regions, and by way of example we will look at the revolution in English land-use systems which was brought about by the process of enclosure of the open fields. This was a change mainly inspired by economic advantage but there were other factors as well, such as population pressure.

In post-medieval England the feudal system was disintegrating and market forces were bearing increasingly upon the lives of ordinary people. The decision-making of farmers, especially those located close to large centres of population, was less motivated by pure subsistence and opportunities opened up for specialized, commercial production. Thus, some time in the mid-fifteenth century the prices of livestock products in England, especially wool, began to outstrip corn. The demand for wool had expanded, both for export and increasingly for manufacture into textiles, but the population had yet to recover its demand for staple grains after the demographic disasters of the fourteenth century, when famine and the bubonic plague under the name of the Black Death (1348) had stalked the land. Inevitably, grass began to look like a better economic proposition and much arable land was either abandoned or actively transferred to pastoralism by aggressive landlords who evicted their tenants. Some of the hamlets deserted in the later medieval period had been subjected to this policy (see Chapter 11).

Large areas were enclosed for sheep pasture, especially in the Midlands where as much as a tenth of the agricultural land was enclosed between 1450 and 1600, but also in the western and south western fringes which were environmentally better suited anyway to a specialization in a grass-based agriculture. About 45 per cent of England was already enclosed in 1450, with a further 2–3 per cent by 1600 (*see* Figure 7.7).

At times enclosure was bitterly contested. The strong-arm tactics of certain manorial lords were crude and relied for their success upon the fact that news travelled slowly and only locally. As communications improved and something

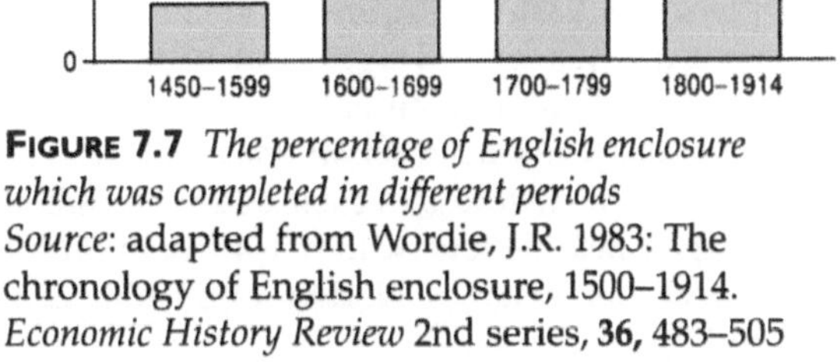

FIGURE 7.7 *The percentage of English enclosure which was completed in different periods*
Source: adapted from Wordie, J.R. 1983: The chronology of English enclosure, 1500–1914. *Economic History Review* 2nd series, **36,** 483–505

approximating to public opinion emerged, so changes in rural areas became political. Such was the uproar in the Tudor period for instance, when Sir Thomas More wrote of 'sheep devouring men', that Parliament instigated enquiries and legislated certain measures of control. The Midland Revolt of 1607 was partly related to enclosure and there was widespread disquiet at the loss of previously collective resources, such as grazing and gathering fuel, which cottagers and smallholders had enjoyed since time immemorial.

From the mid-sixteenth century some of the political temperature was reduced by a wider adoption of the procedure of enclosure by agreement among the landowners of a parish. The incidence of violent coercion declined but the power of large landowners remained a factor. Some theorists argue that enclosure was most popular in regions where the increase in farming profitability was likely to be greatest, and there is logic in that, but there were some fertile areas where open fields remained intact until the eighteenth century. Here it may be that the potential rewards were so great that agreement by persuasion was difficult to achieve.

In the seventeenth century educated opinion came to favour enclosure as a means of improvement. To many it seemed a rational and even a necessary condition for modernizing agriculture. In the early eighteenth century enclosure by Private Act of Parliament became the norm. These were thought important because enclosure by agreement, democratic though it may sound, actually caused endless legal wrangles in the Courts of Chancery and Exchequer, with disputes sometimes dragging on for decades. A Parliamentary Act added the possibility of legal compulsion which over-rode local objections.

On the whole, enclosure seems to have proceeded with least opposition in areas where population pressure was not a problem, where farming had a pastoral bias, and where there was still plenty of waste to improve. Where population pressure was on the verge of outstripping the available waste and common resources, people were obviously not keen on an enclosure which would deny them access to their common rights in the open fields. Riots were not uncommon in such areas, for instance Northamptonshire (*see* Figure 7.8).

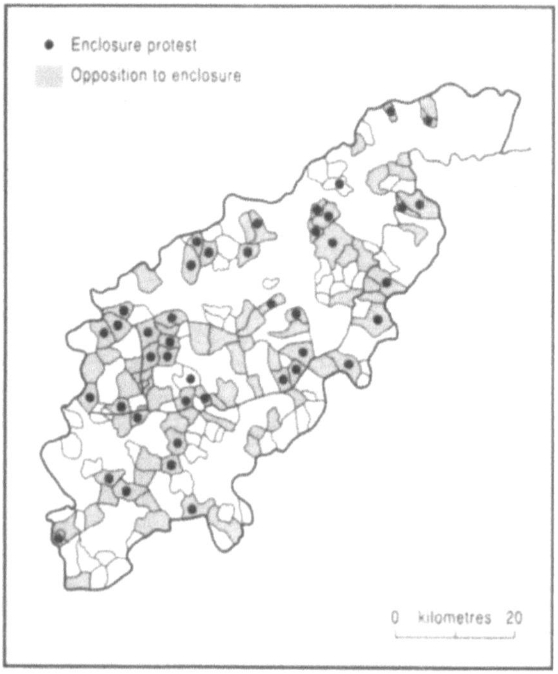

FIGURE 7.8 *Opposition to enclosure in Northamptonshire*
Source: adapted from Neeson, J.M. 1983: Opposition to enclosure in Northamptonshire c. 1760–1800. In Charlesworth, A. (ed.) *An atlas of rural protest in Britain, 1548–1900*. London: Croom Helm

Even where violence was avoided, the impact of enclosure upon the lives of ordinary people was often negative. Of the contemporary writers who described this, John Clare (1793–1864) is perhaps one of the most renowned. He was a Northamptonshire agricultural labourer who put into verse the great changes in the countryside around his native village. He hated enclosure and the changes it had wrought in the beloved landscape of his youth:

> The cow boy with his Green is gone
> And every Bush and tree
> Dire nakedness over all prevails
> yon fallows bare and brown
> Is all beset wi' posts and rails
> And turned upside down . . .
>
> The bawks and Eddings are no more
> The pastures too are gone
> The greens the Meadows and the moors

Are all cut up and done
There's scarce a greensward spot remains
And scarce a single tree
All naked are thy native plains
And yet they're dear to thee.
The Lamentations of Round-Oak Waters (*c*. 1815)

Clare was right in identifying the remarkable modification of the Midland landscape. A more recent commentator is firm that enclosure was one of the most significant events in landscape history:

> Parliamentary enclosure was possibly the largest single aggregate landscape change induced by man in an equivalent period of time. In a number of ways it surely warrants this weight of significance, producing as it did scattered farmsteads where once nucleated villages proliferated, hedgerows or stonewalls and thus a mosaic of geometrically shaped fields, and ordered landownership patterns where once existed the relatively disorderly open fields.
> (Turner, 1980, 33)

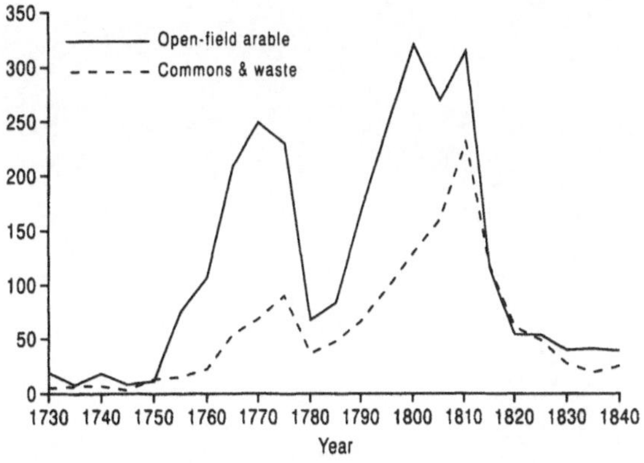

FIGURE 7.9 *The progress of enclosure by Act of Parliament, 1750–1820. The vertical axis shows the number of Acts passed. Note that enclosure from the open fields and from the commons and waste are shown as two separate lines*
Source: Based on Turner, M.E. 1980: *English Parliamentary enclosure: its historical geography and economic history*. Folkestone: Dawson.

During the eighteenth century there was a tremendous upsurge in Enclosure Acts, especially during two periods: 1765–80, and 1793–1815: 80 per cent of all Enclosure Acts were passed during these two periods (*see* Figure 7.9).

An obvious factor was profit potential, especially during the rapid price inflation of food during the Napoleonic Wars which encouraged the expansion of capitalist, market-orientated agriculture. A broader degree of enterprise specialization was possible, with exploitation of comparative advantage causing a shift of the agricultural pendulum further away from regional self-sufficiency. Thus more and more land in the Midlands was laid down to grass (Figure 7.10), so that by 1850 James Caird was able to claim a geographical distinction between the arable east and the pastoral west of England.

As a result of the privatization of farm land accomplished by enclosure, collective risk minimization was replaced by individual enterprise, so innovative farmers felt less inhibited. An example of this was the collective grazing of the stubble which was allowed after the harvest had been gathered in. This made selective breeding of animals difficult and therefore improvements in wool quality, milk yield and carcass weight were delayed until enclosure allowed the separation of beasts with desirable genetic traits.

Enclosure's functional implications for farming were fundamental. Individual decisions about crops and livestock, and the results of those decisions, were no longer dependent upon the agreement of neighbours, although we should not be so naïve as to assume that success or failure now depended only upon the personal qualities of the landlord and tenant. There were still constraints upon both.

The physical nature of the landscape was completely replanned in many villages. Each Act appointed a small group of commissioners and gave them considerable powers. They had to reallocate land held in scattered strips by an individual into a compact block. There were detailed regulations for the construction of roads, fences, gates and hedges. Naturally the whole process was very costly, and expenses were shared out among the new proprietors. Table 7.2 shows the breakdown of costs for the Buckinghamshire parish of Prince's Risborough.

Agreement about the nature of the Act was often difficult to achieve among the potential beneficiaries. In North Thoresby in Lincolnshire, for instance, enclosure was first discussed in 1766, again in 1801–2, 1809–10, and 1823–4. Each

FIGURE 7.10 *The enclosed fields of Padbury, Buckinghamshire, showing the survivial of ridge and furrow from the pre-enclosure period*
Source: Committee for Aerial Photography, University of Cambridge

generation of farmers disagreed amongst themselves until 1836 when the agreement was finally drawn up (Russell and Russell 1987).

TABLE 7.2 *Costs of enclosing Prince's Risborough, 1820–23*

	£	s	d
Soliciting the Act	837	0	10
Clerk	1014	12	5
Commissioners	1400	12	0
Surveyor	1486	8	7
Roads	3930	14	3
Fences	853	9	5
Bridges	209	17	1
Tillage	521	10	5
Other	1467	18	10
Total	11 722	3	10

Source: Turner (1973)

Within a few years of the passing of the Act the landscape of the parish had been transformed by a team of surveyors who carried out the instructions of the commissioners. An open, bare landscape of large fields was divided into compact, regular blocks marked off by straight fences and quickset hawthorn hedgerows. Roads were straight, with wide grass verges. New farms were often created, away from the compact villages, on sites more convenient for individual farming, although inertia maintained the essential structure of the rural settlement pattern.

So there appeared in the eighteenth century a new planned landscape in the English Midlands (*see* Figure 7.11), substantially different from anything seen before. Occasionally relicts of earlier, open-field landscapes survive because some enclosure commissioners, in order to minimize costs, simply hedged individual or groups of strips without properly consolidating and

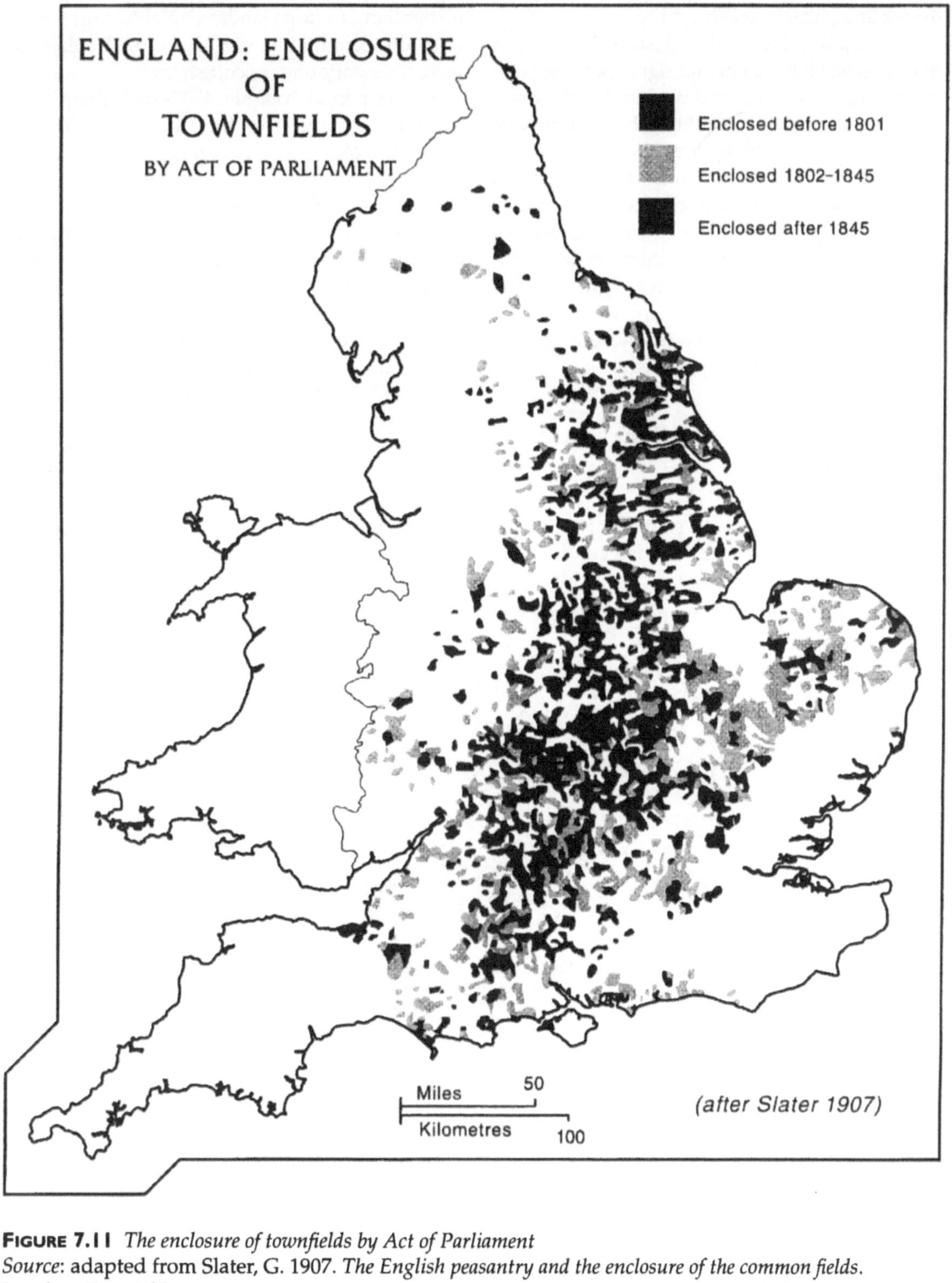

FIGURE 7.11 *The enclosure of townfields by Act of Parliament*
Source: adapted from Slater, G. 1907. *The English peasantry and the enclosure of the common fields.* London: Constable

engrossing them, thus preserving medieval shapes as an anachronistic oddity.

For Johnson (1996) the closure of common resources had a fundamental significance beyond the creation of hedges and ditches. It was a rare moment of shifting mentality when the private interests of individual replaced the collective conscientiousness as a driving force of society. Seen from this perspective it is clear that the public desire for improvement was a smokescreen for social changes in which the interests of smallholders and labourers were downgraded. The physical changes in the landscape were proof of thousands of local skirmishes whose outcomes varied in time but not in symbolism.

Landscape planning in Scotland

In Scotland enclosure was somewhat different. Agricultural improvements were later and slower than in England, and enclosure was usually at the instigation of a large landlord. Neither agreement with the community, nor an Act of Parliament, were necessary under Scottish law.

In the period roughly 1730–1850 there were substantial changes in the rural Scottish landscape. The old open field run rig was replaced by large, square fields. The farming hamlets, *ferm touns*, were replanned and sometimes landlords laid out new towns and villages in controlled, geometric patterns on green field sites with economic activities alternative to agriculture (*see* Figure 7.12). Some were associated with the textile and fishing industries and provided opportunities for tradesmen. Douglas Lockhart (1980) has calculated that over 450 Scottish settlements have this strong impress of the planner.

Traditional customs in Scotland meant that landlords felt obliged to provide for those dispossessed by these changes. But such customs broke down in the nineteenth century as the population increased, and failures of potato crops led to

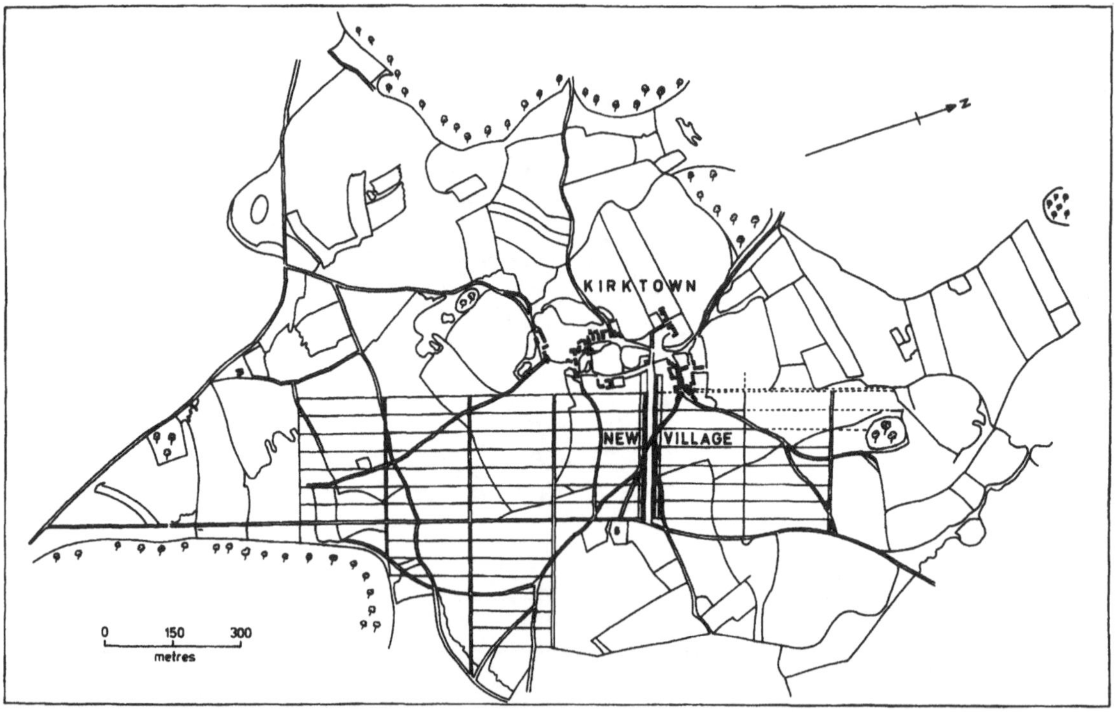

Figure 7.12 *The planned town of Urquhart, Morayshire. Note the regularity of the settlement and of the lotted lands nearby for grazing cows and growing fodder*
Source: D.G. Lockhart

misery. Landlords with an eye to profit from sheep forced their tenants off the land and encouraged them to emigrate, the notorious clearances creating deserts in areas of the Highlands (see Chapter 20).

CASE STUDY: PLANNING A LANDSCAPE FROM SCRATCH: THE US LAND SURVEY

Although the British, French and Spanish colonists in North America imported their unsystematic European traditions of land division, they were nevertheless open to new ideas. This is most obvious in the rectangular street patterns in cities such as Philadelphia and Charleston. In May 1785 a more general system was devised with the passing of the justly famous *Land Ordinance*, a statutory means of allotting land to the west, in the territories which were still the sole realm of native Americans.

In discussion, Thomas Jefferson and others decided that allotment on the basis of a checkerboard of squares was the most appropriate. It would be the cheapest to survey and it also fitted the rational, scientific philosophy of these men of the Enlightenment. Here we have the clearest large-scale expression of ideology in superficial landscape geometry. Functional boundaries from field edges to state lines were affected, creating what Cosgrove (1984) calls Palladian America. (Palladio was an influential Italian architect who was responsible for the revival of many classical ideas of design.)

There were north–south *principal meridians* and intersecting east–west *base lines*. Starting in Ohio, these were carefully surveyed using standardized chains and formed the basis for all of the subsequent land subdivision. *Range lines* were surveyed at 6-mile (9.65 km) intervals and these formed the boundaries of townships of (usually) 36 square miles. In turn, these townships were subdivided into a gridiron of 36 *sections* of 640 acres (260 hectares), each made up of four *quarter sections* or sixteen *forties*. The last of these was the most popular size of plot for the average pioneering farm family.

The American landscape of straight lines was packaged and commodified *de novo*. Unsurprisingly there was no part in it for its aboriginal owners who were unceremoniously dispossessed and herded into reservations. Their lifestyle, especially the nomadism of the plains Indians, was inconvenient for planners and settlers alike. The boundaries were real enough as physical barriers to movement but they were also intended as a symbolic declaration of the independence of each lot holder. The landscape had been privatized and, having invested their sweat in clearing and ploughing it, the new inhabitants were aggressively resistant to any perceived intrusions into their piece of tamed wilderness.

Every landscape evolves and it is interesting to speculate how the land division will influence America in the next millennium. So far forties and quarter sections have been split or amalgamated to suit convenience, leading to a blurring of the primary survey lines (*see* Figure 7.13). Most likely to survive are the transport routes, for their straight lines have become the viscera of American life.

CONCLUSION

Landscape planning is not just a feature of modernism. We have adduced evidence of planned and replanned agricultural landscapes from the Bronze Age to the nineteenth century, with administrative and commercial motives. Sometimes such features have survived in relict form only but there is a strong suspicion that the logic of communal planning underlies much of what may appear to be ordinary landscapes. The specialized skills of historical geographers, landscape historians and archaeologists are required to unravel this particular story about the organizational motivations of our ancestors.

FURTHER READING AND REFERENCES

Fleming (1988) provides a good introduction to the Dartmoor reaves and coaxial field systems in general. Meeus is one of the few writers to address landscapes at a continental scale, but his disciplinary origin is landscape architecture rather than cultural

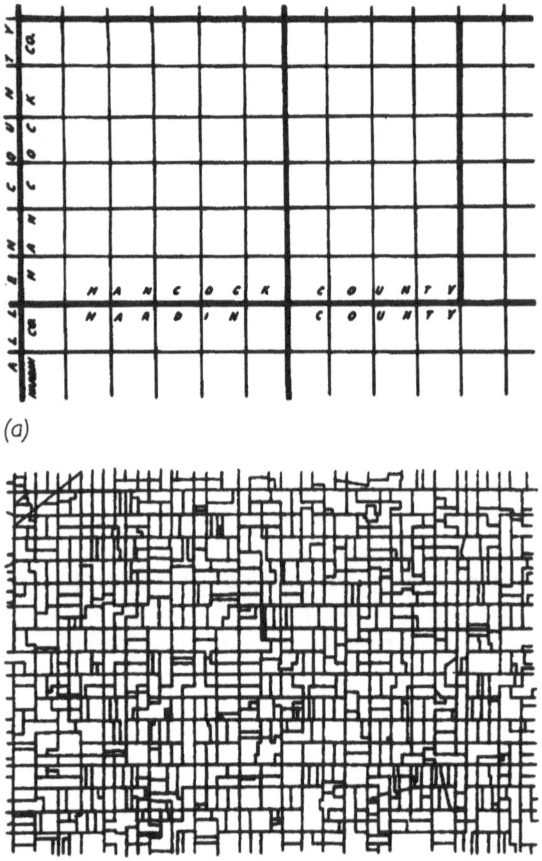

FIGURE 7.13 *The land division of the United States, as exemplified in Ohio.*
(a) County boundaries. Each minor block here is one square mile.
(b) Farm boundaries in the same area.
Source: Thrower, N.J.W. 1966: *Original survey and land subdivision: a comparative study of the form and effect of contrasting cadastral surveys*. Chicago: Rand McNally

and historical research. Rackham's (1986) classic account of the ancient and planned landscapes of England is essential reading, and Bradford (1957) is an authority on centuriation. See Turner (1980) on enclosure and Johnson (1990) on the US land survey.

Bradford, J. 1957: *Ancient landscapes: studies in field archaeology*. London: Bell.

Cosgrove, D. 1984: *Social formation and symbolic landscape*. London: Croom Helm.

Doolittle, W.E. 1984: Agricultural change as incremental process. *Annals of the Association of American Geographers* **74,** 124–37.

Fleming, A. 1988: *The Dartmoor reaves: investigating prehistoric land divisions*. London: Batsford.

Johnson, H.B. 1990: Towards a national landscape. In Conzen, M.P. (ed.) *The making of the American landscape*. Boston: Unwin Hyman, 127–45.

Johnson, M. 1996: *An archaeology of capitalism*. Oxford: Blackwell.

Lockhart, D.G. 1980: Scottish village plans: a preliminary analysis. *Scottish Geographical Magazine* **96,** 141–57.

Meeus, J.H.A. 1995: Pan-European landscapes. *Landscape and Urban Planning* **31,** 57–79.

Meeus, J.H.A., Wijermans, M.P. and Vroom, M.J. 1990: Agricultural landscapes in Europe and their transformation. *Landscape and Urban Planning* **18,** 289–352.

Rackham, O. 1986: *The history of the countryside*. London: Dent.

Russell, E. and Russell, R.C. 1987: *Parliamentary enclosure and new Lincolnshire landscapes*. Lincoln: Lincolnshire Recreational Services.

Turner, M.E. 1973: The cost of Parliamentary enclosure in Buckinghamshire. *Agricultural History Review* **21,** 35–46.

Turner, M.E. 1980: *English Parliamentary enclosure: its historical geography and economic history*. Folkestone: Dawson.

8

CLEARING THE WOOD

Clearing woodland for pasture and cultivation has been the most fundamental and widespread agent of human-induced environmental transformation in world history.

W. Beinart and P. Coates, 1995: (*Environment and history: the taming of nature in the USA and South Africa*. London: Routledge)

INTRODUCTION

Simon Schama's (1995) evocation of the cultural history and mythical context of woodlands is convincing testimony to the power of trees over the human imagination. In particular he shows how German nationalism has its roots in the forests that the Romans failed to conquer, romanticized from the eighteenth century onwards in the work of artists such as Caspar David Friederich (1774–1840) and Anselm Kiefer (b. 1945), and that British oaks were symbols of aristocratic prestige and of the country's naval supremacy. Similarly, French and American interpretations of woodland also made an important contribution to their view of the natural world and of themselves. In Chapter 6 we looked at the sustainable management of woodland, and in Chapters 18 and 20 the theme of forests as wilderness will be picked up. But here we will confine ourselves to the clearance of the primeval woods which figure so strongly in the cultural memory of European and American peoples.

Otto Schlüter's imaginative reconstruction of woodland in Europe in AD 900 and 1900 (*see* Figure 8.1) conveys better than a thousand words an image which is deep within us all. It shows the taming and defeat of wild nature by human action, a deforestation as widespread and as destructive as the current devastation of the Amazon rainforest. Apart from agriculture it was, according to Sir Clifford Darby (1956, 183), 'perhaps the greatest single factor in the evolution of the European landscapes', or indeed the humanized landscapes of the other continents. The mastery of fire and use of the axe have not only changed landscapes visually; there has also been the disruption of ecosystems and their harnessing for agriculture and settlement.

EVIDENCE

The emphasis here will be upon the wide range of evidence which can be called upon to bear witness to the human impact upon the natural world. This will help us to appreciate the necessity for the historian of environmental change to draw upon the skills of cognate disciplines to interpret data as varied as ancient parchments and satellite photographs.

Among these various types of evidence, pollen analysis has proved itself to be especially valuable and effective for reconstructing historical patterns of vegetation. Where pollen is preserved, for instance in the successive layers of a peat bog, it can be recovered by core sampling and analysed microscopically for the relative proportions of pollen grains from different plant and tree species. We know the conditions under which each species thrives and it is therefore possible to hypothesize about changing climatic conditions in the past and the growing impact of humans. In addition, we

(a)

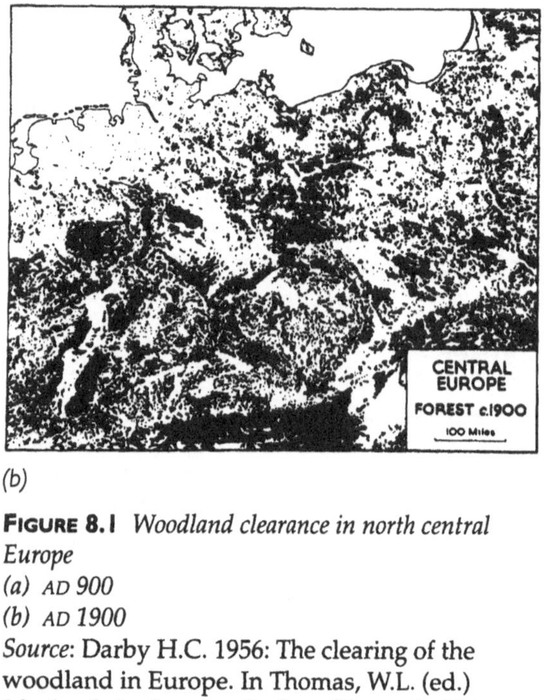

(b)

FIGURE 8.1 *Woodland clearance in north central Europe*
(a) AD *900*
(b) AD *1900*
Source: Darby H.C. 1956: The clearing of the woodland in Europe. In Thomas, W.L. (ed.) *Man's role in changing the face of the earth*. Chicago: University of Chicago Press, 183–216

may be able to infer local environmental conditions, such as the presence of coastal salt marsh conditions due to sea-level transgressions. A greater precision is given to such interpretations if dates can be ascribed to the various levels of a pollen sample, for instance by the carbon 14 isotope dating of wood preserved at different depths. A pollen diagram can then be compiled, showing the species frequencies at each level of the core.

A second, and very different, type of evidence is that of place names. These take many forms but in western Europe they often contain elements which tell us about the time of their origin, and the landscape in which they were created. Thus Figure 8.2 shows a ghostly shadow of former woodland in England by mapping names connected with clearance. Some are of Celtic origin, such as *coed* from Welsh and *cut, quite, coose* and *coys* in Cornish, and *lundr* (small wood), *skogr* (wood), *thveit* or *thwaite* (clearing), and *viothr* (wood) in Norse (Gelling 1984). But the majority are Anglo-Saxon, for example *leah* (also *lea, lee, leigh, ley, lye* – forest, wood, glade, clearing), for example, Durleigh, Somerset (wood with deer); and *hyrst* (also *hurst* – wooded hill), as in Longhirst, Northumberland (long wood). A similar exercise is possible for most European countries.

We can sometimes recover the name of an individual who was responsible for deforestation, as where Kati's *ridding* (clearing) has become today's Kateridden in Yorkshire. Individual species of trees are also commonly included (Elmham, Ashtead, Beech Hill, Ewhurst, Maplebeck), and occasionally the method of clearance, for instance Brindley, Wensleydale (*brende leah* – clearing caused by fire).

Third, there is documentary evidence such as deeds, taxation records and travellers' diaries. For the south of England, the Anglo-Saxon charters (AD 700–1050) are useful because they described territorial boundaries and often used woodland or hedges as descriptive features. But the most impressive early source of environmental information in all Europe is the Domesday Book. Compiled in 1086 by William I's political machinery as a means of assessing the resources of his newly conquered kingdom, this was a large-scale clerical exercise recording a variety of data on land use. It represented the state of the art of surveying in the late eleventh century and, although its accuracy and comprehensiveness have been questioned by modern scholars, it was nevertheless a magnificent achievement.

One question asked in each location was '*quantum silvae*: how much wood?' This elicited a variety of responses by region, with answers in terms

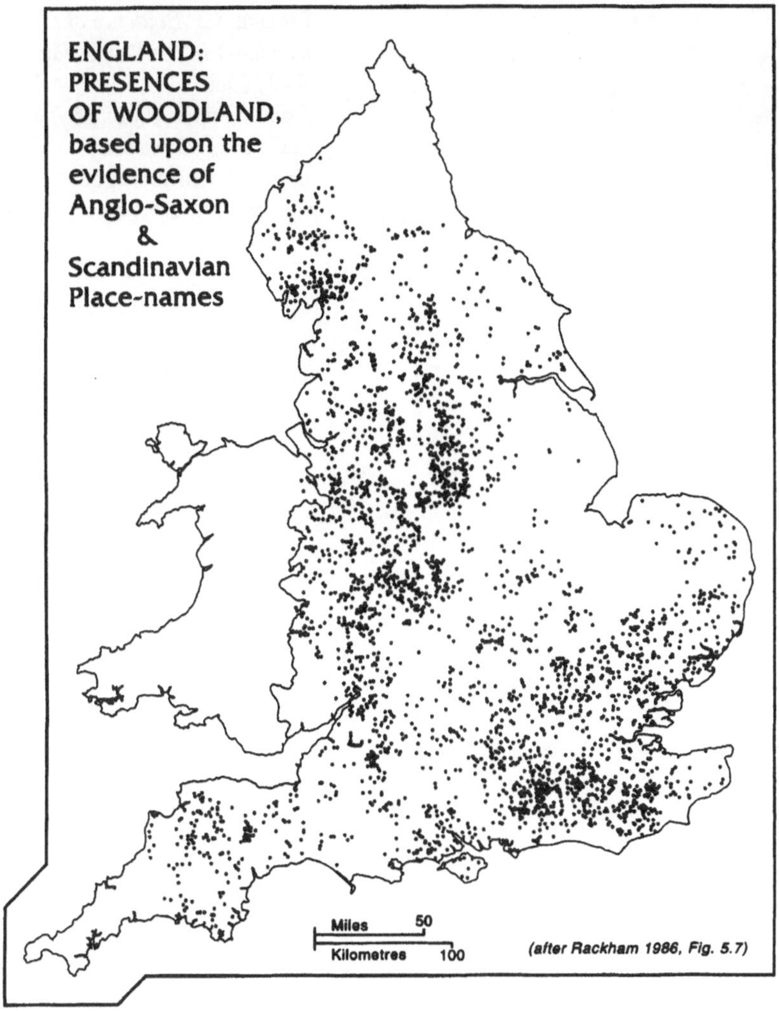

FIGURE 8.2 *Place name evidence of woodland in Dark Age England*
Source: Based on Rackham, O. 1976: *Trees and woodland in the British landscape.* London: Dent

of area (acres), length (leagues) and the number of pigs that could be pastured on the acorns and beechmast (swine renders). Figure 8.3 attempts to map these, and the result, remarkably, is a close reflection of the place name evidence. In other words, there seems to be good confirmation from entirely different directions of the approximate outline of surviving woodland in the Anglo-Saxon/early Norman periods. Only about 15 per cent of the country was wooded at this time, so here we have an indication that the evolution of the settled landscape was well advanced, with little further scope for expansion except perhaps on to marginal land or by the intensification of established farming systems.

Fourth, modern evidence of forest clearance includes maps, aerial photographs and field surveys. In the last 20 years satellite images have revolutionized the monitoring of deforestation, especially the Landsat and SPOT sensors. Figure 8.4 is dramatic evidence from the Rondonia state of Brazil of the process of pioneer settlement next to the new feeder roads, producing the light-coloured herringbone pattern. This scene is approximately 175 km wide.

WOODLAND CLEARANCE

In Europe significant clearance started during the Mesolithic (8000–3500 BC) with hunters burning vegetation, perhaps to help them catch game. But

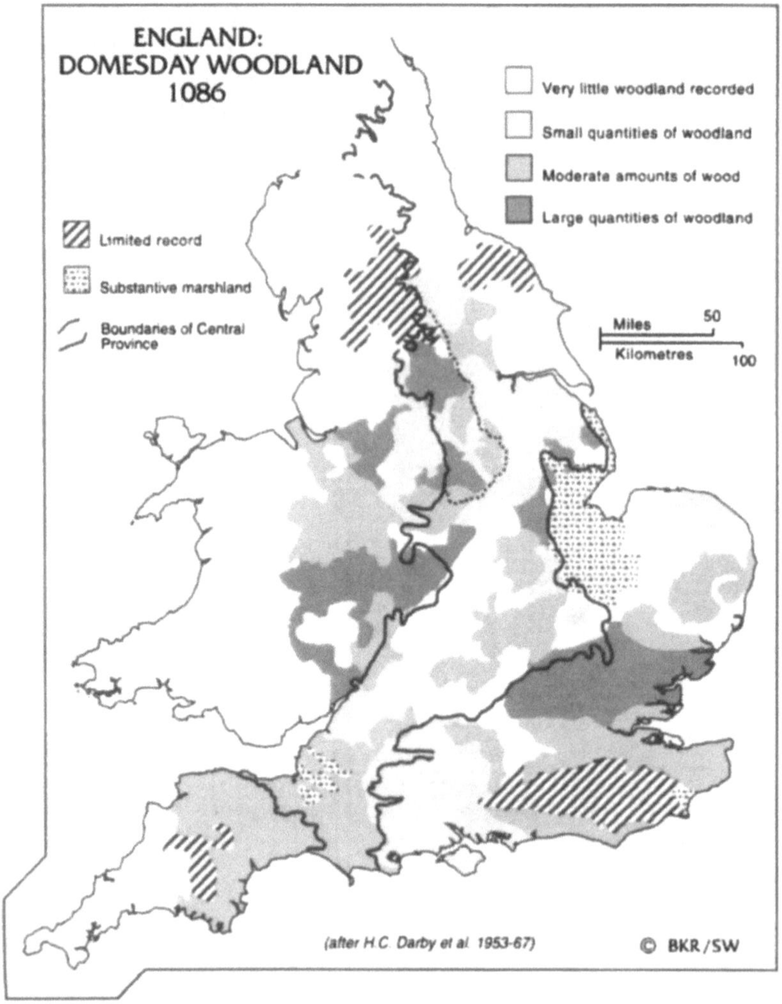

FIGURE 8.3 *Evidence of woodland in the Domesday Book, 1086*
Source: Based on Darby, H.C. 1977: *Domesday England*. Cambridge: Cambridge University Press

their impact was relatively small by comparison with the onslaught of the later Neolithic farmers who used slash and burn techniques to clear land for cultivation. The advent of iron axe technology allowed quick and efficient clearance which, when followed by browsing of young regrowths by stock, meant that woods were often unable to regenerate.

In historic times there have been bursts of activity when trees were felled for timber and fuel and their roots grubbed out in order to make permanent fields. Classical writers such as Strabo provide ample evidence of deforestation in the Greek and Roman spheres, although the impact was certainly localized. More widespread was what Marc Bloch called 'the Great Age of Clearance' which swept across western Europe in the centuries between AD 800 and 1200. This was an era of population growth and economic expansion after the end of the Dark Ages.

Much of this medieval cutting and felling was undertaken by free peasants who colonized a woodland frontier. Their *assarting* or reclamation was often by small groups, even families, who broke away from overfull villages and oppressive manorial control in order to carve out a new life for themselves. Some landlords granted licences to assart because a revenue was created for them where none had existed before, up to one-third of the improved value, and they also gained liegemen who owed a duty to fight when their domains were threatened.

FIGURE 8.4 *Landsat image of Rondonia, Brazil*
Source: NASA Landsat Pathfinder Humid Tropical Forest Project, University of New Hampshire

Religious orders were also involved, especially the Cistercians who spread rapidly in England a few decades after 1120. Their lifestyle was spartan, according to the strict rule of St Benedict, and their philosophy was that of St Bernard, to search out lonely places and tame the wilderness themselves in God's name. Sheep and cattle granges on a large scale were their trademark in places such as the Pennines and the North York Moors (*see* Figure 8.5), initially on land granted by sympathetic landowners. By 1150 the Cistercian abbey of Rievaulx was comprised of 140 monks and 500 lay brothers, who laboured to improve the waste by clearing wood and draining swamps. Their impact, and that of other monastic houses in the region, was to create a landscape looking in places like an open prairie. This was because there were few tenant farmers and therefore no demand for nucleated villages.

Relatively little natural woodland has survived in Europe (*see* Figure 8.6). Most has been managed now for centuries and even the patches of virgin forest require strict conservation measures to prevent erosion and unwanted change.

CASE STUDY: THE UNITED STATES OF AMERICA

Woodland used to cover half of the American landscape. Before the European colonization, the 12 million or so native Indians were active in agriculture and the management of forest ecosystems (Day 1953). As Cronon (1983, 47–51) shows, they created spaces in the forests by taking fuel wood and burning the bark of trees in order to plant maize on mounds in between the leafless trunks. The repeated burning of undergrowth resulted in easy to traverse open forests of large, well-spaced trees, with grass and non-woody plants replacing shrubs. The fires burned quickly and then extinguished themselves, driving game from cover but not destroying timber trees.

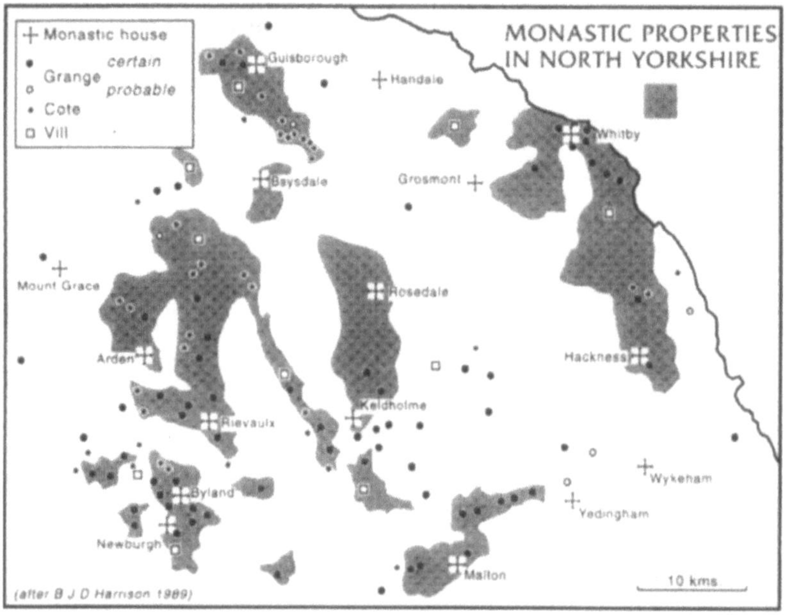

FIGURE 8.5 *Monastic properties in the North York Moors*
Source: Harrison, B.J.D. and Roberts, B.K. 1989: The medieval landscape. In Spratt, D.A. and Harrison, B.J.D. (eds) *The North York Moors: landscape heritage*. Newton Abbot: David & Charles, 72–112

At first the settlers were concerned mainly with self-sufficiency and cleared only as much as they needed for subsistence, winter fuel and a log cabin, perhaps a few hectares at first, expanding later as urbanization generated a demand for food. The labour needed for felling, 80 man days per hectare, was beyond the resources of some families so trees were ring-barked and left to rot. There are no reliable records of the extent of these improvements, but one estimate has 46 million hectares cleared by 1850, with further rapid pulses in the 1850s and 1870s (*see* Figure 8.7).

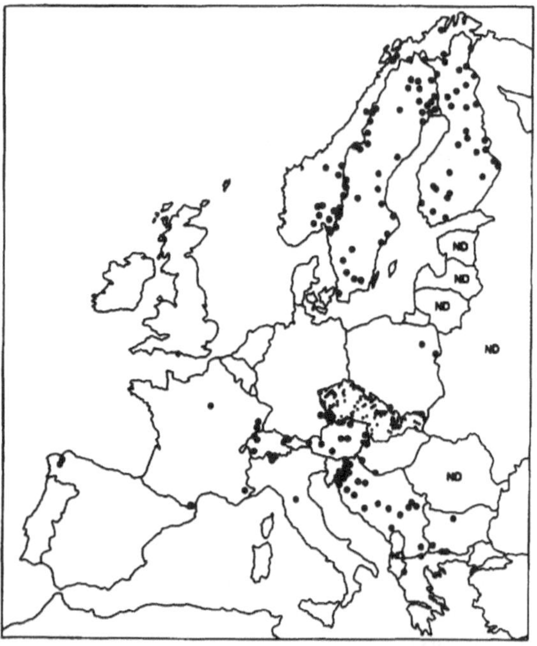

FIGURE 8.6 *Surviving virgin forests in Europe*
Source: Peterken, G.F. 1996: *Natural woodland: ecology and conservation in northern temperate regions*. Cambridge: Cambridge University Press

Opening up the largely treeless prairies in the later nineteenth century was an altogether different exercise, but it still required timber for housing, fencing and fuel (see Williams 1989). The forests of the east, and later of the mountains in the west, remained under the axe in order to provide a steady supply, now as a commercial enterprise rather than a by-product of agricultural clearance. Logging already produced 150 million board metres in 1801, rising to 2.4 billion in 1859, and an all-time peak of 14 billion in 1904 (Figure 8.8). This expansion was made feasible by better cutting and processing equipment, and the transportation facilities of the railways, and was organized by a new breed of industrial capitalism. Much of the cleared land was useless for agriculture, being on poor glacial soils around the Great Lakes or on steep slopes in the Rockies.

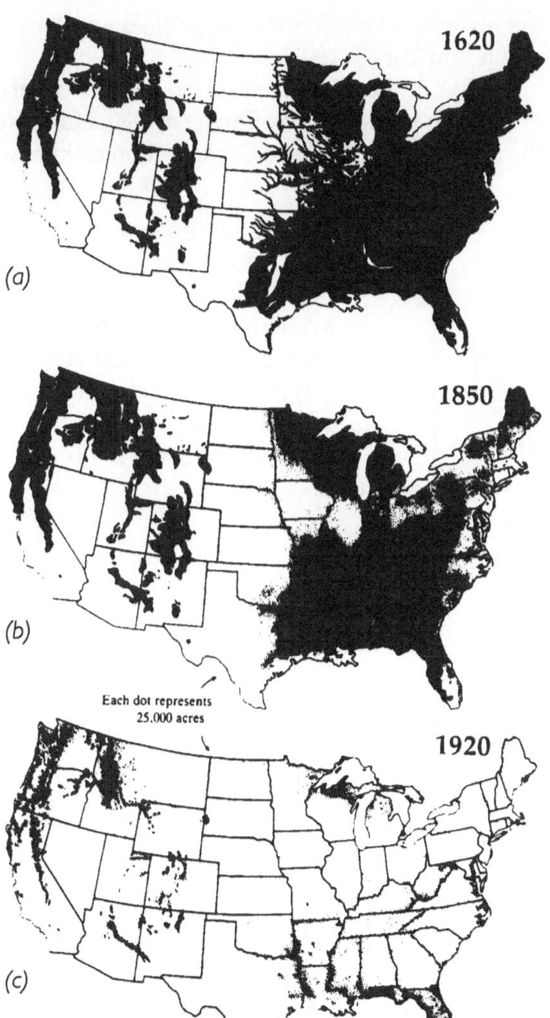

FIGURE 8.7 *Virgin forest in the USA*
(a) 1620
(b) 1850
(c) 1920
Source: Williams, M. 1990: The clearing of the forests. In Cozen, M.P. (ed.) *The making of the American landscape*. Boston: Unwin Hyman, 146–68

MODERN DEFORESTATION

The rate of woodland clearance has accelerated in the twentieth century, especially in the developing countries of Asia, Africa and Central and South America (Table 8.1). Some has been for the creation of farms or the supply of fuelwood to growing populations, but the demand from Europe, North America, and recently Japan, has continued to be insatiable and very difficult for poor countries to resist.

European colonization was especially important in setting the early trends. In Brazil, for instance, clearance was vigorous from the seventeenth century onwards, first in the coastal subtropical forests where hardwoods were the prize and later for the production of export cash crops such as sugar and coffee. Erosion of the extensive inland forests began with mineral extraction and continued in the twentieth century with waves of peasant settlers seeking a new life away from poverty near the urbanized coastal strip or in the drought-stricken North East, often on government-sponsored projects. More recently, there has been the widespread creation of large cattle ranches by the extravagant destruction of forest, and logging by large-scale corporate enterprises has added a new dimension of efficient and ruthless exploitation. Modern technology has exacerbated the problem, the portable chain saw and tractorized machinery in particular allowing a rapid, flexible and often selective harvesting regime.

The last two decades of the twentieth century have seen a hastening of tropical deforestation, balanced by heightened public awareness and an outcry of western opinion. Such indignation is a little ironic coming from societies which have already transformed their own landscapes but the concern is genuine and is rightly focused on the likely reduction of the earth's biodiversity if its most complex ecosystems are violated. The facts are contested, however, with estimates varying alarmingly as to both the extent of forests and their rate of clearance (Williams 1989; Grainger 1993). The most authoritative recent study (FAO 1993) indicates a shrinking of tropical rainforests by 154 million hectares in the 1980s, 8 per cent of the total, with the greatest impact being in Indonesia, Brazil, Malaysia, Zaire and Colombia, in that order (Table 8.2). As a result, there was a loss of approximately 2.7 per cent of species and 7.8 per cent of biomass between 1981 and 1990.

Today 31 per cent of the world's land surface remains forested, but a third of this is open forest. Closed forest covers the whole surface of the land but open forest is degraded so that the canopy is incomplete due to thinning or its fragmentation and isolation into smaller and smaller islands. Inevitably the extension of the forest edge increases the likelihood of human incursions and the invasion of a different mix of flora and fauna (Skole and Tucker 1993).

Table 8.3 illustrates the complex nature of forest change in the tropics. Between 1980 and 1990 the Food and Agriculture Organization estimates

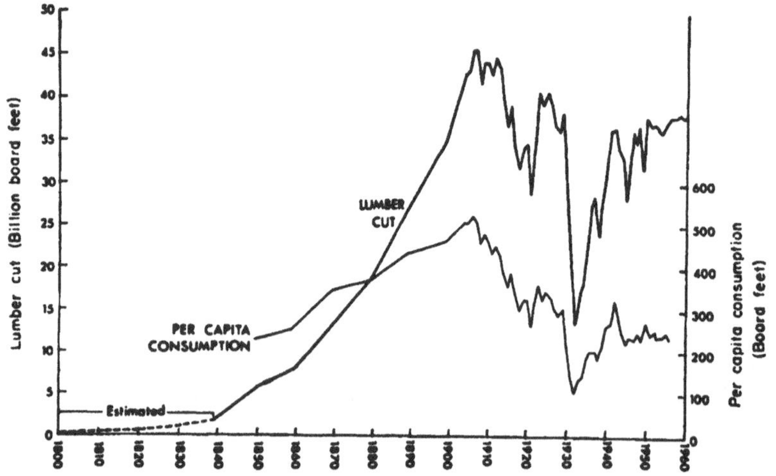

FIGURE 8.8 *Timber cut in the USA, 1800–1960*
Source: Reprinted from Williams, M. 1982: The clearing of the United States forests: the pivotal years, 1810–1860. *Journal of Historical Geography* **8**, 12–28

TABLE 8.1 *Area deforested (000 km²) in selected periods*

	Pre-1650	1650–1749	1750–1849	1850–1978	Total
North America	6	80	380	641	1107
Central America	15	30	40	200	285
South America	15	100	170	637	922
Oceania	4	5	6	362	377
Former USSR	56	155	260	575	1046
Europe	190	60	166	81	497
Asia	807	196	601	1220	2824
Africa	161	52	29	469	711
Total	1254	678	1652	4185	769

Source: mid-range estimates of Williams (1990, 180), based on assumptions about clearances per head of population

TABLE 8.2 *Tropical deforestation, 1981–90*

Forest ecosystem type	Africa	Asia	Oceania	Americas	World
Rainforest	86.4 (0.5)	148.0 (1.2)	29.3 (0.3)	450.2 (0.4)	713.8 (0.6)
Moist deciduous	251.3 (0.8)	41.8 (1.4)	0.1 (0.3)	297.9 (1.0)	591.8 (0.9)
Hill and montane	35.3 (0.8)	41.1 (1.2)	5.4 (0.3)	119.7 (1.2)	201.4 (1.1)
Dry deciduous	92.5 (0.8)	40.7 (1.0)	—	44.9 (1.2)	178.6 (0.9)
Very dry	58.7 (0.5)	—	—	1.0 (1.8)	59.7 (0.5)
Desert	3.4 (0.4)	2.9 (0.9)	—	1.6 (2.0)	8.0 (0.9)
Total	527.6 (0.7)	274.6 (1.2)	36.0 (0.3)	918.1 (0.7)	1756.3 (0.8)

Source: estimates in FAO 1993
Note: (million hectares in 1990 and annual percentage loss 1981–90)

TABLE 8.3 *Changes in tropical forest type 1980–90*

Change from (classes in 1980, km²)	Change to (classes in 1990, km²)										
	Closed forest	Open forest	Long fallow	Fragmented forest	Short fallow	Shrubs	Other land cover	Water	Plantations	Percentage 1980	Percentage 1990
Closed forest	16781	382.1	82.6	291.8	524.3	9.5	247.5	0	0	24.5	22.6
Open forest	23.6	10049.0	48.3	371.2	117.8	12.7	397.3	0.1	1.4	14.8	14.1
Long fallow	7.7	14.6	556.8	1.6	51.7	4.4	28.5	0	0	0.9	0.9
Fragmented forest	24.1	40.0	1.0	8088.8	5.8	7.7	293.5	0	0	11.3	11.8
Short fallow	7.6	10.9	9.6	2.1	2254.8	0	53.3	0.4	0	3.1	4.0
Shrubs	0.8	10.8	0	1.1	0	3877.9	154.3	0.1	0	5.4	5.4
Other land cover	16.9	38.2	11.0	63.1	34.3	86.6	26 452.0	51.2	0	35.8	37.1
Water	0.5	0	0	0.5	3.2	0.1	81.5	2960.1	0	4.1	4.0
Plantations		0	0	0	0.4	0	0.4	0	4.6	0.0	0.0

Source: estimates in FAO (1993, 34)

that 382 100 km^2 of closed forest was converted into open forest, 291 800 km^2 into fragmented forest, 524 300 km^2 into short fallow shifting agriculture and 247 500 km^2 into other land cover. There were changes in the other direction but they paled into insignificance by comparison.

This drain on the world's forest resources is neither inevitable nor irreversible. Sustainable management strategies are possible in tropical environments, as they have proved elsewhere (Steen 1987), although a *forest transition* will be involved with a smaller extent than at present and a significantly modified diversity of species. Such a transition has already taken place in most advanced countries (*see* Figure 8.9), embodying a reafforestation with trees such as pines with convenient economic characteristics.

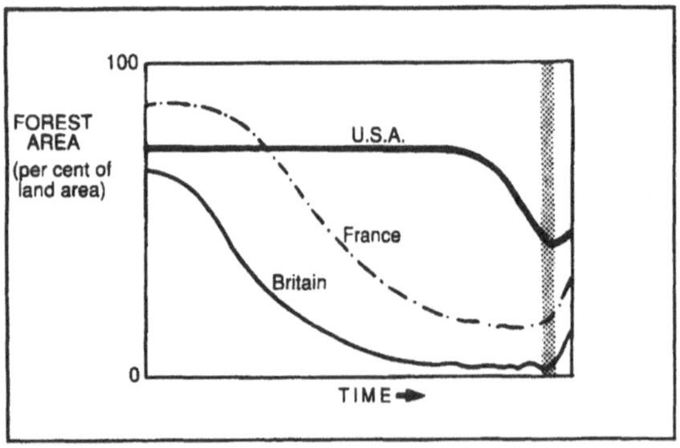

FIGURE 8.9 *The forest transition*
Source: Mather, A.S. 1992: The forest transition. *Area* **24**, 367–79

MODERN FOREST RESOURCES

Most of the world's managed forest resources lie in the coniferous forests of the temperate latitudes (Table 8.4). Here the area of forest lands is stable or increasing slowly as a result of programmes of industrial cropping. In Europe management has become a matter of planting and tending single species stands (see Figure 6.9) but in countries such as Canada and Russia there is still much reliance upon old-growth forest (Sedjo 1986).

The extent of modern timber plantations was approximately 100 million ha in 1985 and increasing. They are effectively industrial enterprises, often highly mechanized and backed up by the advanced technologies of agro-chemicals and tree breeding. Table 8.5 shows the output of industrial timber by continent. The former Soviet Union, Canada, the United States, the Nordic countries, China and Brazil are the main producers of industrial roundwood.

In 1990 about a third, 522 million ha, of tropical forests were managed in some way, including 44 million ha of industrial and non-industrial plantations. The latter is a poor record but is even worse when one considers that 85 per cent of plantations are found in just five countries: India, Indonesia, Brazil, Vietnam and Thailand. The International Tropical Timber Organization in 1988 reported the depressing fact that only 1 million ha outside India could then be said to be under sustainable yield management, out of the 828 million ha of productive tropical forest (FAO 1993).

TABLE 8.4 *World forest resources*

	Coniferous		**Broadleaf**	
	Land area (million ha)	**%**	**Land area (million ha)**	**%**
North America	400	30.5	230	13.4
Central America	20	1.5	40	2.3
South America	10	0.8	550	32.0
Africa	2	0.2	188	10.9
Europe	107	8.2	74	4.3
USSR	697	53.1	233	13.6
Asia	65	5.0	335	19.5
Oceania	11	0.8	69	4.0
World total	1312	100.0	1719	100.0

Source: Sedjo (1987, 9)

TABLE 8.5 *Forest products 1992 (million cubic metres)*

	Fuelwood and charcoal	Industrial roundwood
Africa	480	59
North America	154	593
South America	242	117
Asia	854	265
Europe	53	278
Oceania	9	35
Former Soviet Union	81	256
World	1873	1604

Source: FAO, *Agrostat*

CONCLUSION

Forest clearance has been a major signature of human impact on the earth. The situation has stabilized in the developed countries, where the management of trees now seems to have reached a sustainable level and environmental damage is more or less under control. In poor countries, with the possible exception of India and one or two others, deforestation by peasant colonists, large ranchers and commercial logging companies far outweighs sustainable timber management. This rapacious destruction of common property resources by private and state interests is fuelled by Western demand for forest products and it is therefore incumbant upon the consuming countries to give assistance (financial if necessary) to their suppliers to adopt far-reaching policies which fully account for the cost of environmental degradation. According to this logic, the loss of the tropical rainforest ecosystems with their extraordinary and unique richness of flora and fauna, is a global responsibility and the forests are in a sense therefore global landscapes (see Chapters 21 and 22).

FURTHER READING AND REFERENCES

Darby (1951), Darby (1956) and Williams (1989) provide accounts of woodland clearance in England, Europe and America respectively. For modern deforestation in poor countries, see Williams (1990).

Cronon, W. 1983: *Changes in the land: Indians, colonists and the ecology of New England*. New York: Hill & Wang.

Darby, H.C. 1951: The clearing of the English woodland. *Geography* **36,** 71–83.

Darby, H. C. 1956: The clearing of the woodland in Europe. In Thomas, W.L. (ed.) *Man's Role in Changing the Face of the Earth*. Chicago: University of Chicago Press.

Day, G.M. 1953: The Indian as an ecological factor in the north eastern forest. *Ecology* **34,** 329–46.

Food and Agriculture Organization 1993: Forest resources assessment, 1990: tropical countries. *FAO Forest Resources Paper* 112 Rome: FAO.

Gelling, M. 1984: *Place-names in the landscape: the geographical roots of Britain's place-names*. London: Dent.

Grainger, A. 1993: *Controlling tropical deforestation*. London: Earthscan.

Schama, S. 1995: *Landscape and memory*. London: HarperCollins.

Sedjo, R.A. 1986: Forest plantations of the tropics and southern hemisphere and their implications for the economics of temperate climate forestry. In Kallio, M., Andersson, A.E., Seppala, R. and Morgan, A. (eds) *Systems analysis in forestry and forest industries*. Amsterdam: Elsevier.

Sedjo, R.A. 1987: Forest resources of the world: forests in transition. In Kallio, M., Dykstra, D.P. and Binkley, C.S. (eds) *The global forest sector: an analytical perspective*. Chichester: Wiley.

Skole, D. and Tucker, C. 1993: Tropical deforestation and habitat fragmentation in the Amazon: satellite data from 1978 to 1988. *Science* **260,** 1905–9.

Steen, H.K. (ed.) 1987: *History of sustained yield forestry: a symposium*. Santa Cruz: Forest History Society.

Williams, M. 1989: *The Americans and their forest*. New York, Cambridge University Press.

Williams, M. 1990: Forests. In Turner, B.L., Clark, W.C., Kates, R.W., Richards, I.F., Mathews, I.T. and Meyer W.B. (eds) *The earth as transformed by human action*. Cambridge: Cambridge University Press.

9

THE CONTROL OF WATER

God created the world, but the Dutch created Holland.

(Popular Dutch saying)

INTRODUCTION

Water has always been of major significance in both the material and spiritual lives of people. From the semi-arid thirst of the early civilizations of Mesopotamia to the destructive floods which regularly reshape Bangladesh today, water has been at the top of the list of nature's gifts that must be controlled and exploited. We have seen aspects of this already in Chapter 3 where the hydraulic hypothesis of urbanism was tested, and in Chapter 4 where water's role in agricultural intensification and population growth was explored. In this chapter we will concentrate on wetlands and drainage.

WETLANDS

Wetlands occupy approximately 6 per cent of the earth's surface (*see* Figure 9.1) and account for 24 per cent of primary productivity. They are far from being the dull wastelands of popular imagination, but form an ecologically important part of the natural landscape which has persisted longer than the primeval woodland. This was because they were less easily modified by primitive technologies and they therefore remained largely untouched until the early medieval period. At present, however, they are under threat world-wide.

Wetlands may be divided into *marshes* (fresh or salt water), dominated by reeds, rushes, grasses and sedges; and *swamps*, frequently flooded and dominated by woody plants, sometimes trees. Some wetlands produce more organic matter than is decomposed and as a result they have deposits of peat. They are known as *bogs* when they are comprised of acid peat, *fens* when there is neutral or alkaline peat, and *mires* when the peat is especially thick. Wetlands are most often found in river floodplains where drainage is impeded and in waterlogged coastal situations.

There have been sustainable adaptations of wetland environments (Chapter 6), for instance the Marsh Arabs of Iraq, the Cajuns of Louisiana and the Camarguais of Provence, but the characteristic reaction to wetlands has usually been to drain and utilize them. We can identify seven major types of wetland modification (L'Vovich and White 1990). First there is the use of bog and ill-drained land to create new agricultural resources. This has been very common around the world (see UK case study) and is still continuing.

Second, coastal marshes were used for fish farming, for instance the dyke-pond system of the Pearl River Delta described below, and salt-making in evaporation ponds. Figure 9.2 shows a mound of salt (sodium chloride) produced at Sahline in Tunisia. Here inspiration was provided by natural salt lakes (*sebhkas*) which dry out seasonally leaving deposits of various salts. China is the largest producer of sea salt in the world using traditional evaporation techniques, with an output of 13 million tons a year.

Third, coastal reclamation takes a number of forms. One approach is by embankment in stages, using the natural processes of siltation called

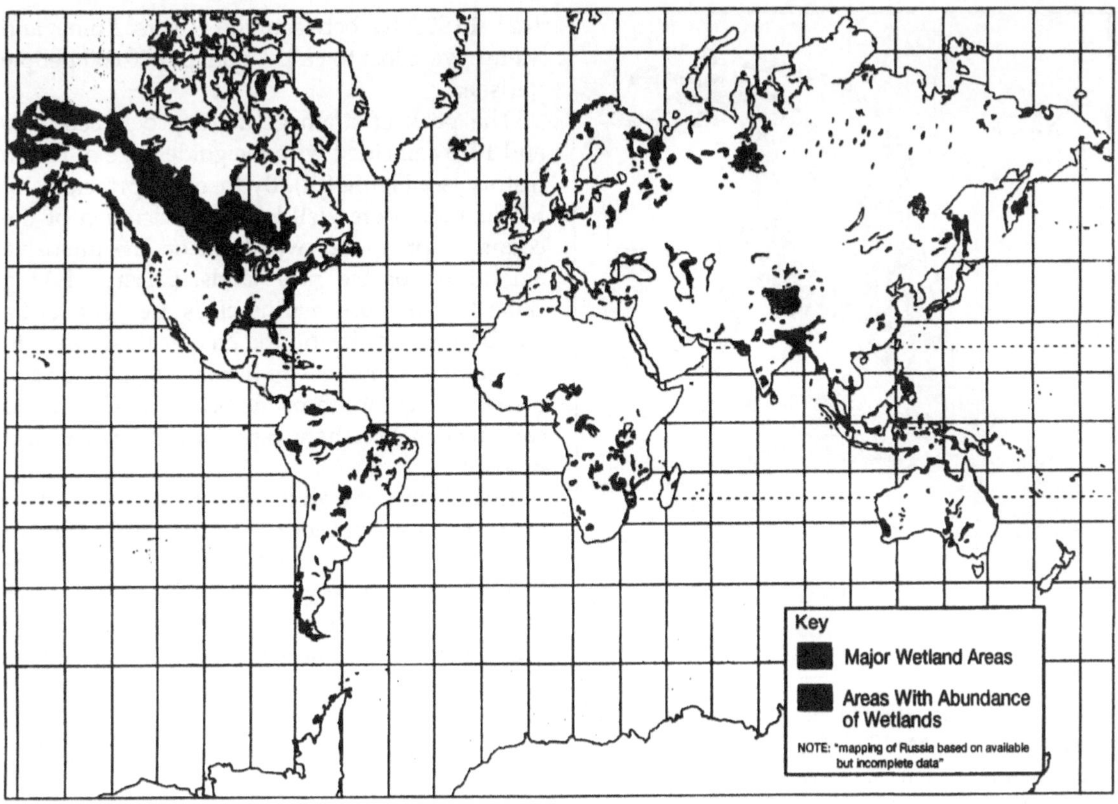

FIGURE 9.1 *Global wetlands*
Source: Reprinted from Mitsch, W.J., Mitsch, R.H. and Turner, R.E. 1994: Wetlands of the Old and New Worlds: ecology and management. In Mitsch, W.J. (ed.) *Global wetlands: old world and new*. Elsevier Science, 3–56

warping. A patch of marsh is identified that is dry at low tide and which is being colonized by salt-tolerant (halophytic) vegetation. A dyke is built to protect this area from storms and sluice gates are positioned so that water can be drained at low tide. Silt will collect gradually and create new ground, especially if helped by artificial silt traps. The freshwater equivalent is where silt-laden river waters are deliberately let on to land by controlled flooding.

Another approach is to use *polders*. These are embanked pieces of land drained in a relatively short time. Nowadays the creation of an artificial salt marsh is not necessary because of the use of advanced technologies such as mechanical diggers and pumps. The polders of the Netherlands are the most impressive. In effect, it is a country created from the sea and the estuaries of the Rhine, Meuse and Scheldt: 40 per cent is below sea level (*see* Figure 9.3).

FIGURE 9.2 *Salt making at Sahline, Tunisia*
Source: P.J. Atkins

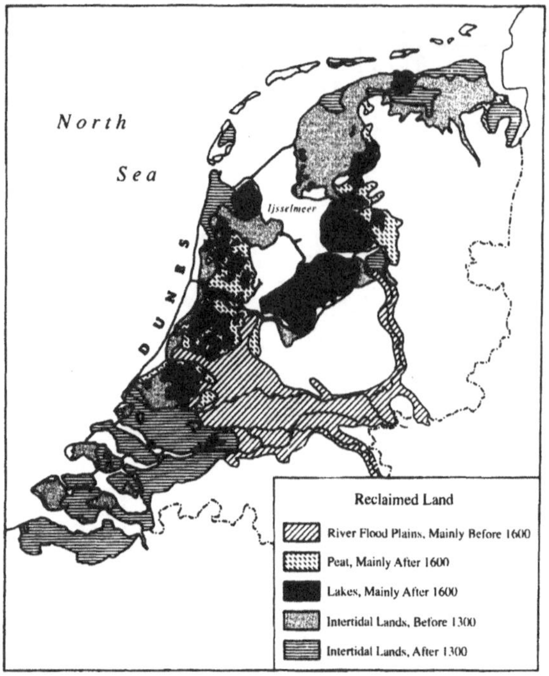

FIGURE 9.3 *Reclamation in the Netherlands*
Source: Williams, M. 1990: Agricultural impacts in temperate wetlands. In Williams, M. (ed.) *Wetlands: a threatened landscape*. Oxford: Blackwell, 181–216

Among the earliest settlers of the coastal marshes of what is now the Netherlands were groups of herders. They built artificial circular or oval mounds, *terpen*, to protect their dwellings and livestock from flooding. These refuges were small at first but were gradually enlarged in diameter and raised up to 10 metres above sea level. Linking terpen by causeways was a logical step and it was these simple dykes which by AD 1000 were beginning to alter the hydrological regime of the marshes.

In the eleventh century groups of people, *wateringues*, were formed to expedite reclamation, encouraged by abbeys, nobles and, later, by wealthy merchants from the flourishing industrial towns. Count William the Dyker, of Holland, achieved a great deal in the thirteenth century, but much of the early progress was set back by the marine floods of the thirteenth and fourteenth centuries when many dykes were swept away. In 1287 50 000 people drowned between the Zuyder Zee and Dollart in a great flood, and in November 1421 18 000 ha between the Rivers Rhine and Meuse were lost to cultivation and 10 000 people perished.

The peak of drainage activity between 1600 and 1625 coincided with the golden age of Dutch history (*see* Figure 9.4). By then the principles of reclamation were well known and much of the country's prosperity was built on the intensive agriculture of the new lands. Coastal dyking came first because marine clays are very fertile and do not shrink, and the greater tidal range in the south (2 metres in north Holland, and 4 metres at the mouth of the Scheldt) encouraged early reclamation there, especially in Zeeland and Brabant.

Inland lakes and wetlands were more of a challenge. The less stable peat material was difficult to drain by gravity. Technological innovation was one solution. Improved windmills, with larger sails which could be rotated into the wind, made lifting water a practicable possibility. Chains of

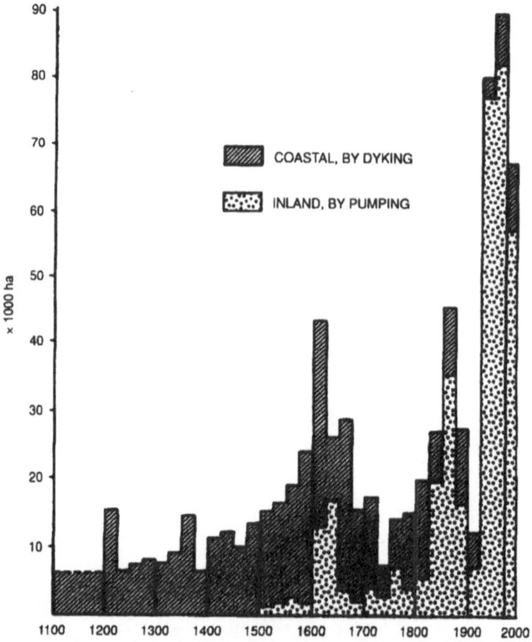

FIGURE 9.4 *Progress of reclamation in the Netherlands*
Source: Williams, M. 1990: Agricultural impacts in temperate wetlands. In Williams, M. (ed.) *Wetlands: a threatened landscape*. Oxford: Blackwell, 181–216

windmills could pump water up slopes into canals and onward to the sea.

In the nineteenth and twentieth centuries mechanical pumps have further improved drainage efficiency and made it possible to contemplate the reclamation of the Zuyder See/IJsselmeer. In all, the total area of the Netherlands reclaimed from 1200 to the present has been 6925 km^2 (*see* Figure 9.5).

Fourth, coastal protection against erosion is common, by the construction of features parallel to the coast, sea walls and embankments, which absorb wave energy, or structures at right angles to the coast, for instance groynes, which deflect or reduce the drift of sediment. Unforeseen consequences may result, however, such as increased erosion, because engineers may not be fully aware of the geomorphological relationships between the forces of erosion and deposition.

Fifth, land is reclaimed for urban and industrial use, for instance in Hong Kong which has a very restricted area. Its rapid economic expansion since the Second World War, particularly the growth of manufacturing industry, required more flat land than was available. Reclamation was an expensive solution but there was really no alternative. A quarter of the urban area is built on sites reclaimed by the dumping of fill.

The other obvious example is Venice. The

FIGURE 9.5 *The land of one thousand islands, Netherlands*
Source: KLM Aerocarto

original inhabitants were probably refugees from the invasions of Alaric the Goth in AD 402, but the advantages of the natural protection of the lagoon and its location with respect to possible long-distance trading routes were not lost on their descendants who drove piles into the mud and began a process of reclamation and building which has created one of the world's great cultural gems at the very edge of early water control technology (Cosgrove 1990).

Sixth, there is agricultural land with impeded drainage that would not, strictly speaking, be included in most definitions of wetlands, but which has nevertheless been the focus of much activity and investment with the objective of bringing it into more productive use. As an example, one of the most important improvements of agricultural land in England between 1840 and 1900 was the underdrainage of about 20 per cent of the cultivated area. Similar adaptations were tried earlier with the ridge and furrow systems of medieval Europe (Chapter 11) and the drained fields of the pre-Colombian Americas.

The seventh type of human activity is the artificial creation of wetlands, either as a by-product of peat cutting as with the Norfolk Broads, the flooding of *water meadows* to increase the productivity of grassland, and the creation of artificial lakes for the purpose of navigation, irrigation or micro-climate modification. A fascinating example of the last category was the nineteenth-century French plan to flood the Shott Djerid in southern Tunisia (Heffernan 1990).

Case study: the drainage of the Fens in eastern England

The Fens around the Wash have formed on a clay plain where the drainage was impeded by the deposition of river and marine silt and organic peat. It is not entirely flat, with some low hills forming islands which became important for settlement, such as the Isle of Ely.

The sea level has risen and fallen many times, for instance in the Roman period it was lower and the Fens were then a prosperous agricultural area. A long channel, the Car Dyke, was even built to reduce the risk of flooding and doubled as a canal for taking produce to the city of Lincoln. But even in its flooded state marshland had its uses. It may have seemed inaccessible, a defensive feature exploited by Hereward the Wake in 1071, but its resources were much appreciated and exploited by neighbouring people (Chapter 6).

Yet there was an incentive to drain the marshes because fen soils are naturally fertile. Reclamation seems to have begun before 1066 and was well under way by the twelfth century. Creating embankments was labour-intensive so it was undertaken most readily by the large monasteries, such as Spalding and Crowland, which had the ability to mobilize the necessary work force. There is some evidence that peasants were also involved, working together as whole villages. By winning new land for themselves they were able to prosper and population increased rapidly.

Large-scale drainage works are evident from the sixteenth century onwards. As Cosgrove (1990) points out, there were more than economic and technological forces at work. The change in cultural orientation that we call modernization at this time was changing people's view of the world, and scientific reason was applied to bringing the earth more extensively under human control. In effect, hydrology as a technical discipline was brought to bear widely in Europe to modify rivers by draining their floodplains, dredging their channels and simplifying their drainage networks, to build embankments and coastal defences against marine encroachment, and to create new tracts of farm land from formerly useless marsh. This modern project has culminated in the construction of the Suez and Panama Canals and water engineering on a large scale in mega projects such as the Aswan Dam.

In the early modern period Fenland waterways were straightened to allow easier trade and better drainage, and also the drainage network was modified with new cuts, but the dissolution of the monasteries by Henry VIII in 1536 (smaller) and 1539 (larger) meant the removal of one of the principal actors, leading to the decay of some banks and flooding. At the same time, however, there were centralized attempts to regulate the use of the Fens, such as an Act of Parliament restricting fowling (1534), and a code of fens laws (1548).

There were some experiments with draining, but rapid and large-scale drainage was delayed until the early seventeenth century, prompted by the worst floods for 400 years (1607, 1613, 1614). It had to be done by groups of adventurers who clubbed together and invested the large sums necessary. In 1630–1, for instance, the Earl of Bedford and thirteen co-adventurers were approached by local landowners to drain the peat of the southern fenland. They employed a Dutch engineer, Cornelius Vermuyden who, despite some disruption during the Civil War, cut many drains, some of substantial size such as the Old (1637) and the New (1651) Bedford Rivers (*see* Figures 9.6 and 9.7).

There were problems. First, Vermuyden underestimated the technical challenge of his gravity-based system. For instance, it was soon found that the Old Bedford River could not cope with the flow of water and had to be supplemented by a parallel cut, the New Bedford River. Second, no one anticipated the fact that peat shrinks when it is dry. By 1700 the new drains were above the level of the peat, increasing the risk of floods. Third, many of the locals objected to any drainage. The livelihoods of the fishermen and fowlers were ruined and sluice gates interfered with the commercial waterways. There were frequent riots, with people breaking down the banks and filling in channels.

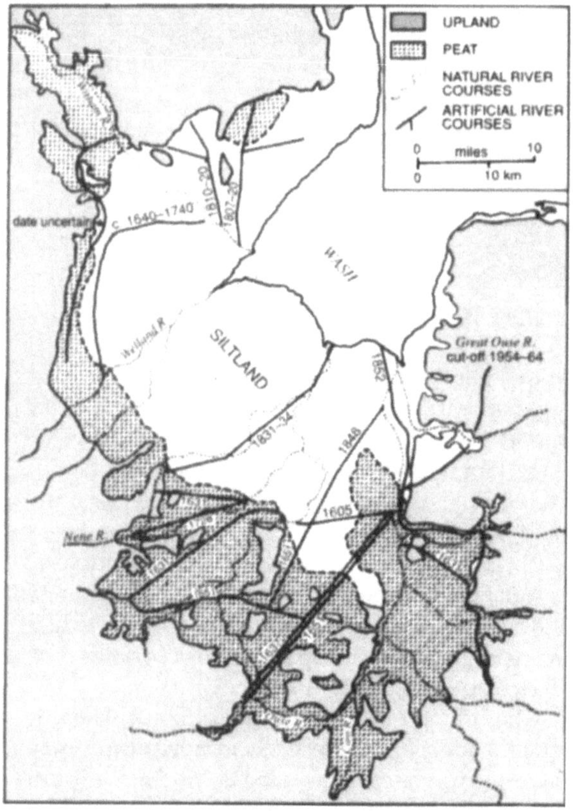

FIGURE 9.6 *The course of major artificial drains in the Fenlands created from 1600 onwards*
Source: Williams, M. 1990: Agricultural impacts in temperate wetlands. In Williams, M. (ed.) *Wetlands: a threatened landscape*. Oxford: Blackwell, 181–216

The problems caused by shrinking peat was partially solved by the use of windmills for pumping from the eighteenth century, and later by steam pumps (1819) (Butlin 1990). The greatest single project was the drainage of Whittlesey Mere (Cambridgeshire) in the mid-nineteenth century. At the time this was heralded as a great technical achievement, but today it is seen as one of the greatest disasters to have befallen British nature conservation. Much new land was created by drainage and the extent of the Fens gradually dwindled from the original 280 000 hectares to only 855 hectares now left in their natural state.

In the twentieth century further damage to wetlands has been caused by nitrogen enrichment from fertilizers, leading to eutrophication. Because of incentives from the UK government and the subsidies of the European Union's Common Agricultural Policy, a lot of marshland has been drained in recent decades. The peak period was the 1970s and 1980s, with 104 000 hectares of drainage in 1974 alone.

We can still see the old water courses (roddons) from the air (*see* Figure 9.8). Their silt stands above the level of the shrunken peat. There remains a danger of flooding from storm surges in the North Sea, such as that which caused loss of life in 1953. The current rise in sea levels is an additional threat.

The modernist discourse of wetland drainage and modification has in the last two decades come under sustained fire, however, from the environmentalist movement who have pointed to the loss of precious habitats and their biodiversity. This green lobby has been supplemented by the argument from activists in developing countries that large dams and their associated lakes

FIGURE 9.7 *The Old and New Bedford Rivers (1978)*
Source: Committee for Aerial Photography, University of Cambridge

displace poor farmers and that the newly created stretches of stagnant or slow-flowing water are ideal for the breeding of disease vectors such as the schistosomiasis-carrying snail and the biting flies that spread river blindness. In recent years a number of hydrological mega projects have been cancelled or shelved because of political opposition and there is a crisis of confidence now among the water engineers and their political masters (Cosgrove 1990).

CASE STUDY: RAISED, DRAINED AND SUNKEN FIELDS

An exceptionally interesting method of moisture control amongst those pre-industrial peoples who possessed no plough was the use of raised, drained and sunken fields. These have recently received attention from researchers and we now know that they were widespread in the Americas, sub-Saharan Africa and Oceania. Figure 9.9 shows the current state of our knowledge in the New World, and this will no doubt be refined as archaeologists and geographers undertake more field work. The total area identified so far is 270 000 ha but this is very approximate, not least because of the hotly disputed identification of 128 500 ha in the Maya region by a group of scholars using remote sensing techniques (Adams *et al.* 1981; Pope and Dahlin 1989; 1993).

The dating of these fields is difficult. There is some evidence that raised field agriculture may have begun as early as 1850 BC in Surinam but they certainly flourished most widely from about AD 600–1500. The Spanish conquistadores did not find them in every location because local circumstances occasionally led to their abandonment.

Most of the raised field systems have been found in areas where pre-industrial population densities were high, at or above 300 per km^2. In

FIGURE 9.8 *A roddon near Littleport, Cambridgeshire (1969)*
Source: Committee for Aerial Photography, University of Cambridge

each case they had been preceded in time by dryland systems, so it seems reasonable to speculate that the construction of raised fields was a response to an increased demand for food, requiring intensification. Certainly, the remarkable levels of productivity were sufficient to support brilliant civilizations such as the Tiwanaku of the Titicaca Basin (Kolata 1993).

Could such systems have grown organically? Perhaps in some locations they were the result of the efforts of individual small farmers, but there is evidence of large-scale planning in the Basin of Mexico, and at Pulltrouser Swamp, Belize, the hydraulic regime suggests that most fields could not have functioned until the whole system was completed.

The fact that such a successful indigenous agricultural technology has decayed is a matter of regret and efforts are in hand in several countries to revive and adapt it to modern needs. Again, planning will be needed.

THE VARIETY OF RAISED FIELD FORMS

Riverine situations, including backswamps and other low-lying floodplain terrain where water tables are high, are relatively common. They are not enclosed, and experience regular surface flow and periodic inundations. Here raised fields are usually used for drainage purposes.

In contrast, basin depressions, both palustrine and lacustrine, are closed systems with little flow. Water tables fluctuate in response to the changing balance between evaporation and subsurface outflow. This sort of environment is found on lake peripheries, and in seasonally inundated depressions, swamps, marshes and bogs. Here raised fields are used for water retention.

The third relevant biotope is the seasonally flooded open savanna found in regions such as the Llanos de Mojos, Bolivia (*see* Figure 9.10).

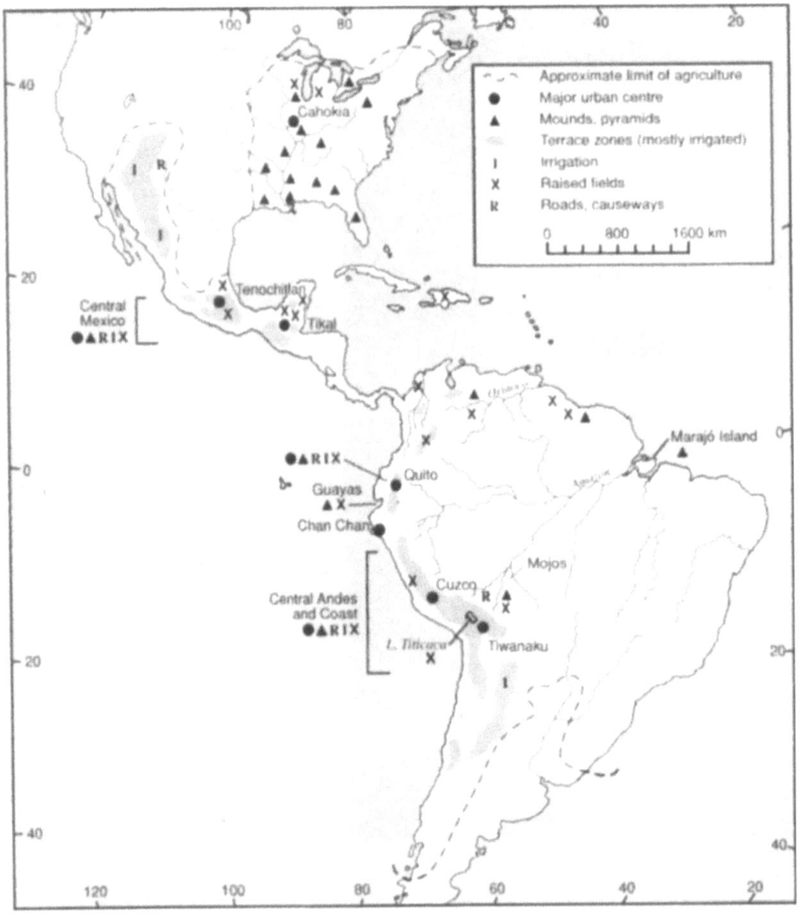

FIGURE 9.9 *The major sites of raised fields in the Americas*
Source: Reprinted from Denevan, W.M. 1992: The pristine myth: the landscape of the Americas in 1492. *Annals of the Association of American Geographers* **82,** 369–385

Between them, these three settings yield evidence of at least nine types of field systems.

1. Water recessional. Here the crop is planted in the dry season as the waters recede, and harvested before the floods return. There is an advantage that the subsoil is moist and the flood waters keep the weeds down. Water recessional agriculture is a simple technique requiring little technology. Modern examples include the cultivation of playas, fertile alluvial soils on sand bars, shores and islands exposed in dry season, and levées (river bank backslopes).
2. Drained fields. In South America there are drained fields near Lake Titicaca and on the savannas of western Venezuela. There are also raised fields and ditches running down natural levées adjacent to rivers channelling overflow water into back swamps, such as at San Jorge, Colombia.
3. Hydrophytic agro-ecosystems. The use of water-loving plants such as rice, taro is an obvious wetland adaptation. Many Pacific islands have taro mounds.
4. Ridged fields. Here earth is heaped into linear ridges which have a number of functions. Among the best examples of ridged fields are those near Lake Titicaca.
5. Chinampas. Mexican chinampas have water held in canals between linear fields. They are more advanced versions of ridged fields.
6. Planting platforms. The *tablones* of highland Guatemala are raised planting beds separated by ditches which are used for both drainage and irrigation. Platform fields are common in the Pacific.

FIGURE 9.10 *Ridged fields in the Llanos de Mojos, Bolivia*
Source: Denevan, W.M. 1978: Latin America. In Klee, G.A. (ed.) *World systems of traditional resource management*. London: Arnold, 217–44

7. Sunken fields. Examples of one type of sunken field, *mahamaes*, are found in at least 10 coastal Peruvian valleys. Their origin is uncertain but water management was certainly involved, either to improve access to the water table or to collect run-off. The latter function is more likely with the depression fields, *Qochas*, of the altiplano. Flores Ochoa (1987) has identified 20 000 qochas of various shapes and sizes, each up to 4 metres deep and 100 hundred metres or more in diameter. As the water recedes they are planted with potatoes, rotated with quinoa, cañiwa, oats and barley.
8. Linear fields. Andean linear fields have a drainage function. They are found principally in northern Ecuador and central Colombia, and run downslope at 10 to 20 metre intervals, alternating with ditches. Soil erosion is not a major problem in this region but the accelerated drainage of those soils with a high clay content would have prevented slumping. Similar fields are found in Tanzania.
9. Mounds and ridges are also common field types for the control of water, mainly for the control of upper soil waterlogging in the cultivation of maize, manioc and potatoes. They are common in Africa and archaeological evidence indicates that they were extensively used by native North American peoples.

The likely functions of raised fields

An obvious function of a raised field is to facilitate the drainage of excess moisture. There are other considerations:

- As a result of the proximity of the planting surface to water, a perched water table may be induced if the soil is sufficiently porous. This allows sub-irrigation of the plant roots, but the depth of root penetration will be limited by the saturated layer. Periodically, soil must be removed to prevent surface height building up so that sub-irrigation becomes inefficient.
- Raised beds are currently popular with modern gardeners because careful soil preparation means that a well-aerated soil warms more easily, and suffers less from compaction during cultivation, so plant roots can flourish, and weeding and harvesting of root crops involve less effort and drainage is improved.
- Raised field hydraulic systems allow management of salts and pH levels by soil modification and the partitioning of fresh and salt water, for instance by using water from a river or spring.
- Soil erosion is reduced or reversed as sediments are trapped. The soil builds up a high organic content which facilitates water retention.
- Microclimate modification results because the wet soil and water in the swales absorb heat and release it at night, increasing the surface temperature by 0.5–1.5°C. Frost hazard is reduced and growing seasons are lengthened. The surface geometry of the fields may also have an effect, depending on the orientation of the ridges, because of exposure to solar radiation.
- The soils of raised fields have a high organic content because of mulching with plant material, human and animal manure, and fertility can be raised with regular applications of nutrient-rich canal muck. Green manure is available in the form of aquatic plants growing in the ditches.
- The ditches are used for fish farming.

CASE STUDY: THE CHINAMPAS OF MEXICO

The swamp gardens of Mexico are justly famous for their high productivity and an agricultural regime which benefits from the close proximity of water. In the eyes of many they are the archetypal raised fields, although they were relative latecomers. The archaeological evidence suggests that they did not become significant until after AD 900 and did not achieve maturity as a system until *c.* 1350 in association with urbanization and careful planning by the Aztecs (*see* Figures 9.11 and

FIGURE 9.11 *An Aztec map of Tenochtitlán–Tlateloloco showing chinampa holdings*
Source: Reprinted from Coe, M.E. 1964: The chinampas of Mexico. *Scientific American* **211,** 1, 90–8 by permission of Eric H. Mose

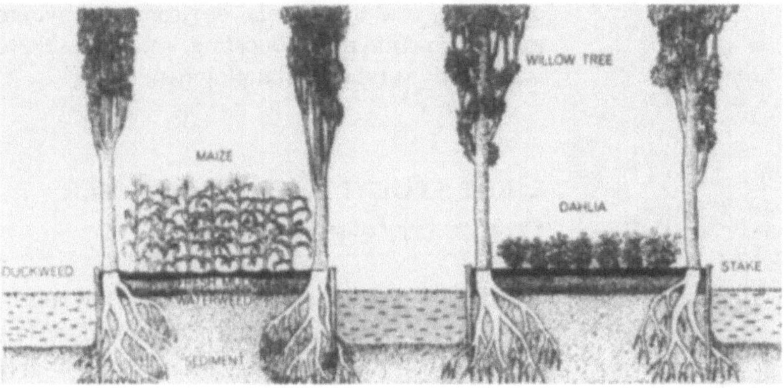

FIGURE 9.12 *Cross-sectional diagram of chinampas*
Source: Reprinted from Coe, M.E. 1964: The chinampas of Mexico. *Scientific American* **211,** 1, 90–8 by permission of Eric H. Mose

9.12). Cultivation was tied into a large-scale water-control system constructed, maintained and managed by the state.

Research by Armillas (1971) has identified about 120 000 ha of raised fields in the Basin of Mexico, clearly planned in rectangular blocks and generally aligned SSW-NNE. They would have produced perhaps half of the food needed for the 200 000 population of the Aztec capital of Tenochtitlán in the early sixteenth century. The Xochimilco area, 15 km to the south, has the greatest concentration of surviving chinampas.

There were never any floating gardens in Mexico, as sometimes reported. They were made of mud scooped up from the lake bed, mixed with organic matter such as plants growing in the canals, held together in a rectangular cane framework and bordered by willow (*Ahuejote*) trees (*see* Figures 9.13 and 9.14). Chinampa sizes vary from 2–10 metres wide, and 10–200 metres long (some are as much as 900 metres in length). Plot height is generally 1 metre above the canal level. Fertility was acquired from canal mud, household refuse, bat dung and human manure. With transplanting of seedlings, it is possible to raise two maize crops per year, and many other crops are grown intensively, such as beans, chili peppers, tomatoes and grain amaranth. In the wet season (June to October) the full canals boost the water table sufficiently for sub-irrigation of the roots, but water has to be added to the field surface in the dry season.

There may be some modern applicability of chinampa agriculture in other parts of Mexico. In the late 1970s experiments began in the tropical lowland states of Tabasco and Veracruz, in a very different environment from the altiplano of the Basin of Mexico. Results have been generally positive, with crops being produced on swampy land. It is conceivable that this type of traditional,

FIGURE 9.13 *Chinampa plots and canal, Tlahuac, Mexico*
Source: Armillas, P. 1971: Gardens on swamps. *Science* **174,** 653–61

FIGURE 9.14 *Chinamperos scooping mud from a canal*
Source: Coe, M.E. 1964: The chinampas of Mexico. *Scientific American* **211**, 1, 90–8

labour-intensive, and low-tech agriculture might be appropriate to the needs of people in a variety of tropical and subtropical environments where food production needs boosting, and where there is currently rural underemployment.

CASE STUDY: THE PEARL RIVER DELTA DYKE-POND SYSTEM

The delta of the Zhujiang (Pearl River) in southern China provides soils which are rich, deep but ill-drained. The region was a swamp but has gradually been reclaimed by a piecemeal process of embanking which has given it a unique landscape (*see* Figures 9.15, 9.16 and 9.17). An area of 800 km^2 has been made into a dyke-pond system which has an integrated economy of agriculture and aquaculture unlike anywhere else in the world supporting 1.2 million people.

At first many local farmers used their land for rice paddy but over the last 600 years much has been converted to fish ponds, with sugar cane and mulberry trees growing on the intervening dykes. The silt which accumulates in the ponds is scooped out two or three times a year and used as fertilizer on the dykes and in return the pond

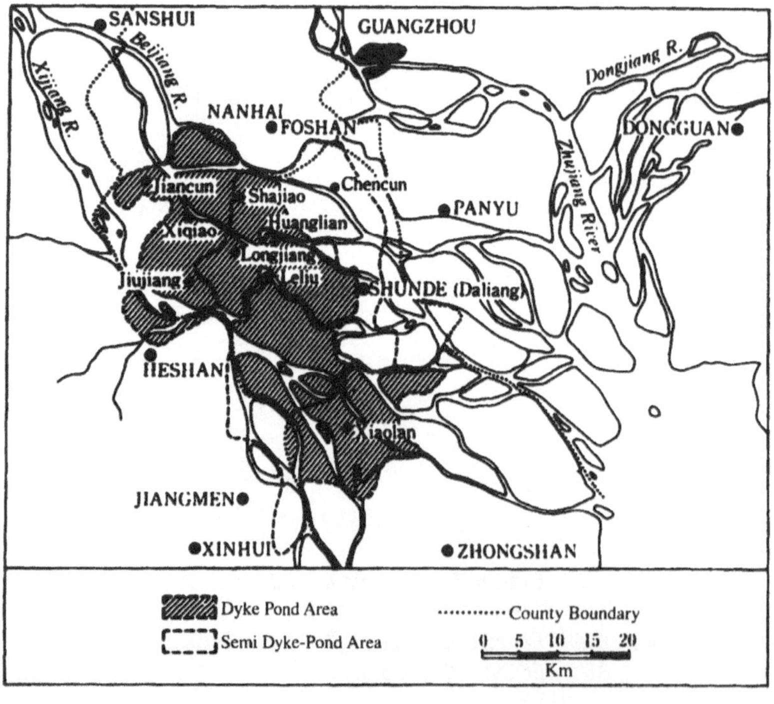

FIGURE 9.15 *The Pearl River delta*
Source: Ruddle, K. and Zhong, G.F. 1988: *Integrated agriculture-aquaculture in south China: the dike-pond system of the Zhuijiang delta.* Cambridge: Cambridge University Press

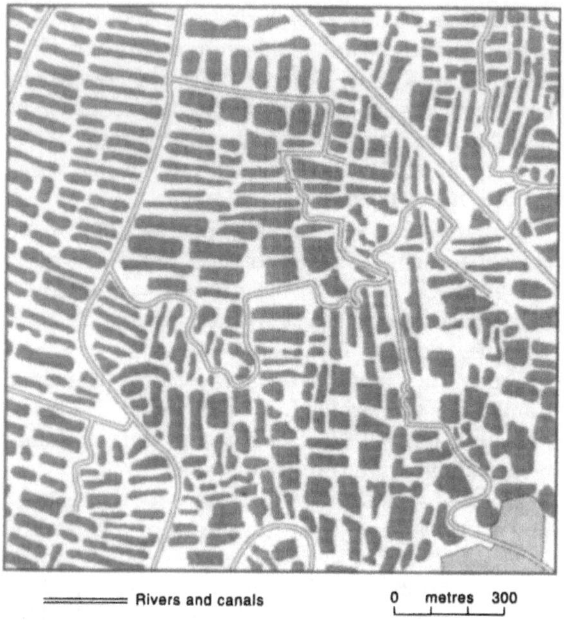

FIGURE 9.16 *A dyke-pond landscape, Leliu Commune, central Zhuijang delta*
Source: Based on Ruddle, K. and Zhong, G.F. 1988: *Integrated agriculture-aquaculture in south China: the dike-pond system of the Zhuijiang delta.* Cambridge: Cambridge University Press

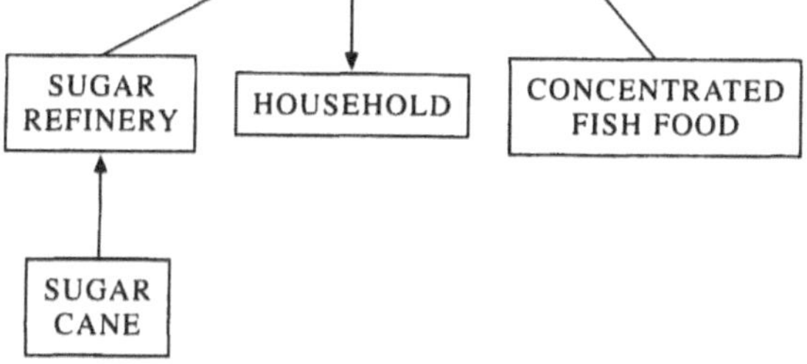

FIGURE 9.17 *A flow diagram of the dyke-pond system*
Source: After Ruddle, K. and Zhong, G.F. 1988: *Integrated agriculture-aquaculture in south China: the dike-pond system of the Zhuijiang delta.* Cambridge: Cambridge University Press

receives various inputs. Most of the fish are sold into the market and this one region provides half of the total production of Guangdong Province and 80 per cent of China's live exports. The sugar cane and mulberry trees are also cash crops, the leaves of the latter being used to feed silk worms. Pigs are fed on sugar cane waste and water hyacinth and their manure is recycled as pond fertilizer. Vegetables are also grown on the 6–10-metre wide dykes, and mushrooms are cultivated during the off-season in the silk worm sheds (on pond mud).

In the pond the Grass carp is perhaps the most valued fish. It is voracious, consuming between 100–174 per cent of its own body weight each day in grass, aquatic plants and crop residues. The other species are complementary feeders, the Big-head carp preferring zooplankton, the Silver carp phytoplankton, and the Black, Mud and Common carp eat decomposing matter on the bottom. This fresh water ecosystem is not self-regulating. It needs careful management to maintain the balance of nutrients and to prevent the build-up of the toxic products of decomposition and other chemical reactions. Skilled farmers are able to keep the ponds aerobic, with acceptable levels of pH, and with a good mix of microflora and microfauna. The fish are fed daily on bean cake, peanut cake, elephant grass, household and crop residues, and sugar cane waste from the refinery. Pond fertilizer includes pig and human excrement, and silkworm waste. Tea-seed cake and quicklime are added to adjust the water's chemical composition.

This very successful system is in fact rather delicate. The market price of the various components will shift through time now that the government has liberalized the economy, and there is no guarantee that all of the sub-systems will remain profitable. The price of silk, for instance, plummeted during the depression of the 1930s. At present there is prosperity and the dyke-pond system looks a like beautifully adapted and sustainable wetland ecosystem. It could be replicated in similar environments to good effect, especially since over-fishing is threatening offshore stocks. But well-meaning planners intervene at their peril. Recent expansion in the Zhujiang delta has involved reduction of the land–water ratio to 1:4 and this is insufficient to supply the necessary inputs for the ponds without imports from other parts of China. Outward linkages imply a reduction in local levels of self-sufficiency and sustainability.

CONCLUSION

The management of wetland ecosystems has produced some dramatic landscape changes visually and in environmental terms. Much additional agricultural land has been created, some of it producing high crop yields because of the fertility of marine and riverine sediments, but a negative by-product has been the reduction of biological productivity in wetlands, which are second only to tropical rainforests in their variety and complexity of flora and fauna. We must ask whether this has been worthwhile, especially when one considers that there are sustainable and productive wetland land-uses of ancient pedigree, such as the dyke-pond system or the raised/drained field systems, which do not demand the elimination of standing water. The modernist approach to producing yet more dryland arable has been very powerful over the last 500 years but the turn of the millennium may coincide with its re-evaluation and decline.

FURTHER READING AND REFERENCES

Audrey Lambert's book on drainage in Holland is still the best account in English, and Darby (1956, 1983) is the authority on the English Fenland. For wetland fields in the Americas, see Coe (1964).

Adams, R.E.W., Brown, W.E. and Culbert, T.P. 1981: Radar mapping, archaeology, and ancient Maya land use. *Science* **213,** 1457–63.

Armillas, P. 1971: Gardens on swamps. *Science* **174,** 653–61.

Butlin, R.A. 1990: Drainage and land use in the Fenlands and fen-edge of northeast Cambridgeshire in the seventeenth and eighteenth centuries. In Cosgrove, D. and Petts, G. (eds) *Water, engineering and landscape: water control and landscape transformation in the modern period*. London: Belhaven, 54–76.

Coe, M.E. 1964: The chinampas of Mexico. *Scientific American* **211,** 90–8.

Cosgrove, D. 1990: An elemental division: water control and engineered landscape. In Cosgrove, D. and Petts, G. (eds) *Water, engineering and landscape: water control and landscape transformation in the modern period*. London: Belhaven, 1–11.

Darby, H.C. 1956: *The draining of the fens*. 2nd edn, Cambridge: Cambridge University Press.

Darby, H.C. 1983: *The changing fenland*. Cambridge: Cambridge University Press.

Flores Ochoa, J.A. 1987: Evidence for the cultivation of *Qochas* in the Peruvian altiplano. In Denevan, W.M., Mathewson, K. and Knapp, G. (eds) Pre-hispanic agricultural fields in the Andean region, *British Archaeological Reports, International Series* **359**: 399–402.

Heffernan, M.J. 1990: Bringing the desert to bloom: French ambitions in the Sahara desert during the nineteenth century – the strange case of 'la mer intérieure'. In Cosgrove, D. and Petts, G. (eds) *Water, engineering and landscape: water control and landscape transformation in the modern period*. London: Belhaven, 94–114.

Kolata, A. 1993: *The Tiwanaku: portrait of an Andean civilization*. Cambridge, MA: Blackwell.

Lambert, A. 1985: *The making of the Dutch landscape: an historical geography of the Netherlands*. 2nd edn, London: Academic Press.

L'Vovich, M.I. and White, G.F. 1990: Use and transformation of terrestrial water systems. In Turner, B.L. *et al.* (eds). *The earth as transformed by human action*. Cambridge: Cambridge University Press, 235–52.

Pope, K.O. and Dahlin, B.H. 1989: Ancient Maya wetland agriculture: new insights from ecological and remote sensing research. *Journal of Field Archaeology* **16,** 87–106.

Pope, K.O. and Dahlin, B.H. 1993: Radar detection and ecology of ancient Maya canal systems – reply. *Journal of Field Archaeology* **20,** 379–83.

10

LANDSCAPES ON THE MARGIN

DESERTS, HILLSLOPES, HEATH, MOOR AND GRASSLAND

> The great inviolate place had an ancient permanence which the sea cannot claim ... The sea changed, the fields changed, the rivers, the villages, and the people changed, yet Egdon [Heath] remained.
>
> (T. Hardy 1891: *Tess of the D'Urbervilles*)

INTRODUCTION

The discussion in Chapters 8 and 9 concerned human efforts to bring under control land with a potentially high productivity. The technical challenge was often severe, especially in the case of the reclamation of extensive wetlands, but the rewards were significant because of the high yields that could be won from fertile soils. This chapter is rather about the marginal lands that were distinctly less inviting to human improvement. They were severely constrained physically by aridity, infertility, or steepness of slope, and were often only tackled when all other available areas had been settled. Of course the process of colonization on the margin is still going on as the human ecumene expands to accommodate population growth.

ARID LANDS

We have already seen (Chapter 3) how several early civilizations challenged the desert and succeeded by the control of water supplies for crop irrigation. These hydraulic skills continued to be exercised in a broad belt from the Mediterranean to China but it was not really until the nineteenth century that the West European powers began to take an interest. Their colonial endeavours had brought them into contact with frontiers of aridity in the Americas, Africa, Asia and Australia and they sought solutions which would allow the expansion of cultivation and settlement.

At first the desert was a puzzle to its would-be modern exploiters. They had difficulty in defining where it began and ended due to climatic misperceptions, and settlers were often drawn in at the desert margins who were far too optimistic about the risk of drought (*see* Figure 10.1). At first it seemed that their fragile economies would be little match for the nomadic pastoralists and pre-industrial cultivators, such as the Hohokam or Pima Indians of North America, who had exploited these regions for millennia. But the learning experience of the last 150 years has eventually yielded improved technologies such as dry farming, and irrigation using reservoir storage and ground water reserves.

The resulting landscapes of modernism imposed on the desert are remarkable. The stark boundary line between the watered and unwatered is especially noticeable in those areas where the large centre-pivot method of irrigation is

FIGURE 10.1 *Dust Drift near Lamar, Colorado*
Source: Nebraska State Historical Society

used, and newly planned desert settlements are often also unusual because they require innovative architecture and ground planning to allow their populations a chance of a normal life. The kibbutzim in the Negev desert are among the best examples of this conquest of the arid environment.

AGRICULTURAL TERRACING

Among the human-made features of the landscape which cannot be ignored, terracing must rate highly. Its visual appearance can be stunning, as with the wet rice terraced fields of Banaue in the Philippines, a world-famous icon of the human impact. Terracing's economic importance has been crucial at times to various civilizations, mainly pre-industrial. Modern terrace construction is rare and usually confined to countries such as China where there is the political will to mobilize the necessary labour.

The origins of terracing are obscure. Figure 10.2 suggests hearths of innovation in the Middle East, South-east Asia, Peru and Mexico but these and the routes of dispersal are highly speculative. Williams (1990) prefers an explanation in terms of what he calls 'incremental slope levelling', in other words, farmers trap soil eroded by natural geomorphological processes and thus create terraces which do not require elaborate architecture. The appeal to cultural diffusion is then unnecessary.

The purposes of terracing are straightforward. First, level surfaces are important for irrigation and plough agriculture. Levelling is not necessary for rain-fed agriculture or the use of digging stick or foot plough but the sophisticated hydrological regime of wet rice cultivation does require flat fields.

Erosion control is a second objective, to intercept downslope drainage and encourage the deposition rather than removal of its sediment load. Spores (1969) has found in the Nochixtlan Valley in Oaxaca, that purposive erosion of hillsides was practised by cutting through caliche to expose underlying softer red soils. These were eroded by gullying and soil collected by constructing stone and rubble dykes (10–200m long) across stream beds. New terrace fields (*lamabordo*) were thus created. Donkin (1979) does not see erosion control as a major function of terraces, however, and it is true that small-scale, piecemeal terracing is not particularly effective in soil conservation. Modern contour ploughing has replaced terracing as a means of erosion control.

A third purpose of terracing is soil moisture retention. Even on non-irrigated terraces the construction of a stone retaining wall will create a localized water table which may retain sufficient moisture for the sustenance of plant roots. Finally, terraces create a microclimate which retains humidity in the dry season and in winter resists frosts. Their rock faces retain heat over the course of the night, keeping temperatures relatively high.

The many variations of terrace type can be summarized in three main categories with subdivisions (based on Spencer and Hale 1961, and Donkin 1979).

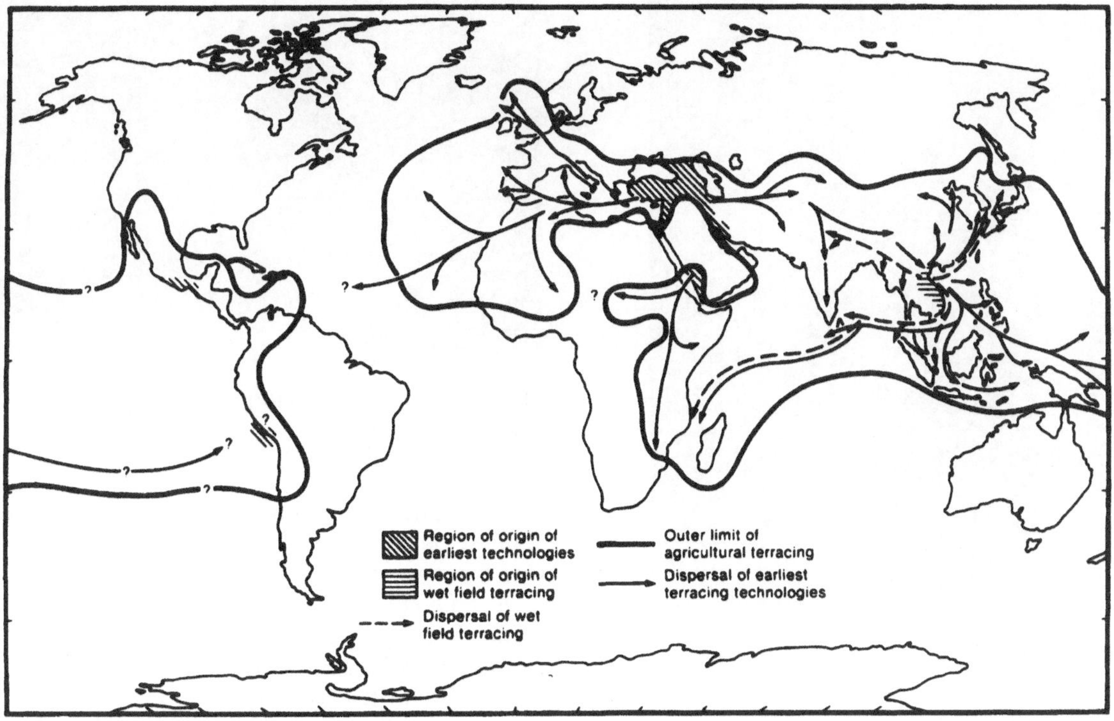

FIGURE 10.2 *The origins and dispersal of agricultural terracing*
Source: Spencer, J.E. and Hale, G.A. 1961: The origin, nature and distribution of agricultural terracing. *Pacific Viewpoint* **2**, 1–40

1. Cross-channel terraces.
2. Lateral, bench or contour terraces.
3. Valley-flzoor terraces.

Cross-channel terraces are of a simple design and may have been the earliest form of terrace. The *channel-bottom weir terrace* involves a low barrage constructed across a stream bed to divert water and encourage deposition of silt. Such terraces were often across a broad embayment. By contrast, the *narrow channel, barrage terrace* is a feature of constricted valleys originally occupied by an intermittent stream in an arid or semi-arid area. Its catchment is much larger than the area cultivated. The fields are level, with water draining from surrounding slopes.

Lateral, bench or contour terraces are popularly associated with *linear contour, irrigable terraces* (*see* Figures 10.3 and 10.4), but in the *linear sloping, dry field terrace* the natural slope is little modified and the plants are rainfed.

To produce level fields on a steep hillside is a major engineering task requiring high, thick retaining walls. The steeper the slope, the deeper the foundations have to be. A double skin of masonry with rubble fill may be desirable for the most massive walling, although the construction skills were not universal. The limiting specification of the wall is to hold the soil under high moisture conditions, perhaps with built-in drains to release pressure. The terraced wet rice field systems of Asia require a level of hydraulic sophistication which demonstrates the extraordinary landscape-modifying skills of pre-industrial peoples.

Many contour terraces require to be backfilled with sub-soil and topsoil. This was often carefully graded to keep the most fertile and friable material at the surface. In areas of thin soils it was necessary to move earth from the valley or, in the case of Malta, to import it as ballast from overseas.

Contour terraces are often found stacked in combinations like staircases, in Latin America

Figure 10.3 *Staircase between Inca terraces, Machu Picchu, Peru*
Source: P.J. Atkins

known as *andenes*, up the slope. Spencer and Hale (1961) suggest that this is by no means universal and propose a category of *isolated, short, sloping dry field terraces* which are likely to have been the work of small groups of farmers or individuals.

The *tree-crop, buttress terrace* is a type of bench terrace very common in countries flanking the Mediterranean Sea (*see* Figure 10.5). Walling may not be regular or continuous and the field surface is rarely perfectly horizontal, but this type of terrace is serviceable for perennial tree crops such as olive and citrus. A sufficient depth of soil is retained for the tree roots on dry hillsides that would otherwise be of little economic value.

Figure 10.4 *A staircase of Inca terraces, now abandoned, Pisac, Peru*
Source: P.J. Atkins

Figure 10.5 *Tree-crop, buttress terracing in Greece*
Source: P.J. Atkins

Valley-floor terraces may be formed by a simple mud bund, as in the *stream bed, wet field, mud terrace*, or have more formal stone walls set at right angles to a stream, as in the *wet field terrace*. Both are usually irrigated.

Case study: agricultural terracing in Latin America

Central and South America have a significant concentration of terraces. Peru alone has an estimated one million hectares. Perhaps one ought to use the past tense, however, because a high proportion have been abandoned for centuries. Although desperately short of agricultural land, Peruvians do not cultivate 75 per cent of their terraces. The proportion of disuse in neighbouring northern Chile is 80 per cent. Why? Denevan (1992) estimates a Latin American population of 53.9 million at the time of contact with the outside world in 1492. This declined to 5.6 million in 1650 due to the ravages of war and disease. It is hardly surprising therefore that terraces were abandoned.

In the 1980s an intensive study of terracing in the Colca Valley, Peru, produced some interesting results. Four zones of terraces were identified:

valley bottom fields; bench terraces on the lower hillslopes; sloping field terraces (which have never been irrigated) on very steep slopes above the current level of cultivation; and upland walled terraces above the bench zone on gently sloping terrain. The crops recorded in the sixteenth century were mainly maize and quinoa, with some potatoes. Denevan (1987) found that 62 per cent have been abandoned, including 91 per cent of the upland terraces abandoned, 61 per cent of the bench terraces, and only 7 per cent of the valley bottom field terraces.

The population of the Colca Valley was about 62 500–71 000 in 1530, falling to 35 000 in 1570, and never since recovering to its former levels. The Spanish conquistadores moved the Indians from the hillsides into *reducciones* in the valleys, while cattle were put to graze the higher terraces.

The research found fertile soil on both cultivated and abandoned terraces and it seems possible that the present area of arable agriculture could be expanded. We could argue that terrace rehabilitation is appropriate for developing country situations because it is labour-intensive and employs traditional technology and inexpensive local materials, but Peruvian farmers will only respond if water is provided for irrigation.

Case study: the hanging gardens of Yemen

Arabia is not entirely dry. The 3000-metre high mountains in Yemen receive some 500 mm of rainfall every year, the result of erratic, high-intensity storms and this is used by the people to sustain an agricultural system of ancient pedigree and fascinating holistic complexity, all of this in a country on the same latitude as the Sudan and surrounded by the deserts of Saudi Arabia and south Yemen.

Hillslope terracing and careful management of moisture have made Yemen a model of agricultural sustainability. The location of terraces is designed to catch the maximum amount of precipitation and in each micro watershed they are linked together by overflow channels. Each field is carefully levelled, with a masonry lip to hold the water until enough has soaked into the soil. The water then passes downslope, being led by a system of channels from field to field, ensuring that every farmer benefits. To prevent soil erosion from flash floods the farmers maintain a system of subterranean conduits which carry excess water away without damage. Irrigation from permanent and seasonal springs is an additional source of moisture for the crops.

This sophisticated regime of water harvesting is coupled with a range of husbandry practices whose origins go back hundreds of years. The soil is ploughed at certain times to conserve soil moisture and household and farmyard wastes are added to replenish fertility. The main subsistence crops are barley, wheat and sorghum.

Mechanical weathering is extensive, especially among the volcanic rocks in parts of the highlands. The resultant particles are worked by the rains and become suspended silt in small highland streams which are utilized in a gravity system of irrigation. As a result, most of the terrace soils are artificially accumulated colluvials, often highly fertile. Sediment is also carried downstream to the 20–40 km-wide coastal lowland region of the Tihama where rich irrigated farming also exists. In a sense this valley floor farming is the complement to the upland terraces in that it depends on water and soil which has already passed through the upper part of the system and can now be re-used. The release of too much or too little water from the mountains would ruin the agriculture of the Tihama, which is in balance with its environment.

Traditionally the Yemeni government supported terrace farming through the provision of credit, the control of market prices to give the farmers an incentive to produce, and the control of disputes over water and land. In recent times, however, the ancient system has begun to collapse. Two major issues have arisen since the 1960s. First, the chronic shortage of fuel in highland Yemen has encouraged the cutting of many of the trees which were a means of binding the soil on some terraces while at the same time yielding a crop of fruit and wood from controlled of whose coppicing. Second, as Yemen has become more integrated into the world economic system, a greater emphasis has been put on the cultivation of cash crops such as cotton, sesame, fruit and vegetables, coffee and a local narcotic *qat* (*Catha edulis*). The small farm sector has been neglected and disadvantaged by the import of subsidized grain from the USA. As a result, many people have migrated to cities such as San'a or abroad, with less effort now being devoted to repairing the terrace walls. Many have collapsed and this is not just a problem for the farmer of a particular field; it may be a disaster for those further down the system. Broken walls and untended fields mean greater, unmanaged runoff, with the result that lower fields may be in danger of destruction by gullying. An estimated 50–250 tonnes of soil per hectare is lost annually on recently abandoned terraces. More sediment enters the system and even the farmers of the Tihama have suffered as some of their fields have been buried by boulders.

The delicate balance of Yemeni farming has been disrupted and it is difficult to see how it can be recovered. After 2500 years of sustainable agriculture, a period of 20 years of modernization and policy changes has been sufficient to threaten ruin.

Reclaiming the Heath

Heathland is a landscape of low-growing, woody shrubs such as heather (or ling), gorse, bracken, erica, broom, which is reminiscent in some ways of the Mediterranean *garrigue* vegetation. There is little natural heathland in Britain; most is the result of human action, starting in the Mesolithic and accelerating in the Bronze Age. Commonly the original brown forest soils have degenerated to podzols, soils which are low in fertility and poorly drained. Most heath was once covered by woodland and after clearance the heathland vegetation has been maintained by grazing and other forms of management.

Most British heath is in the lowlands, usually in dry areas which are subject to periodic droughts, having acid soils such as podzols, or they are on chalk, limestone or sandy soils. The heath is normally imagined as an *empty* wilder-

Figure 10.6 *Squatter settlement on heathland, Oreton Common, Shropshire*
Source: Committee for Aerial Photography, University of Cambridge

ness, such as the *blasted heath* of Shakespeare, but in prehistoric times some heathland areas were comparatively densely populated. Thus the Breckland with its neolithic flint mines was an early centre of industrialization.

In the medieval period heath was legally protected as common land and there were several forms of management:

1. Rabbit warrens for producing meat. Rabbits were introduced in about AD 1100.
2. The use of heathland plants. Gorse was used as a fuel because it burns readily and quickly. Heather was used as a low grade thatch. Bracken also has potential as a fuel, as litter for livestock, or as a thatching material. In the eighteenth and nineteenth centuries it was burned for potash. The ashes were used in glassmaking, soap manufacture and as a detergent.
3. Grazing in areas of heath was maintained and renewed by burning.

Traditional heath management practices have disappeared and most are now returning to woodland. In Surrey in the 1790s 20 per cent was heath and 4 per cent woodland. This has now been reversed with 3 per cent heath and 16 per cent woodland.

Enclosure and reclamation for agriculture have been popular for over 200 years (Figure 10.6). The Dorset heath of Thomas Hardy was 40 000 ha in 1750, shrinking to only 7900 ha at present (Figure 10.7), with threats of further encroachments. There are obvious implications for flora and fauna such as the rare Dartford Warbler. In the Breckland, 11 800 hectares were afforested between 1920 and 1967, with a further 1215 hectares converted to arable farmland. There remain approximately 70 000 hectares of heathland, of which 37 000 hectares is in the Breckland, Surrey, Hampshire, Dorset and the Lizard Peninsula (*see* Figure 10.8). Most heath survives because it is commonland or because it is in military use, such as Lakenheath in Norfolk.

CASE STUDY: THE DANISH HEATH

The relationship between heathland and society is most aptly illustrated by the Jutland heath of

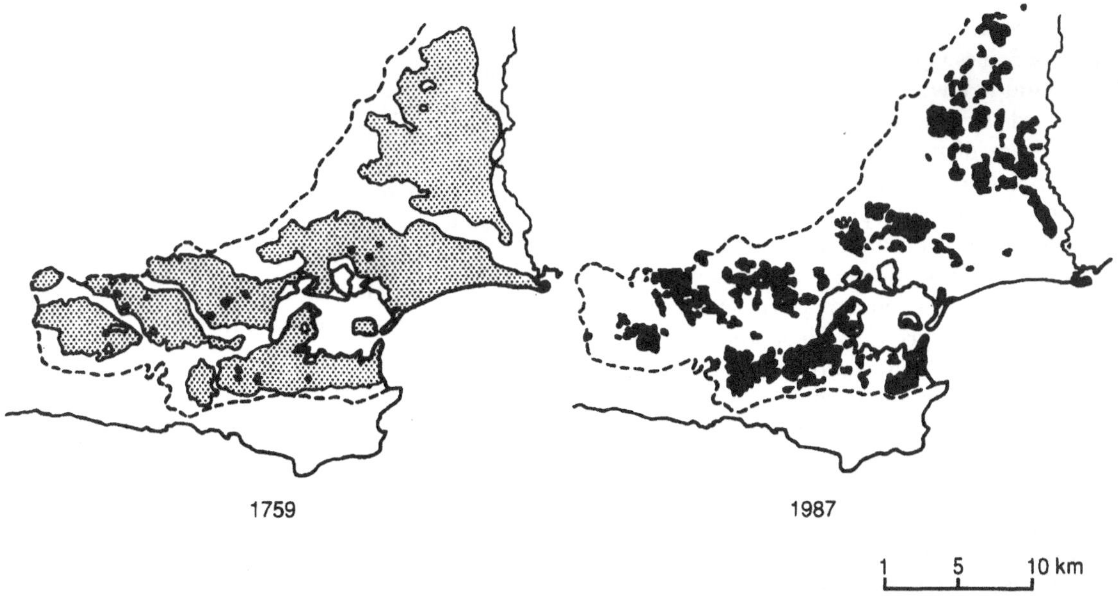

FIGURE 10.7 *The reduction of heathland in Dorset*
Source: Reprinted from Webb, N.R. 1989: Studies on the invertebrate fauna of fragmented heathland in Dorset, UK, and the implications for conservation. *Biological Conservation* **47**, 153–65

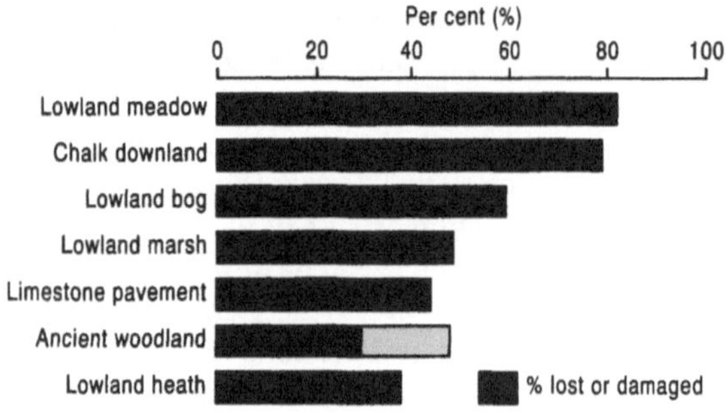

FIGURE 10.8 *The loss of wildlife habitat in England and Wales, 1945–90*
Source: Adapted from Bowers, J. and O'Riordan, T. 1991: Changing landscapes and land-use patterns and the quality of the rural environment in the United Kingdom. In Young, M.D. (ed.) *Towards sustainable agricultural development*. London: Belhaven, 253–92

Denmark. Heath dominated the landscape (Figure 10.9) in 1800 and it also dominated the mind of many nineteenth-century intellectuals. To them it was a negative pastoral, the ruin of a golden age landscape. The heath became a symbol of disgrace for a country which had suffered two heavy military set-backs: the first during the Napoleonic Wars when Denmark had sided with France, and the second at the hands of Germany over the territory of Schleswig-Holstein. To have a portion of one's land as a wilderness was taken as a sign of backward thinking and an inability to grasp the possibilities of improvement. The heath was 'a symbol of the potential for development buried in the Danish soil and its people' (Olwig 1984, 84).

In 1866, two years after the second débâcle, the Danish Heath Society was formed. It was hoped to unite Danes in a common purpose and recover a little national pride. A clever publicity campaign caught the imagination of the public and a programme of drainage, irrigation and afforestation was soon underway. Agricultural production was expanded and intensified and the 1.2 million hectares of heath in the early nineteenth century shrank to 260 000 hectares in 1950.

After some time there was nostalgia for the former glories of the heather-clad heath. In 1904 Jeppe Akkjaer used a military metaphor:

> 'Those who describe the Heath Society as a conqueror are correct: West Jutland is, in my eyes, a conquered country. Any Jutlander who thinks about what has occurred must regard the Heath Society and its so-called cultivation of our forefather's land with the same eyes as a conquered people looks upon the monuments to victory which the enemy raises upon the land of the conquered.
>
> (Olwig 1984, 86)

Today much of the remaining heath has been preserved but, ironically, its appearance is changing. Without the heather burning and grazing of the traditional management system, the heath is degrading into scrub.

MOORLAND

Moorland, with its open aspect, is the upland equivalent of heath. It is found in cold areas with a high rainfall and peaty soil, and has features in common with the tundra: low vegetation, impeded drainage, generally treeless. The moors are the most extensive area of vegetation left in the British Isles which has been relatively untouched by humans in the last 500 years, but they are not *natural* landscapes. The fossil soils preserved under Bronze Age burial mounds are often brown earths, indicating a former vegetation of mixed deciduous woodland, at least in some areas. This woodland was disturbed by Mesolithic people and climatic change to wetter conditions contributed to landscape modification (Simmons 1996). The present vegetation is heather, bog myrtle, crowberry, cotton grass, and sphagnum moss where drainage is impeded. In drier areas bracken is common; in wetter areas blanket peat bogs may have formed.

Moors are among the most valued of landscapes and, as pointed out by Marion Shoard (1982), all 10 national parks have been selected principally to preserve remote moorland. Tourists visit Howarth in Yorkshire to visualize the windswept moors of Cathy and Heathcliffe in *Wuthering Heights*, but their main place of pilgrimage is to the muirs of the Scottish Highlands (*see* Figure 10.10).

The principal activity, apart from peat cutting and occasional mineral workings, has been rough grazing. Quite a lot of moorland has been

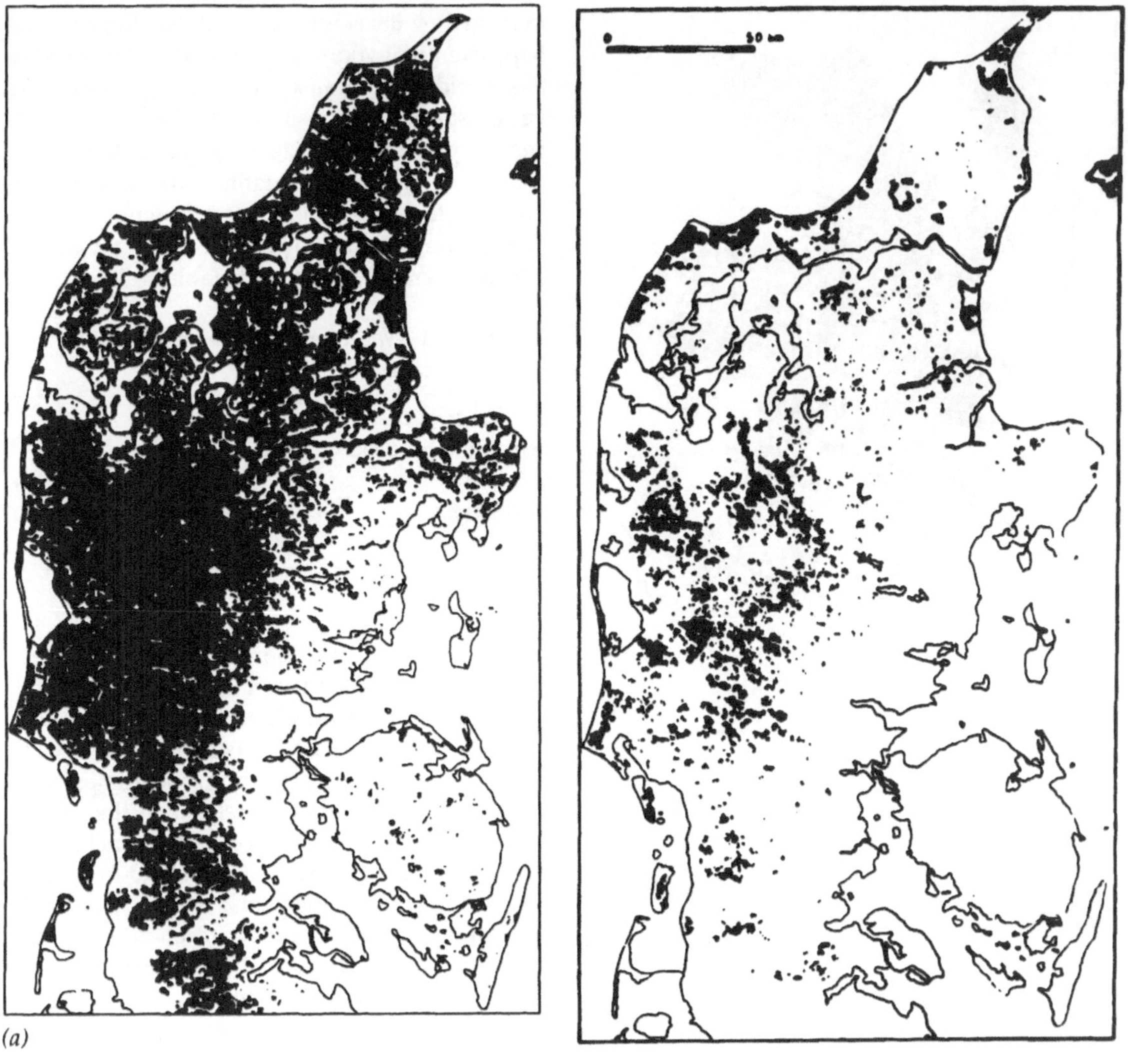

(a)

(b)

FIGURE 10.9 *The extent of heathland in Denmark in (a) 1800 and (b) 1950*
Source: Reproduced courtesy of Dr K.R. Olwig and The Geographical Institute, University of Copenhagen

reclaimed, especially during periods of population pressure. In the eighteenth and nineteenth centuries peat was removed by *paring and burning*, with the ashes used as fertilizer. Stones were cleared and used as boundary walls. The soil was then ploughed to improve drainage and sown with improved grasses. The creation of sheep pasture was the main objective, for instance during the Highland Clearances in Scotland from 1782 to 1854.

Modern moorland management is multifaceted. Apart from sheep and more recently deer husbandry, estate owners can make a profit from providing facilities for shooting grouse. The other competitors for space range from water gathering, military training, forestry, mining and quarrying, to tourism and wildlife conservation in National Parks and Areas of Outstanding Natural Beauty.

In the twentieth century much moorland has been lost to afforestation and agricultural reclamation, and there have been ecological changes

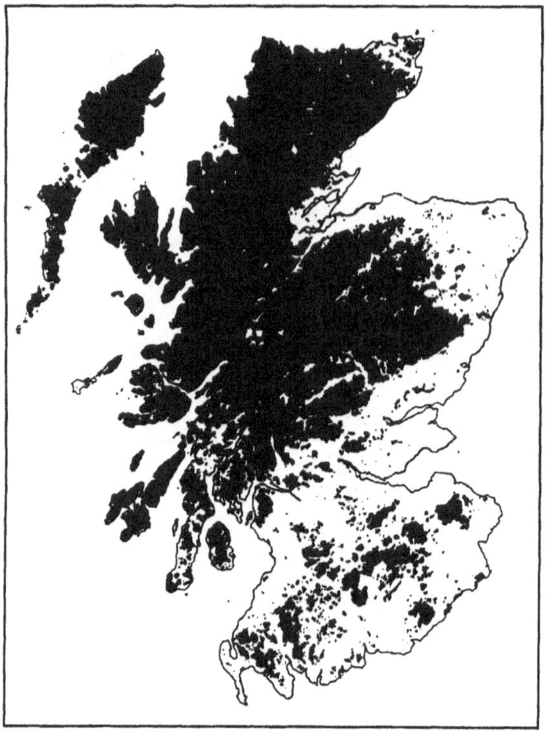

FIGURE 10.10 *Heather in Scotland*
Source: Miles J. 1994: The soil resource and problems today: an ecologist's perspective. In Foster S. and Smout, T.C. (eds) *The history of soils and field systems*. Aberdeen: Scottish Cultural Press, 145–58

such as the spread of bracken. Satellite surveillance and field sampling suggest that about 14 150 km² of moorland survives in England and Wales, of which about 80 per cent is dominated by heather (Bardgett *et al.* 1995). Some moors are threatened, for instance, in Scotland which has lost about a fifth of its heather moorland since the late 1940s. On Exmoor piecemeal enclosure began in 1962, and by 1976 the 24 000 hectares of moorland were reduced to 19 000 hectares. A voluntary scheme of notification of improvement had failed, but there has been better protection under recent legislation.

GRASSLANDS

Studying the changing area of grassland utilized is problematic. The extent of improved pastures, managed by the sowing of seed, fencing and perhaps even irrigation, is quantifiable, but most of the world's pastoralists use wild grasslands of varied species composition. The latter vary from year to year in their extent according to climatic conditions and the fluctuating numbers of grazing livestock and their market value. Graetz (1994) estimates that 25 per cent of the global land area is covered by grassland (Table 10.1).

TABLE 10.1 *World grassland area (in millions of square kilometres)*

Tropical grassland	2.1
Temperate grassland	10.5
Savannas	10.7
Arid shrubland	12.0
Total	35.3

Source: Graetz (1994, 131)

Fire has been an important factor in the creation of the grassland biome, coupled with the discouragement of woody species by grazing animals. Much new pasture has been created in recent decades by deforestation, sometimes inadvertently but often deliberately, as in the case of the subsidized extension of cattle ranching in Brazil. But the conversion of grassland to cultivation has far outweighed this, with one estimate suggesting a 20 per cent reduction since the beginning of agriculture. Other factors are at work here, including desiccation due to global warming and desertification as a result of overgrazing in marginal areas such as the Sahel in Africa.

The impact of humans on grassland ecosystems has been profound. The introduction of alien species has been especially important in the New World, where Blumler (1995) estimates that California's valley grasslands are now comprised of 80–90 per cent of exotics. Here the transfer of biota adapted to pseudo-Mediterranean conditions was initially by the Spanish, and the successful colonizing plants were more resistant to the pressure of grazing than native species. In North America this pressure came from the cattle, sheep and horses that replaced the bison and other indigenous mammals that were decimated by hunting.

By far the most significant modification of

grasslands, however, has been its ploughing for cultivation. In America the extensive grasslands of the mid-west offered an important opportunity for expansion, once the settlers had crossed the Appalachians in numbers in the nineteenth century. The prairies were generally the most fertile and least hilly tracts in the country, and were colonized by wheat, maize and cattle farmers. Prosperity was not always secure in the drier regions, however, and the search for the most suitable methods of cultivation has included some notable disasters such as the so-called Dust Bowl of the 1930s (Figure 10.1) when much soil was eroded by wind. The mid-west is in many ways a global landscape because, not only have its large, open cereal farms been reproduced in other areas of capitalist agriculture, especially in western Europe, but it has also been called upon to supply food-deficient countries with its immense output of grain surplus to the needs of American consumers.

Conclusion

The marginal lands were a challenge to human ingenuity and a number of adaptations have resulted. Simple but effective water management techniques among settled cultivators and nomadic lifestyles were traditional means of using otherwise harsh arid environments, and the modern era has seen its own landscapes imposed with three main aims: (a) using the limitless solar energy for intensive agriculture; (b) extracting oil and minerals; and, (c) establishing border settlements to prove territorial control.

Terraced fields can be found in a wide range of topography, from hilly country to steep mountain sides, and also in areas of varied soils and climate. Population pressure seems to have been at least as important as environmental considerations, and the need must have been great in view of the amount of labour input required for construction and maintenance. New terraces are being created in some parts of the world but the overall acreage has fallen over the last few hundred years. The lesson seems to be that, as traditional bonds and obligations are dissolved by the economic and cultural forces of the modern world, so terraced agricultural systems degrade and collapse.

Heathland and moorland in Europe are essentially artificially created ecosystems, but they are so valued by the public that recent threats to their integrity have brought a response by conservationists eager to preserve their unique combination of flora and fauna. There is less concern about grasslands around the world, although the last 30 years have seen worries about overgrazing and the desertification of the rangelands.

Further reading and references

On terracing, see Spencer and Hale (1961) and the work of Conklin (1980). The Yemen is described by Vogel (1987, 1988a, 1988b) and Olwig (1984) offers some penetrating insights on the Danish heathlands. Graetz (1994) is a good up-to-date source on world grasslands.

Bardgett, R.D., Marsden, J.H. and Howard, D.C. 1995: The extent and condition of heather on moorland in the uplands of England and Wales. *Biological Conservation* **71,** 155–61.

Blumler, M.A. 1995: Invasion and transformation of California's valley grassland, a Mediterranean analogue ecosystem. In Butlin, R.A. and Roberts, N. (eds) *Ecological relations in historical times: human impact and adaptation*. Oxford: Blackwell, 308–32.

Conklin, H.C. 1980: *Ethnographic atlas of Ifugao*. New Haven: Yale University Press.

Denevan, W.M. 1987: Terrace abandonment in the Colca Valley, Peru. In Denevan, W.M., Mathewson, K. and Knapp, G. (eds) Pre-hispanic agricultural fields in the Andean region. *British Archaeological Reports, International Series* **359:** 1–43.

Denevan, W.M. 1992: The pristine myth: the landscape of the Americas in 1492. *Annals of the Association of American Geographers* **82,** 369–85.

Donkin, R.A. 1979: *Agricultural terracing in the aboriginal new world*. Tucson: University of Arizona Press.

Graetz, D. 1994: Grasslands. In Meyer, W.B. and Turner, B.L. (eds) *Changes in land use and land cover: a global perspective*. Cambridge: Cambridge University Press, 125–47.

Muir, R. 1983: *History from the air*. London: Michael Joseph.

Olwig, K. 1984: *Nature's ideological landscape*. London: Allen & Unwin.

Shoard, M. 1982: The lure of the moors. In Gold, J.R. and Burgess, J. (eds) *Valued environments*. London: Allen & Unwin, 55–73.

Simmons, I.G. 1996: *The environmental impact of later Mesolithic cultures: the creation of moorland landscape in England and Wales*. Edinburgh: Edinburgh University Press.

Spencer, J.E. and Hale, G.A. 1961: The origin, nature and distribution of agricultural terracing, *Pacific Viewpoint* **2,** 1–40.

Spores, R. 1969: Settlement, farming technology, and environment in the Nochixtlan Valley. *Science* **166,** 557–74.

Vogel, H. 1987: Terrace farming in the Yemen. *Journal of Soil and Water Conservation* **42,** 18–21.

Vogel, H. 1988a: Impoundment-type bench terracing with underground conduits in Jibal Haraz, Yemen Arab Republic. *Transactions of the Institute of British Geographers* N.S. **13,** 29–38.

Vogel, H. 1988b: Deterioration of a mountainous agroecosystem in the third world due to emigration of rural labour *Mountain Research and Development* **8,** 321–29.

Williams, L.S. 1990: Agricultural terrace evolution in Latin America. In Kent, R.B. (ed.) *Yearbook, Conference of Latin Americanist Geographers* **16:** 82–93.

Part 2

The transition to modernity

INTRODUCTION

In Part 1 we looked at the pre-industrial world through discussions of prehistoric landscapes, the relationships between population, resources and sustainability, and also the role of large-scale planning. This brought us to the brink of the modern age, and indeed in some chapters we found ourselves looking at modern issues such as the land division of the United States and the current status of moorland and heathland in Britain.

It is time to press forward to address the impact of modern economic, social and cultural systems upon landscape and environment, but first we must look at the transitional period between traditional, pre-industrial, and pre-capitalist organizational forms and those which have emerged in more recent times. Chapter 11 will begin this by analysing a number of patterns and trends in medieval feudalism in Europe, and Chapter 12 is devoted to the emergence of urban and industrial landscapes.

In essence, the argument follows that of Dodgshon (1987), that the transition is essentially the story of the expression of newly formed hierarchies and nodes in a spatial form that can be called *landscape*. In a complex and tightly argued book, to which we can do little justice here, he finds that the territorial phenomenon of tribal societies, including the sort of Bronze Age field system planning that we encountered in Chapter 7, was later regulated and deepened in the nascent state-forming process. Dodgshon sees the state as 'a revolution in spatial order' and with it there was a revolution in its impact upon the surface of the earth. Urban centres emerged, with intensified, irrigated agriculture producing food surpluses, and social mobilization reached the level where permanent modifications could be made to ecosystems and unwitting damage could be inflicted upon the environment. These points have already been covered in Part 1.

The argument then encompasses feudalism, which is really a clever device for concentrating power in the hands of the few at a time when technology remained so primitive that it was difficult for the authorities to reach out and enforce their will in each locality. All levels of society voluntarily surrendered a degree of their personal power to their feudal superiors in return for the use of certain assets and other, vague promises about protection. This was a preliminary step along the evolutionary road of individual–state relations which has led to modern forms of governance with their highly efficient forms of surveillance and enforcement.

It was the stirring and eventual acceleration of trade and markets that at the same time conditioned and structured feudalism but also ensured its eventual demise, once its temporary usefulness was exhausted. The timing of the transition varied substantially and significantly, as we will see in Chapter 11. The outcomes were not homogeneous either, and a great deal of ink has been spilt describing the phases and incarnations of mercantilism (trading by merchants) and capitalism.

The transition to modernism was mediated through a number of channels. There were economic channels, of course, such as the concentration of capital that comes with industrialization and the controls over exchange, information and political power that are facilitated by urbanization. But part at least of the shift was in the minds of people. New psychological and philosophical attitudes began to emerge which facilitated change, and which were manifested in modern cultural forms such as art, architecture, literature, and the rational planning and use of landscapes.

The greater integration of national space which came with the modern era introduced or elaborated geographical forms which were remarkable in their scale and increasing functional specialization. Core and peripheral landscapes emerged from the restructuring encouraged by capitalism and it is possible even to identify international linkages as a result of trade, exploration and colonization. Later, in Part 4, we will see that world economies and world systems have their origins in this era of the transition to capitalism.

11

FEUDAL LANDSCAPES

[Feudalism was] a deep seated cause of many effects, a principle which once introduced is capable of transfiguring a nation.
(F.W. Maitland 1897: *Domesday Book and beyond*. Cambridge: Cambridge University Press)

INTRODUCTION

Class-stratified agricultural societies are identifiable from the Neolithic period onwards (Chapter 2), and they were conventional in many parts of the world by the time of what in Europe are called the *Middle Ages*. Social organization varied of course but networks of power focusing upon a king as personification of the embryonic nation–state (*see* Figure 11.1) were common, often materialized through the payment of tribute. Frequently found also was the virtual slavery of an underclass of ultra-poor individuals who were expected to undertake much of the heavy manual labour associated with agriculture.

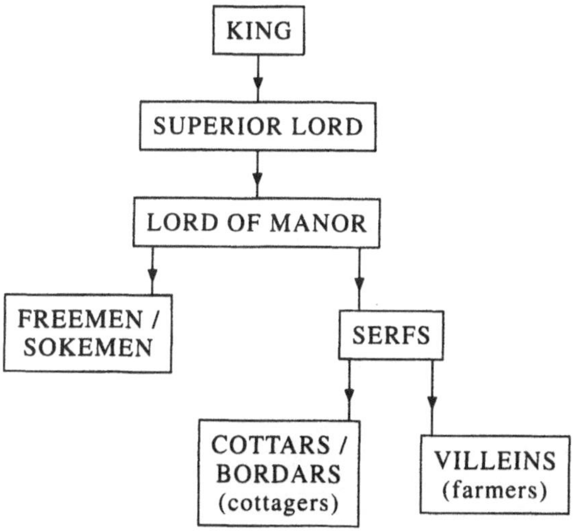

FIGURE 11.1 *The feudal pyramid*

In Europe feudalism had an impact on the landscape which was the result of its organization of landholding. The locally powerful magnates (*barons*) held land (*fiefs*) from the king in return for their loyalty and service; and they in turn had vassals, and so on down the hierarchy of patronage until the poor serf was reached at the very bottom. In theory, everyone gained something from the system, even the downtrodden peasant farmer who received protection from his lord, but the balance of privileges and responsibilities was certainly tilted in favour of the superior person so that inferiors were always indebted.

The classic model of the feudal mode of production has rural life centred around the *manor*, and a village-scale farming system in which the landlord ('Lord of the Manor') farmed some land himself directly in a *demesne* and let out the rest to his peasants in return for labour service and money payments (*fines*) of various kinds. Village resources, including both fields and waste, were regulated collectively. Not all of Europe was subject to this mode of production, farmers in parts of the Mediterranean having greater independence, yet there were similar *seigneurial* and slave-based systems in India and China (Blaut 1993).

So what is special about feudalism that brought about distinctive landscapes in northern Europe, other than the well-known agricultural features such as strip fields? The deployment of military might was certainly one factor. Although we have plenty of evidence of warfare during the early empires and through into the Dark Ages, the technology of aggression and defence was brought to a new, deadly level in the early medieval period. In particular, stone-clad castles, perfected by the Arabs, were introduced into Europe. Affordable at first only by the wealthiest of magnates, they gradually spread, controlling the major route ways and choke points such as river crossings. Their ability to withstand siege made them a rallying point for people from the surrounding area, and many became the initial kernel of urban settlements which grew in the comforting shadow of their curtain walls. Eventually the development of explosives gave a greater edge to attackers and the role of the castle changed from defensive bastion to symbol of privilege. They remained the homes of the wealthy and powerful, at least until a desire for greater domestic comfort made them redundant. Many castles were so well built that they have survived and now serve as icons of a feudal age that has been harnessed in the name of *heritage* (*see* Figure 11.2). Tourists pay to walk along battlements now drained of their original meaning and refilled with a leisure value.

FIGURE 11.2 *The keep, Newcastle-upon-Tyne*
Source: P.J. Atkins

A second anchor of the medieval landscape was the Church. It was an integral part of the feudal hierarchy, with bishops holding temporal political power and controlling large estates of farmland. The Bishop of Durham, for instance, ruled a palatinate or principality and was a considerable figure of semi-independent status, with his own castle and army. This was tolerated by the English King because Durham was an important buffer against a possible Scottish invasion. The Church's greatest impact, however, lay in its widespread presence in every town and in the majority of villages. The predominance of religious buildings of a medieval origin in England is obvious to anyone who cares to look around (Figure 11.3), and this is not an accident. The Church at this time was phenomenally successful in extracting wealth from the community, sufficient to engage in a building spree that seems more extraordinary the more one considers it. Here was a relatively undeveloped agricultural economy which somehow supported the construction of 8500 stone churches and many monasteries in England alone (Rodwell 1981), at a time when most people lived in flimsy huts. Architectural design and the building trade were also in the early stages of evolution, yet Europe as a whole has tens of thousands of medieval stone churches which have survived for 500 years or more. A similar story might be told of other religious realms, for instance the Hindu temple landscapes of south India (Figure 2.9) or the influence of geomancy (locating and aligning buildings to

FIGURE 11.3 *Durham Cathedral*
Source: P.J. Atkins

maximize positive spiritual energy) on town planning in China.

Religious institutions have long been manufacturers of *place* in the sense of tangible locations with which whole communities could associate. Cathedrals, churches, chapels, mosques, temples and shrines, resonate with a universally understood message of significance, and have become a basic building block of landscapes, both urban and rural. In most countries they are still the most important meeting places and the network of churches remains a template within which settlements have been planned. Only in the twentieth century have rivals arisen in the shape of the shopping centre and the location of leisure.

The feudal system, and the medieval landscape with its fields, villages, castles and churches, was not invincible or immutable. Feudalism never took on the monolithic rigidity that is suggested by the simple model outlined above, and there were national and regional variations that suggest a fascinating complexity and sophistication that are often not appreciated. Yet the system had its in-built contradictions that led to crises and eventually to its gradual decay and demise. In this and the next Chapter we will review the transition from feudalism to capitalism, the successor system which remains ascendant five centuries later. This was one of the most important thresholds in human history because, in the words of Cosgrove (1984, 61) it involved:

> the dissolution of existing relations between human beings and the material conditions of their lives, relations often termed 'natural'. These are the dissolution of the relation of human beings to the earth as the natural basis of production; of those in which humans appear as proprietors of the instruments of production; of ownership as a means of consumption of prior significance to production; and of the labourer as living labour power, part of the objective conditions of production.

Under feudalism the close relations of humans and the land had engendered a dependence which affected even those not directly involved in cultivation. Everyone had an interest in the harvest and the links with environmental factors were plain for all to see. The context of capitalism evolved rather differently. Now humans were increasingly alienated from the natural and stood outside environmental concerns. They saw land as an object which could be owned and used for profit and therefore had a different set of land *values* that were concerned with *commodities* for marketing and sale rather than the immediate life-giving consumption of food. In the jargon of economics, land now had an *exchange value* as well as a *use value*. Table 11.1 shows some of the other changes which took place.

TABLE 11.1. *Some differences between feudalism and capitalism*

Feudalism	Capitalism
Land 'held'	Land owned
Labour service	Rents
Collective organization	Individual decision-making
Labour-intensive	Capital-intensive
Subsistence	Markets
Use value	Exchange value
Prescribed status	Social differentiation into classes

FACTORS IN THE TRANSITION FROM FEUDALISM

A number of writers have highlighted the continuity of long-distance trade in Europe after the decline of the Roman empire, gradually turning to an intensification of local and regional trade. This affected even areas inland, away from navigable waterways, which had previously been economic backwaters (Dodgshon 1987). By the ninth and tenth centuries there seems to have been an urban resurgence in Europe based upon trade and marketing. These new or revived urban places were responsible, on the one hand, for the gathering in of raw materials and foodstuffs for consumption and their exchange with other towns and, on the other hand, for the dispersal of manufactures such as cloth and iron goods, and raw materials such as salt, corn and spices. Craft industries were urban-based and their controlling organizations (*guilds*) came to be important in the political management of towns and cities.

There is evidence also that agriculture was not entirely subsistent in its objectives. In the thirteenth century a commercial form of farming was beginning to make headway, especially on the large and efficiently run estates of the church and the monasteries. But the period 1100 to 1300 was one of demographic growth across Europe, with the expansion of the agricultural area on to new land in order to produce sufficient food. In England this seems to have caused population pressure because most of the best farmland was already in cultivation and the newly cleared lands were marginal in their productivity and environmentally fragile. Some writers argue that this led to a widespread decline in soil fertility as a result of inappropriate intensification, and that the famines and epidemics of the first half of the fourteenth century (including the Black Death of 1347–9) were a demographic check of the kind later famously predicted for pre-industrial societies by Thomas Malthus. Climatic change may also have been implicated in this demographic crisis.

In England the population fell from roughly 6 million in 1300 to only 1.5 million in 1450. Large areas went out of cultivation and there was a drop in demand for grain and a rise in wages. To attract tenants many landlords felt they had no choice but to rent out their own land and generally reduce feudal restrictions. There was a realization that the ruthless exploitation of the peasantry in some parts of Europe had actually been counterproductive, leading to reduced incentives to think innovatively or conserve resources, and there were even peasant revolts in some countries (Brenner 1976; 1982).

Leasing land for rent implies a monetization of the economy which, while not incompatible with feudalism, introduced a new dynamic into people's lives. It also implies the emergence of three new social classes: the rentier landlord; the peasant who might gradually accumulate land through good husbandry or marriage and eventually become a large farmer; and the landless labourer who could not afford to rent land, although earlier he might have been a small tenant farmer (*villein*). Such a polarization of wealth was now possible, as was a greater mobility beyond the previously established status hierarchy.

The production of grain and other arable and livestock products remained largely for the local market but it seems that large cities certainly did exert an influence on their hinterlands. In 1300 London had a population of approximately 70 000 and was already drawing food and fuel from 30 kms by road and 100 kms by water (Campbell *et al.* 1993). Such market opportunities were enough to influence farmers to intensify their methods and perhaps to specialize. Subsistence was no longer their sole objective.

The importance of the feudal period to an inhabitant of Western Europe today is at once simple and complex. In simple terms much of the structural framework within which our lives are lived was laid down in the centuries between 1000 to 1500: most of the major towns and ports, most villages and hamlets, many field arrangements, many single farmsteads; and while we are only fully aware of these through the great cathedrals – Chartres and Compostella – and castles, heritage sites, many key landscape foundations are also medieval. The matter becomes much more complex when we think of social structures and trade routes.

A GEOGRAPHICAL PERSPECTIVE: THE EXAMPLE OF ENGLAND

Dodgshon (1987) argues that the transition to capitalism had a varied impact upon the landscape according to region. It seems that feudalism, as represented by the continued presence of labour services, lasted longest in certain areas:

- in the south east of England, in the core rather than the periphery;
- on large, grain-producing manors, rather than on smaller livestock farms;
- in old settled areas rather those newly cleared;
- in areas of strongest feudal control.

This was due to the geographical differences which were becoming clearly evident between the more developed south and east of England and the backward north and west. Both population and wealth were already spatially concentrated in the eleventh century and became even more so by the fourteenth (*see* Figure 11.4), a trend which was reinforced by new departures in the economy. One of the most important was the new, commercially orientated farming which had the

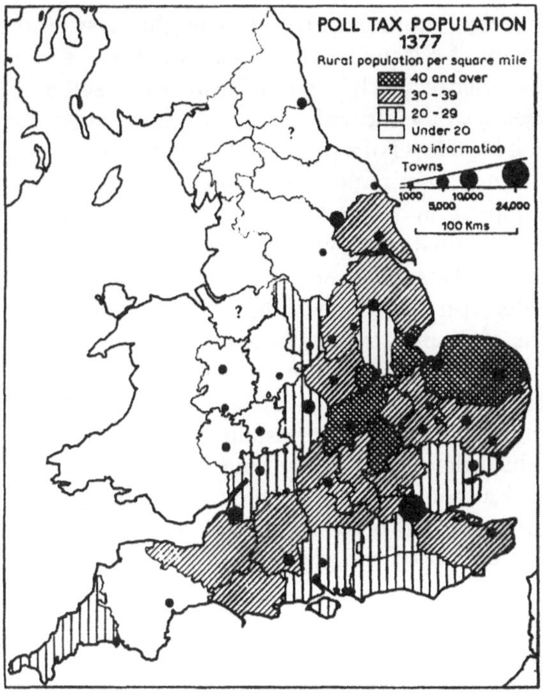

FIGURE 11.4 *A map of population in England, based upon the Poll Tax of 1377*
Source: Darby, H.C. 1973: *A new historical geography of England*. Cambridge: Cambridge University Press

most convenient marketing outlets either close to large cities or in areas where towns had grown during their period of resurgence from the tenth to the twelfth centuries. Greater population density increased the likelihood of a farmer's proximity to a market in the south of England and made possible the first small steps towards an economy dedicated to the matching of supply and demand. Thus urbanization and food marketing encouraged each other in a slow, virtuous circle of growth.

The advantage of the south was mainly one of ecological potential. Climate, soil and topography all played a part in enabling a greater density of population here, although the historical accident of in-migrations from the continent was another factor. Any superficial resemblance to the present-day structure of uneven development, however, should not be taken as an indication of a fully articulated modern economy. Regions remained only loosely connected, self-sufficient units. Even in political terms the concept of a unified and coherent kingdom was some way off, the give and take of the feudal system allowing for a considerable degree of local autonomy.

Although regional economies were still largely autonomous, there were forces of integration which had profound consequences. During the period of the unified Saxon kingdom from the tenth century, and especially after the Norman Conquest of 1066, many estates were accumulated by barons with a scattering of lands in different parts of the country. The Earl of Chester and Count Mortain, for instance, each held land in twenty different counties. Others developed later, such as the Duchy of Lancaster. Such holdings covered a wide range of ecological niches and provided for the household of the lord all of his possible needs for food and other supplies.

Private inter-regional connexions of this sort, coupled with long-distance trade for the open market, would in a small way have given a kick start to economic integration. Most important, it seems to have encouraged a move towards greater specialization according to the comparative advantage of each region, for instance in livestock products in the north and west, and in grain in the south and east. Holders of widely spread estates had investment strategies which mirrored this logic. The peripheral areas of the north and west tended to have farms with lower yields, some of which were colonized late from the waste lands. Here lords, in the hope of attracting and keeping tenants, were more ready to commute labour service to rents or to grant rights to *assart* (clear wood). They encouraged the exploitation of resources such as wool, leather, timber, livestock and game, which were not produced in sufficient quantity in the south and east. In addition new markets were established here and boroughs created to stimulate a system of trade.

Meanwhile, in the core area of the south and east, there was nothing to gain from switching from labour services to rent and so this aspect of feudalism lasted longest in this, the most developed region. Guaranteeing a labour supply for the crucial periods in the arable farming calendar was a priority after the Black Death, but gradually the switch to rents came as the commercial economy took hold.

The implications of this account are twofold. First, an increasingly wide range of resources was required by medieval society and at every level of the economy, from the village to the nation, steps were taken to get access to the basic needs of everyday life. Self-sufficiency remained a goal but the trickle of intra-regional and later, inter-regional trade, suggests that the advantages were well established of moving goods from areas of surplus to those of deficit, either between ecological zones or to feed urban consumers.

Flowing from this is the equally important recognition of the need for complementarity. Simply put, this was the realization that prosperity could be maximized through trade if farmers could increasingly specialize in the products for which they had an advantage. This advantage might arise out of the soil or perhaps a location close to a market. It might even be the result of a farmer's particular skill after years of experience. Such a move away from self-sufficiency at the individual farm level, or perhaps at the local level of a group of parishes, represented a very significant risk, and will have been adopted most readily where a powerful lord could initiate his own private trade or where public markets grew steadily, without the sort of interruptions which would destroy the confidence of specialist producers.

The geographical expression of complementarity was the emergence of agricultural specialist regions. These were sustained by the markets which developed at the boundary between ecological zones, and often they were quite starkly dichotomized between hill and vale, arable and pasture, wood and plain, chalk and cheese. Close

FIGURE 11.5 *South Field, Laxton, Nottinghamshire (1949)*
Source: Committee for Aerial Photography, University of Cambridge

economic links between neighbouring regions, which were often reinforced by social processes such as cross-boundary labour- and marriage-migration, gradually grew into a mutual dependence. Ties which had been very localized and weak at first were deepened with time and the geographical scale broadened.

Since landscape is to an extent the expression of economic specialization, the emergence of complementary regions was a very fundamental factor in its changing appearance. At first the impact was fine-grained but the gradual expansion of trade would have increased the size of the regional mesh as groups of farmers came to appreciate their common cause. National, international, and now global forces have in modern times introduced readjustments of the scale at which economic and cultural processes operate, but landscapes are slow to adjust and in old-settled European and Asian countries they may continue to exhibit the realities of previous centuries. This is why the feudal economy and its transition to capitalism are still important to us today. The landscapes of parts of Britain, for instance, especially those in rural areas, were precipitated out of the ever-changing cultural brew at a time when micro- and meso-scale regional economic specialisms were important and the texture of the countryside now is therefore an indirect expression of a medieval or early modern logic.

Thought of in this way, the regional geography of feudalism is not merely a historical curiosity. The livestock economies, with their landscapes of open pasture and wood pasture, were needed for their specialist resources. These were best encouraged by a liberal form of feudalism and so it was that the north and west had a freer atmosphere. In the arable areas feudalism hung on tenaciously, not out of the class interests of the estate owners so much as the need to secure the labour to sustain

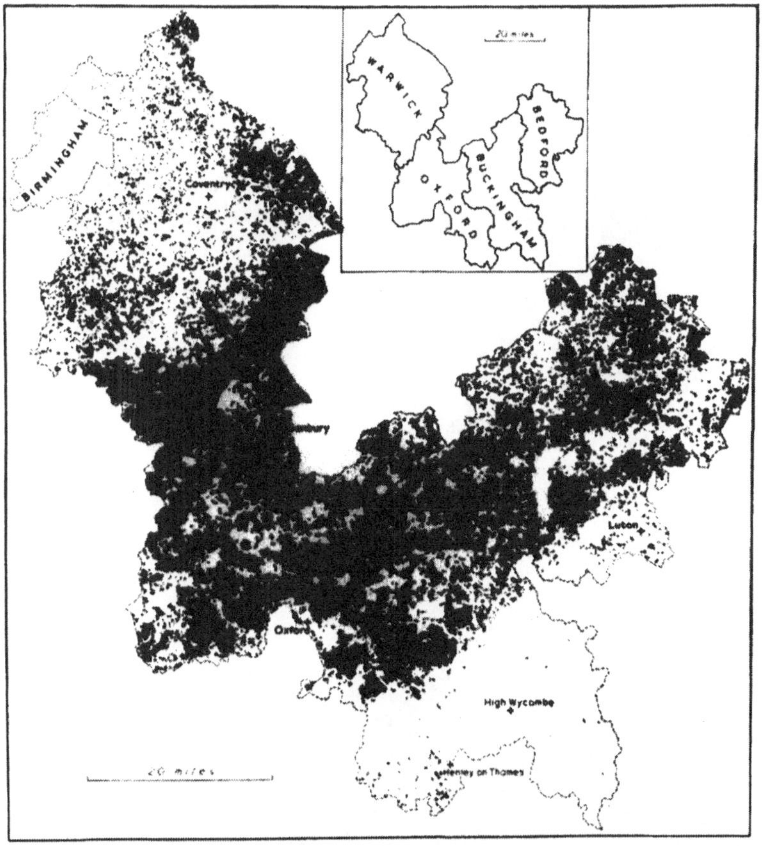

FIGURE 11.6 *The shading shows the distribution of ridge and furrow in the English Midlands*
Source: Harrison, M.J., Mead, W.R. and Pannett, D.J. 1965: A Midland ridge-and-furrow map; *Geographical Journal* **131**, 365–69

grain production. In a sense, both were complementary and both were essential transitional phases in the shift towards the monetized, commercialized era to come.

FIELDS: THE EXAMPLE OF MEDIEVAL EUROPE

There is a problem that the map of European agrarian landscapes, although it has had an astonishing in-built inertia for at least 500 years and probably much longer, has nevertheless evolved under the influence of a number of factors, some of them contradictory. On the one hand, there were good reasons why people, especially in areas where grain cultivation had the greatest potential, should want to collaborate in various aspects of their agriculture. On the other hand, such cooperation was never without its tensions and the logic of private farming was never far from the surface.

Among the reasons for collective effort, a key factor was the sharing of risk. Having strips and scattered holdings reduces the danger of insect pests spreading, as they do most readily in large blocks of a single crop, and it also minimizes the probability of total crop failure because each farmer has a mixture of soil and moisture conditions. The disadvantages were that time was wasted walking between holdings and that disputes were likely to break out between neighbours simply because there are so many boundary edges in this type of farming.

Second, in situations where there was competition for scarce resources, especially in areas where the population was growing and pressing on the land, it made sense to regulate the activities of villagers through a system of collective responsibility. This prevented a few people degrading the common property by their selfish actions for short-term gain.

FIGURE 11.7 *Curvilinear ridge and furrow, Brassington, Derbyshire (1967)*
Source: Committee for Aerial Photography, University of Cambridge

A third reason for joint effort was that lumpy investments in expensive ploughs and other tools could be shared. It is not surprising, therefore, that many villages in medieval Europe displayed evidence of communalism. The arable fields were divided into *selions* (plough strips) (*see* Figure 11.5) and barriers such as the few hedges, walls and ditches were more for functional reasons than to divide properties in the sense that we understand them today. The appearance of the landscape would have been open (*see* Box 11.1), but with a complex patchwork of land uses. The stubble and weeds in the arable fields were communally grazed after the harvest, thus overcoming the problem of a shortage of fodder, and meadows were set aside for making hay. Usually some common pasture was deliberately kept for stock rearing, but it was usually in the fields most distant from the village.

Box 11.1 *Definitions*

> **Open field:** an open agricultural landscape, with few hedges, fences or walls.
> **Common field:** a field system organized communally.
> **Subdivided field:** a field divided into strips.

One strip did not necessarily equal one person's holding. Strips were often held in contiguous bundles but were also frequently scattered, and a peasant could build up a bundle of strips by swapping, marriage settlements, leasing, or purchase. Subdivision by inheritance and in the normal process of the land market was a strong factor in creating the fragmented texture of medieval fields.

There was usually an agreed system of rotation of crops by the *furlong* (amount of land that could be ploughed in a day), and the practice of having three large fields in many Midland villages therefore did not necessarily mean a three course rotation. Furlongs would be left fallow every second, third or fourth year. The whole system was administered by a formal meeting of farmers in the manorial court. There were by-laws to regulate all aspects of agriculture: arable, pasture, meadow, harvesting, storage, fencing, boundaries, roads, and many other matters.

Plough technology was a very important influence on this arable landscape. The heavy ox-drawn plough was used in some areas to pile up the soil into ridges and furrows (*see* Figure 11.6) and headland ridges and, because it was

FIGURE 11.8 *Rural settlement types in Germany.*
A. Irregular forms
1. Irregular villages (Haufendörfer, Kettendörfer)
2. Hamlets (Weiler, Drubbel)
3. Scattered farmsteads
B. Regular forms of medieval eastern colonization
4. Small villages and hamlets, including Gassendörfer, Rundlinge *and* Gutssiedlungen
5. Areas with particularly strong development of Rundling *forms*
6. Large street villages and villages with elongated greens (Angerdörfer *and* Strassendörfer)
7. Forest clearing and heath villages (Waldhufendörfer *and* Hagenhufendörfer)
8. Marsh villages (Marschhufendörfer)
C. Post-medieval planned forms
9. Geometrically laid-out villages of the eighteenth century
10. Moorkolonien
Source: Reprinted from Smith, C.T. 1967: *An historical geography of western Europe before 1800.* London: Longman

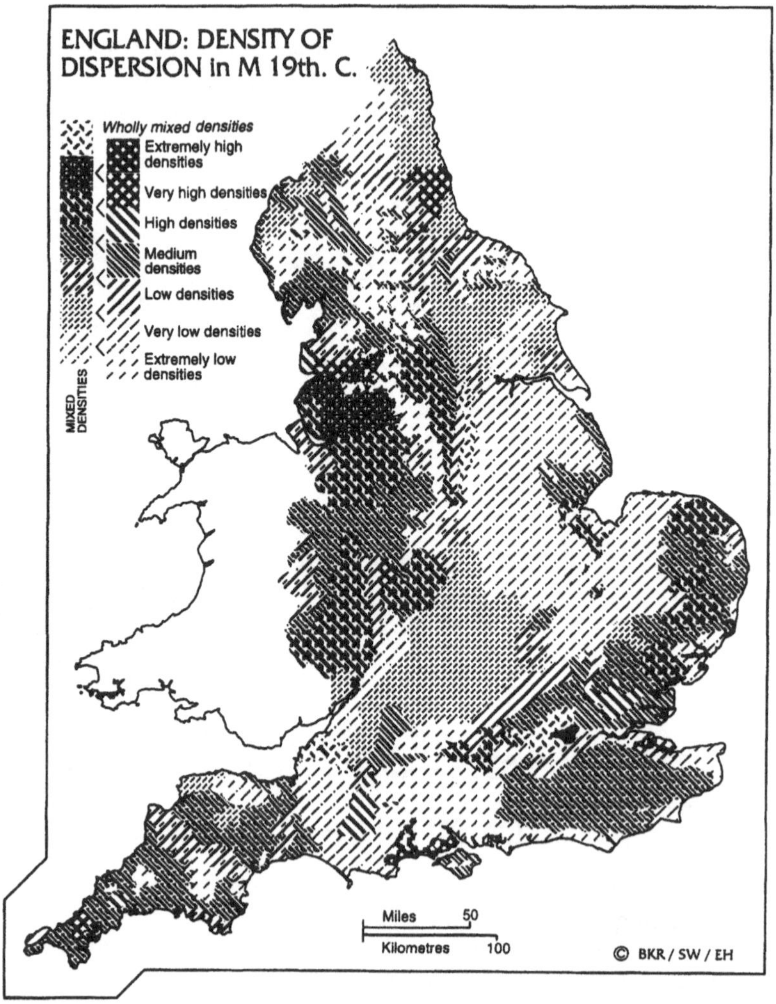

FIGURE 11.9 *Village dispersion in England*
Source: B.K. Roberts

difficult to turn, it also left its traces in the reverse-S shaped strips which are clearly visible from the air (Figure 11.7). The crown of the ridge could be as much as 1 to 2 metres above the bottom of the furrow. The ideal strip was 220 yards long in old English measure by 22 yards wide, making a traditional acre in extent. The shorter of these dimensions was adopted as the length of a cricket pitch and thus medieval ploughing measures have been exported to the wide world.

There were important regional variations on this arable theme (Baker and Butlin 1973). Kent had no communal organization but did have subdivided open fields that were probably the result of partible inheritance, and there were many other complex local adaptations according to circumstances such as regional ecology, population pressure and the strength of feudal control as outlined above. Thus in areas of ancient or early enclosure and in the marginal uplands, broadly corresponding to Rackham's Ancient Landscape (Chapter 7), open fields were less evident. This was mainly because among pastoral farmers the need to cooperate was never strong. There was no need to share risk or purchase tools collectively, and hedged or walled fields made sense if stock were to be prevented from wandering.

The mature Midland system of open field agriculture seems to have evolved, probably during the Dark Ages, from earlier forms of agriculture such as the infield–outfield system. This survived into the eighteenth century in parts of

FIGURE 11.10 *Milburn, Cumbria, a classic green village*
Source: Committee for Aerial Photography, University of Cambridge

Scotland (Dodgshon 1973; Whyte 1979), where an infield was intensively cultivated on a permanent basis and manured with animal and household waste to preserve its fertility, and the outfield was mainly used for grazing, but would also be cultivated intermittently. If there was an increase in the number of mouths to feed, the outfield could be made permanent, for instance by the addition of lime and manure to upgrade its soil status. Other evolutionary factors may have been more agronomic: as soil fertility fell, for instance, the introduction of a fallow period and crop rotations would have encouraged the division of fields.

Evolution continued during the Middle Ages. There is evidence of experimentation with new crops and new cultivation practices and in some parishes the open fields were enclosed at an early date. Chapter 7 gives an outline of the later phase of enclosure, which marked the transition to conditions where capitalist agriculture could flourish.

RURAL SETTLEMENT

The pattern of villages and hamlets in Europe was very much the product of its agrarian history. Nucleated villages make sense where collective effort in arable agriculture encouraged the sharing of effort and information, and scattered homesteads were feasible in the more independent context of pastoralism. Maps of settlement types (Figures 11.8 and 11.9) are not a perfect match with that of agricultural specialisms, however. There are a number of other considerations:

- Large villages sometimes spawn outlying daughter hamlets which exploit pockets of marginal resources but which are unable to grow beyond their nascent stage. Their distribution may indicate areas of late woodland clearance.
- In some areas scattered populations were gathered together, either forcibly for the sake of control by a powerful lord, or for ease of

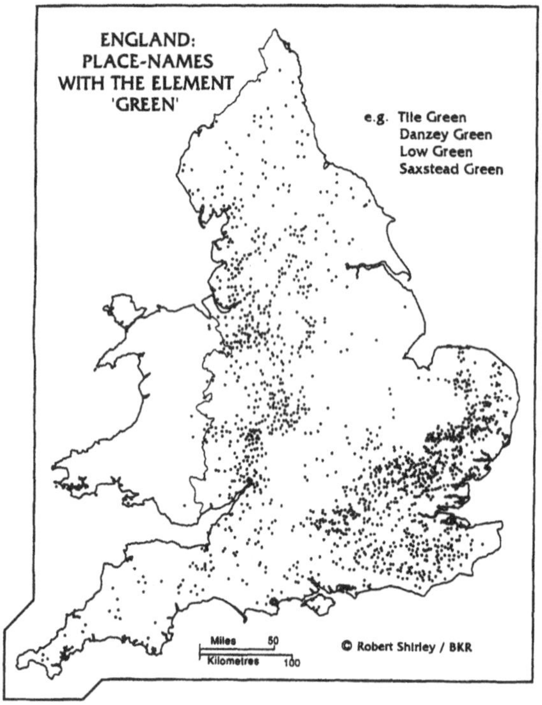

FIGURE 11.11 *Place names with the element 'green', e.g. Tile Green, Danzey Green, Low Green, Saxstead Green*
Source: Shirley, R. 1994: *Village greens of England: a study in historical geography*. Durham: unpublished PhD thesis, University of Durham

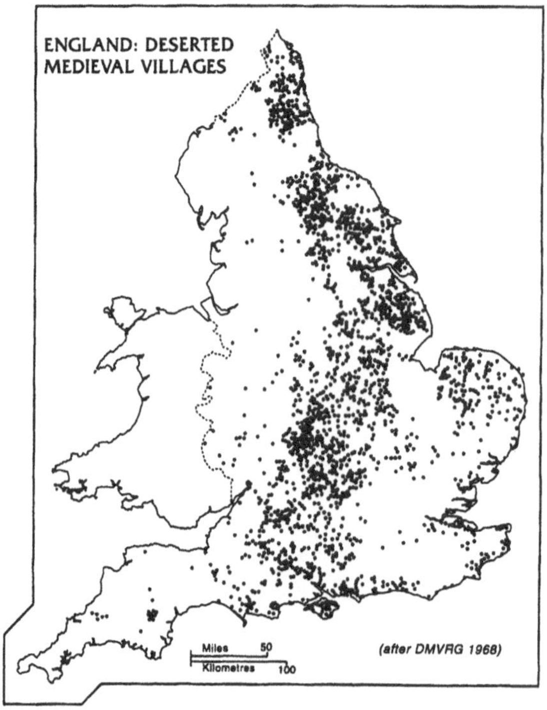

FIGURE 11.12 *Villages deserted in medieval times*
Source: adapted from Beresford, M.W. and Hurst, J.G. (eds) (1971). *Deserted medieval villages: studies*. London: Lutterworth Press

mutual protection in times of military threat. The hilltop village so typical of Mediterranean landscapes was born of this need for defence. The green villages of County Durham and other northern counties (*see* Figures 11.10 and 11.11) are probably also the result of a deliberate reorganization, probably following the devastating 'harrying of the North' by William I in 1068–70.

- Industrial development in areas of 'dual economy' (Chapter 12) encouraged the inmigration of people seeking work and confounded any simplistic notion of rural settlement drawing its sustenance from agriculture alone. The same is true of villages with a marketing function, although some of these might be reclassified as 'towns'.
- A true map of medieval rural settlement should allow for the large number, 2000 or so, of villages and hamlets that have been deserted, and have left only earthworks in the fields, and also the villages which have shrunk to hamlets or single farmsteads (*see* Figures 11.12 and 11.13). Lack of economic competitiveness is important here, but so are the visitations of the bubonic plague, and the deliberate depopulation of the countryside by Lords keen to create extensive sheep ranches (Chapter 7).

Regional variations in the spatial patterns of rural settlement are matched by an equally complex diversity of village plans (*see* Figure 11.14). These are the result of the differing balance between public and private spaces but they are made up of essentially the same elements, including houses, farmsteads, streets, church, green, pond, and fields. The degree of regularity or organic accumulation in the plan is one dimension of possible

FIGURE 11.13 *Deserted village of Ulnaby Hall, Co. Durham*
Source: Committee for Aerial Photography, University of Cambridge

classification and there are others: dispersion/agglomeration, with/without green, mono/poly-focal. Such forms evolve through time under the influence of economic and demographic forces and it is not uncommon to find villages that have changed their size and shape significantly over the centuries.

Village evolution accelerated as the communalism of the feudal era dissolved into an increasingly privatized landscape where capitalism eventually became a dominant force. The pull of the core features of church and manor house weakened and even the logic of nucleation dimmed with the creation of new blocks of land by enclosure. The regularity of rows imposed by green and street was increasingly enhanced by the geometry of drawing-board design, especially where new mining and industrial settlements were established or commuter villages extended in metropolitan regions to accommodate a rapidly growing influx of residents.

CONCLUSION

The crisis of feudalism in the fourteenth and fifteenth centuries undermined both its communal and economic logic. It also presented a

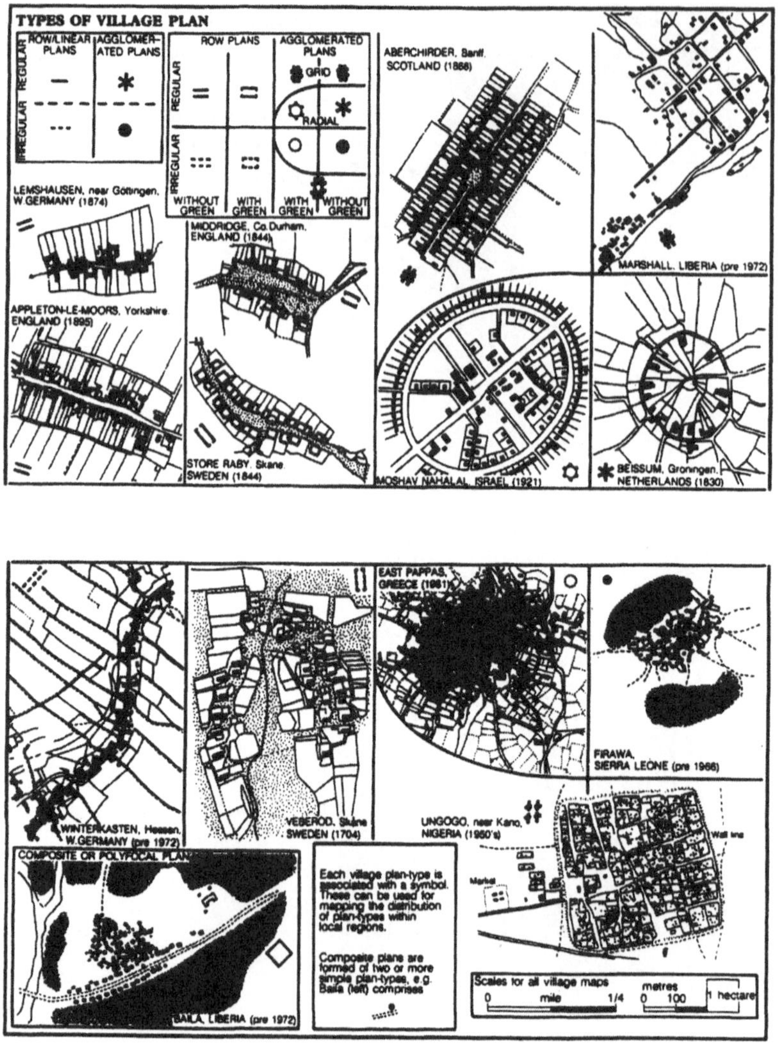

FIGURE 11.14 *Village plan types*
Source: Roberts, B.K.

series of reasons why people would want to reshape their landscape to accommodate the new realities of market forces that gradually crept in. The transition to capitalism took many centuries because the pressure for change had very different local outcomes in the highly complex spatial matrix of feudalism and mercantilism. In Britain there were open fields and common lands still being enclosed in the nineteenth century, and in Russia serfdom was abolished as late as the 1860s.

Industry and towns were catalysts in the transition from feudalism in Europe. We will look at their impact in Chapter 12.

FURTHER READING AND REFERENCES

Dodgshon's (1987) is still the best spatially contextualized account of the economic and social changes that have affected Europe in the last 2000 years. A simpler but useful collection of papers was published by Baker and Harley (1973). Baker and Butlin (1973) is a reference text of enduring value on field systems. For rural settlement see Roberts (1987).

Baker, A.R.H. and Butlin, R.A. (eds) 1973: *Studies of the field systems in the British Isles*. Cambridge: Cambridge University Press.

Baker, A.R.H. and Harley, J.B. (eds) 1973: *Man made the land*. Newton Abbot: David and Charles.

Blaut, J.M. 1993: *The colonizer's model of the world: geographical diffusionism and Eurocentric history*. New York: Guilford Press.

Brenner, R. 1976: Agrarian class structure and economic development in pre-industrial Europe. *Past and Present* **70**, 30–75.

Brenner, R. 1982: The agrarian roots of European capitalism. *Past and Present* **97**, 16–113.

Campbell, B.M.S., Galloway, J.A., Keene, D. and Murphy, M. 1993: A medieval capital and its grain supply: agrarian production and distribution in the London region *c*. 1300. *Historical Geography Research Series* No. 30.

Cosgrove, D. 1984: *Social formation and symbolic landscape*. London: Croom Helm.

Dodgshon, R.A. 1973: The nature and development of infield-outfield in Scotland. *Transactions of the Institute of British Geographers* **59**, 1–23.

Dodgshon, R.A. 1987: *The European past: social evolution and spatial order*. Basingstoke: Macmillan.

Roberts, B.K. 1987: *Rural settlement*. London: Macmillan.

Rodwell, W. 1981: *The archaeology of the English church: the study of historic churches and churchyards*. London: Batsford.

Whyte, I. 1979: Infield-outfield farming on a seventeenth century Scottish estate. *Journal of Historical Geography* **5**, 391–401.

12

URBANIZATION AND PROTO-INDUSTRIALIZATION

> A . . . very long-term process of urban network development creation is a necessary preparation for entry to the modern industrial world.
> (J. De Vries, 1984: *European urbanization 1500–1800*. London: Methuen)

INTRODUCTION

An important aspect of the transition from feudalism to capitalism was the increasing proportion of agricultural and industrial produce that was marketed. The feudal economy was not anti-market as such, but its organization was not sufficiently flexible and supportive to allow the accumulation of capital, nor did it encourage the development of the complex series of market outlets, associated urban hierarchies and integrating flows of long-distance trade, that characterize mercantilist and capitalist modes of economy.

MARKETING, URBANIZATION AND THE SETTLEMENT HIERARCHY IN EUROPE

Most medieval markets were local affairs. At first they were means of small producers swapping occasional surpluses, perhaps by barter. Trading grain for cheese or meat for fruit would have been on the basis of customary exchange values, but this would have been informal. Eventually the advantage of using money and standard weights and measures would have been obvious to all concerned, not least to the market authorities who could charge a fee. One of the reasons why markets were encouraged by estate owners and the Crown was that the potential for taxes and other revenues was very attractive.

Each market operated perhaps only once or twice a week because the business would not have been sufficient to justify permanent pitches. One of the biggest steps came when professional merchants grew in numbers. They acted as intermediaries between informal markets, organizing transport and buying up quantities of items that they knew they could sell on at a profit. Most were itinerant, moving around from market to market. For their convenience the market cross was often supplemented by a small market hall and the village or town elders chose a fixed market day when interaction could be maximized. Interestingly, neighbouring villages seem to have realized that they should have different market days so that they could visit each other's markets (*see* Figure 12.1) and this represents a degree of economic planning that illustrates the move to greater integration.

Not all markets were successful. Three-quarters of those founded between 1200 and 1350 failed, especially those in small villages. Survivors tended to be in towns or in favoured locations on transport routes. Through the sorting process of competitive trial and error there emerged a pattern of market towns and a hierarchy of settlement size (city–town–village–hamlet) that by the sixteenth and seventeenth centuries was geared to a commercial economy that was crystallized into larger and larger regional units, with the influence of London growing rapidly.

On a broader scale, higher order goods were traded through fairs which would happen perhaps once or only a few times a year, and merchants would come from longer distances to display their wares. Networks of fairs evolved regionally but there were also fairs in the Low Countries and in northern Italy that exchanged goods between northern and southern Europe. Rather than dealing just in grain, salt and iron, these fairs specialized in luxury goods such as high quality textiles and other manufactured goods. They were outlets for industrial products.

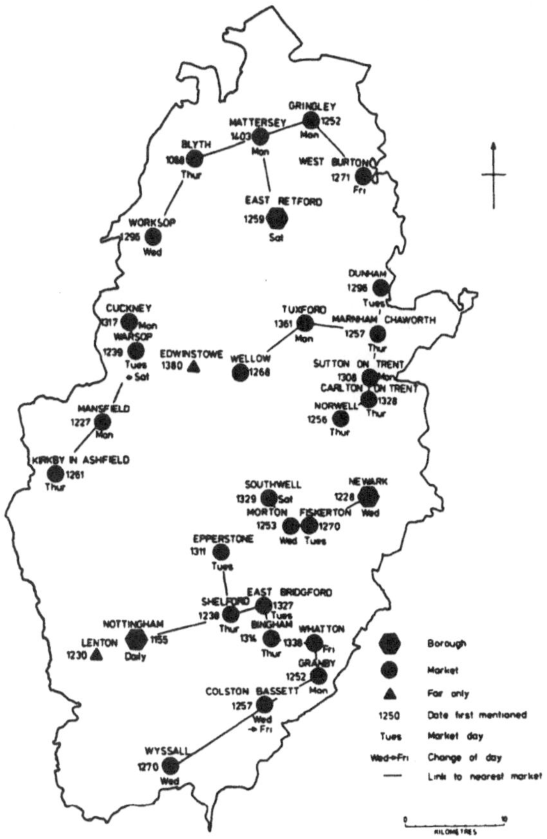

FIGURE 12.1 *Medieval markets, boroughs and fairs in Nottinghamshire*

Source: Reprinted from Unwin, P.T.H. 1981: Rural marketing in medieval Nottinghamshire. *Journal of Historical Geography* **7**, 231–51

PROTO-INDUSTRIALIZATION

In the twelfth century most industry was urban-based. Weavers, iron and lead workers, coopers, leather workers, potters and other craftsmen used local raw materials and served the local market. From about 1300 to 1700 the link between towns and industry became more tenuous as entrepreneurs took up opportunities in the countryside, and gradually regions developed specialisms such as a particular type of cloth or metalwork. One factor was organization. In urban areas the craft guilds regulated production and sought to protect the interests of their members, but they also had a stifling effect upon innovation and acted to keep out newcomers. In contrast, in the countryside there were few restrictions. Such was the shift in the centre of gravity of manufacturing that the prosperity of some towns was seriously threatened.

Technology was also important. Human muscle power was the main input in urban workshops but from the mid-fourteenth century on there was a greater use of water power and wood charcoal as fuel. In the textile industry, for instance, the water-driven fulling mill was adopted from the thirteenth century as a labour-saving device for cleaning and finishing woollens.

Mendels (1972) and other writers have argued that a set of conditions existed in the early phase of industrialization that were very similar in the establishment of rural centres of manufacturing throughout western Europe. These may be conveniently summarized in three points:

- Small-scale domestic production, using patriarchally directed family labour, was cheap and flexible. It made use of the time available between tasks on the farm. The fit with the daily rhythms of livestock farming was better for this than arable areas where the seasonal peaks of ploughing, sowing, weeding, harvesting and threshing left little spare time.
- Such dual economies were characteristic of areas of harsh environment where the industry provided a welcome supplement to agricultural income. They were also encouraged where there were limited resources for expansion, for instance where commons were stinted.

- Rural industry also seems to have been popular in regions with a rapidly growing population, for instance where inheritance practices stipulated the division of a father's properties equally among his sons. This *partible inheritance* anchored population to the land but impoverished successive generations as the average size of farms declined. Population also grew where manorial control was weak and unable to prevent immigration. In these sorts of areas industry was a means of providing work for the underemployed.

Since many rural industrial economies were distant from both their raw materials and their markets, they relied upon intermediaries, often urban-based merchants who organized channels of assembly and distribution and who could deal with the transport of bulky commodities. Some merchants came to control the quality and quantity of output by their clients in a sophisticated putting-out system. This involved an increasing investment by the merchant until he or she owned the raw materials and machinery used in the rural workshop, reducing the craftsmen and their families almost to wage labour. The concentration of economic power in the hands of a few wealthy merchants was sometimes a precursor to the industrial capitalism of a later age, but there were exceptions.

Not all of the proto-industrial regions fulfilled the conditions listed by Mendels. Neither of the textile-producing regions of Suffolk or south Yorkshire, for instance, could be called marginal agricultural economies. Nor can we say that all proto-industrial regions were long-term successes leading to the development of factory-based industrialization and large-scale urban growth. Some did, as in the Lille district of north eastern France and the lower Rhineland of Germany (Pounds 1985). Others waned and returned to agriculture, leaving only traces of their former glory. The textile villages of the Suffolk–Essex border are the classic example of this latter phenomenon, where it is frankly difficult to believe on present inspection that they were once hives of industry, and other areas of vigorous proto-industry have declined, such as the textile districts in northern Italy, Catalonia (Spain) and Lódź (Poland). Some hearths of the Industrial Revolution, such as the North East of England or South Wales, had little or no tradition of rural industry and started, as it were, from scratch in the eighteenth and nineteenth centuries.

The landscape impact of proto-industry was minimal. The energy needed was derived from wind or water, or from human or animal muscle power. Where wood or charcoal were employed, for instance in iron smelting, renewable resources were drawn upon in the shape of managed woodlands (Chapter 6). Mining technology was primitive, leaving only the remains of bell pits or adits, and the miners operated in small groups so that workings were localized and superficial. The dual economies of farming and textile industries would have merged into their rural background, with only minor architectural

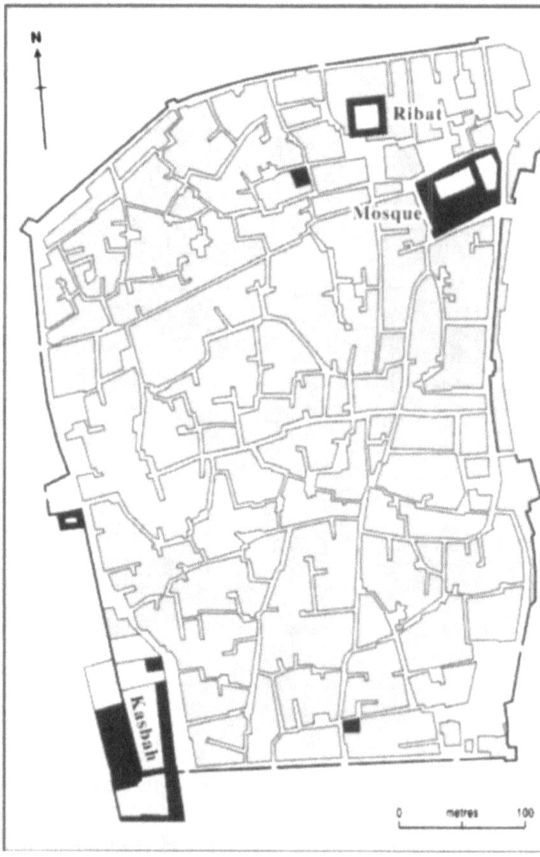

FIGURE 12.2 *A street plan of the medina (old Arab town) of Sousse, Tunisia*

modifications such as spinning galleries or well-lit weaving sheds to show. The major landscape modifications came in those districts which went on to participate in the Industrial Revolution, and the evidence of their early beginnings have often been obliterated.

CASE STUDY: SUFFOLK TEXTILES

Medieval and early modern Suffolk was a wood-pasture region. In the sixteenth century about two-thirds of the county was devoted to dairy farming and both cheese and butter were exported by sea to London. Although in the Lowland Zone, agriculture was not organized into common fields, but had been enclosed at an early date and held in severalty by individual farmers. Partible inheritance was practised and manorial organization was generally weak. Immigration and local population growth meant that by the mid-sixteenth century central Suffolk was suffering from a shortage of commons and resources from the waste land.

The local cloth industry reached its peak in the late fifteenth and early sixteenth centuries, when its coloured cloth was much sought after. Many of the villages on the Suffolk–Essex border were comparatively wealthy, for instance, Sudbury, Long Melford, Clare and Lavenham. Some individuals were spectacularly rich, for instance the daughter and widow of Thomas Spring of Lavenham had the highest assessed tax bill outside London in 1524–5.

Suffolk did not maintain its economic advantage, however. Much capital was invested in large churches, the so-called *wool churches*, which today seem significantly more lavish than is warranted in a small town or large village. The merchant capital was dissipated in buildings and land, with their associated status and was not reinvested in the spiral towards factory industry that was characteristic of other textile regions.

PRE-INDUSTRIAL URBAN LANDSCAPES

Several attempts have been made to theorize the physical and social morphology of cities which might be termed *pre-industrial*. Sjoberg (1960) in particular argued that they share certain characteristics which lead to predictable outcomes. First, there is a low level of technology which influences all aspects of life, from the poor transportation which makes it a walking city, to the primitive energy sources (fuelwood, human

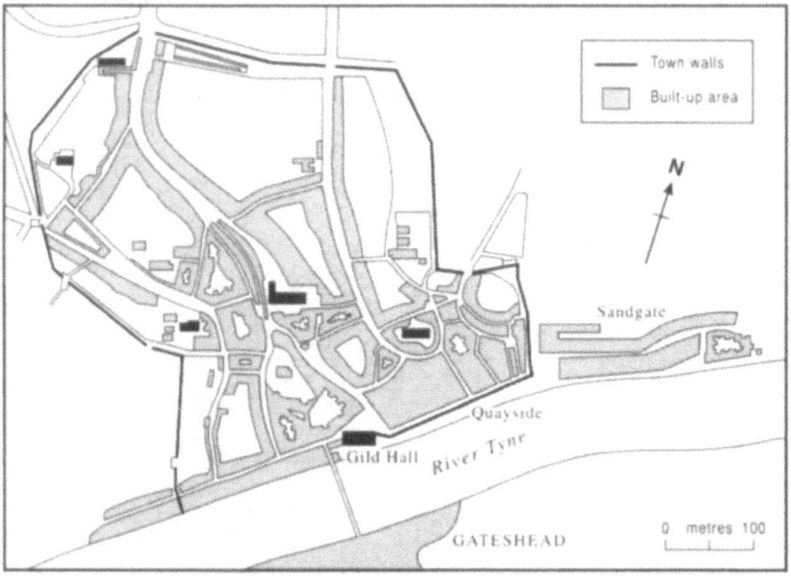

FIGURE 12.3 *Newcastle-upon-Tyne*
Source: James Corbridge's map of Newcastle (1723)

and animal power) which limit manufacturing to a small-scale workshop activity.

Second, the lack of public works means pot-holed and rubbish-filled streets, an unreliable and polluted water supply, poor sanitation, and inevitable consequences for the ill-health of the citizens. This lack of formal planning, as understood in the modern western tradition, is also evidenced in the narrow, winding streets and apparently chaotic ground plan (*see* Figure 12.2). The high density of population is another source of concerns about health because of the rapid spread of infectious disease.

There is, third, a spatial outcome of social processes, with élite groups monopolizing land close to the city centre, in order to dominate the levers of economic, political, cultural and religious power. The high-status core is surrounded by lower order groups, including a ghetto for immigrants, with the ultra-poor and social outcasts living beyond the defended city wall, along with noxious trades such as tanning and soap boiling. There is a lack of functional specialization of land-use and no separation between the place of residence and the place of work.

Sjoberg has been roundly criticized for his tendency to over-simplify and subsequent writers have elaborated upon the fascinating variety of pre-industrial cities. Wheatley's (1971) account of Chinese urbanism reveals a complex process of urban planning, with cosmo-magical undertones. James Vance (1977) has hypothesized two variants on a *pre-capitalist* (rather than pre-industrial) European city. The first, found in northern Europe, was dominated by craft and commercial guilds, the most powerful contemporary social and economic forces. Prosperity was derived from holding and using land rather than owning it, and in consequence there was no land market organized in terms of rent-paying ability, access to the city centre, or even social class. Instead, the social ecology of the city was based upon occupational zoning, as witnessed by the street names such as Butcher Bank and Milk Street.

In contrast, the southern European city, according to Vance, was a *factional city* in which space was fragmented into areas dominated by rival social groupings. These factions were based upon family and commercial ties, and it was they who constructed the fortified houses with tall towers still visible in many Italian cities. The Montagus and the Capulets of Shakespeare's *Romeo and Juliet* were just such factions.

Other scholars have identified categories of urbanism associated with particular cultural or religious associations, such as the Islamic city and with more or less homogeneous regional characteristics, for example the Latin American city. Suffice to say here that we must be alert to these apparent regularities of form and function, yet wary of over-generalizing. Similar forms may have arisen from a disparate variety of processes and do not necessarily give clear insights into the underlying framework of power and social process.

CASE STUDY: NEWCASTLE-UPON-TYNE

At first Newcastle, as the name suggests, was a fortified town, originally at the eastern end of the Roman Hadrian's Wall and later focused on a Norman keep. Its strategic location on the River Tyne (Figure 12.3) was a bastion against invasion from the north but its medieval prosperity was founded rather upon commerce. From the sixteenth century it was one of Europe's most

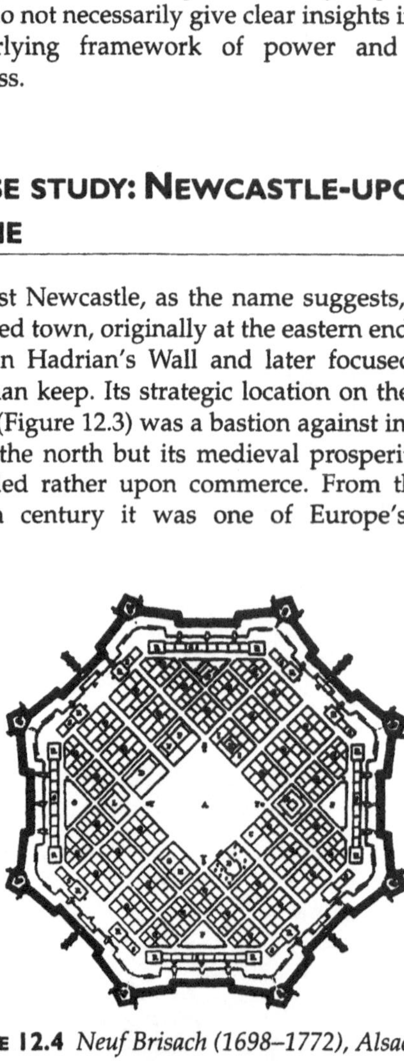

FIGURE 12.4 *Neuf Brisach (1698–1772), Alsace, France*
Source: Reprinted from Morris, A.E.J. 1979: *History of urban form before the industrial revolutions*. 2nd edn, London: Godwin

important centres of the coal trade and the River Tyne was crowded with *keels*, barges for carrying the coal from staithes on the banks upstream down to the waiting coal ships at the estuary mouth.

The urban landscape of historic Newcastle shows many of the characteristics predicted by Sjoberg. Its physical site, with building constrained to the south by the river, to the east by Pandon Dean and the Ouseburn, to the north by the Town Moor, and to the West by a number of large estates (Elswick, South Benwell, Blackett-Ord) which were held off the housing market until the nineteenth century. Most growth until the 1860s therefore came from infill of the city's interstices or the subdivision of existing properties. Over-crowding was very bad, with 40 per cent of the population living in single rooms in 1854. The squalid conditions in the narrow *chares* (lanes) made Newcastle one of the unhealthiest cities in Britain.

The medieval city was peppered with religious institutions, but the central focus of pre-industrial Newcastle was on the river around the Guildhall on the Sandhill and Side. Here were the headquarters of the craft and trading guilds and the city Corporation, surrounded by a cluster of houses owned by wealthy burghers.

Langton (1975) has shown that occupations of freemen in the seventeenth century varied in the wealth of their practitioners, in a clear hierarchy (Table 12.1) which was reflected on the ground in their location with respect to the Guildhall. The Hostmen, who controlled the coal trade, owned large houses on Sandhill, while the shipping and building trades were peripheral. There was occupational zoning as expected by Vance, but overall the city does not fit any single model of social geography.

TABLE 12.1 *House size in Newcastle, 1665, as measured by the average number of hearths*

Occupation	No. of hearths
Mayors	8.4
Hostmen	5.7
Other merchants	4.3
Bakers	3.2
Master mariners	3.1
Mariners	3.0
Barber-surgeons	2.5
Joiners	2.5
Cordwainers	2.4
Weavers	2.2
Shipwrights	2.0
Butchers	2.0
Coopers	2.0
Tailors	2.0
Tanners	2.0
House carpenters	1.6
Blacksmiths	1.4

Source: Langton (1975, 15)

THE RENAISSANCE AND THE URBAN SPECTACLE

Apart from their defensive and trading functions, Italian cities such as Florence, Venice, Siena, Milan and Rome by the sixteenth century had developed institutional and cultural contexts that were highly sophisticated. The way of thinking about towns developed in a rather different direction from previous eras, with a design philosophy that owed much to new military technologies and absorbed the need for a perfect defensive design (*see* Figure 12.4), but also acknowledged the discoveries of artists and engineers about perspective, space and form. Notions about order and good government were extended to plans of ideal cities drawn up on rational criteria (Morris 1979).

One of the most famous events of sixteenth-century European history was the Turkish invasion of Malta in 1565 and their attempt to crush the resistance of the defenders led by the Knights of St John. Voltaire was later to say that no other single military action was so well known. Malta was symbolic because it was small but strategic defensively against the expansion of the Ottoman Empire and represented a last stand for the Knights, who had been expelled from their previous strongholds in Rhodes and Tripoli. After a short but violent siege, the Turks were defeated, but military analysts predicted their return and the Knights therefore called upon the architectural and military engineering skills of Europe to construct a series of fortifications around a new capital city to be called Valletta after Grand Master Jean de la Valette.

The renewed onslaught did not occur but the city of Valletta, with its magnificent position overlooking the Grand Harbour (*see* Figure 12.5), was typical of Renaissance town planning. Valletta exhibits a street plan of rectangular blocks imposed upon a far from flat plateau surface. There were splendid *auberges*, one for each of the constituent nationalities making up the Order, and many other public buildings with a baroque grandeur that established Malta in the collective mind of the Christian world as one of the achievements of contemporary civilization. The defences were massive, employing the latest provisions for enfilade fire and more traditional features such as deep moats and drawbridges. According to Blouet (1987, 88) Valletta became

> a microcosm of the great baroque, absolutist capitals which developed in western Europe. In the latter part of the eighteenth century an English visitor to the island, Patrick Brydone F.R.S., having seen the building and society of Valletta described the city as 'an epitome of all Europe'.

PUBLIC AND PRIVATE SPACE

The balance between public and private space in a city is an important measure of its texture and cultural reference. Despite the classical examples

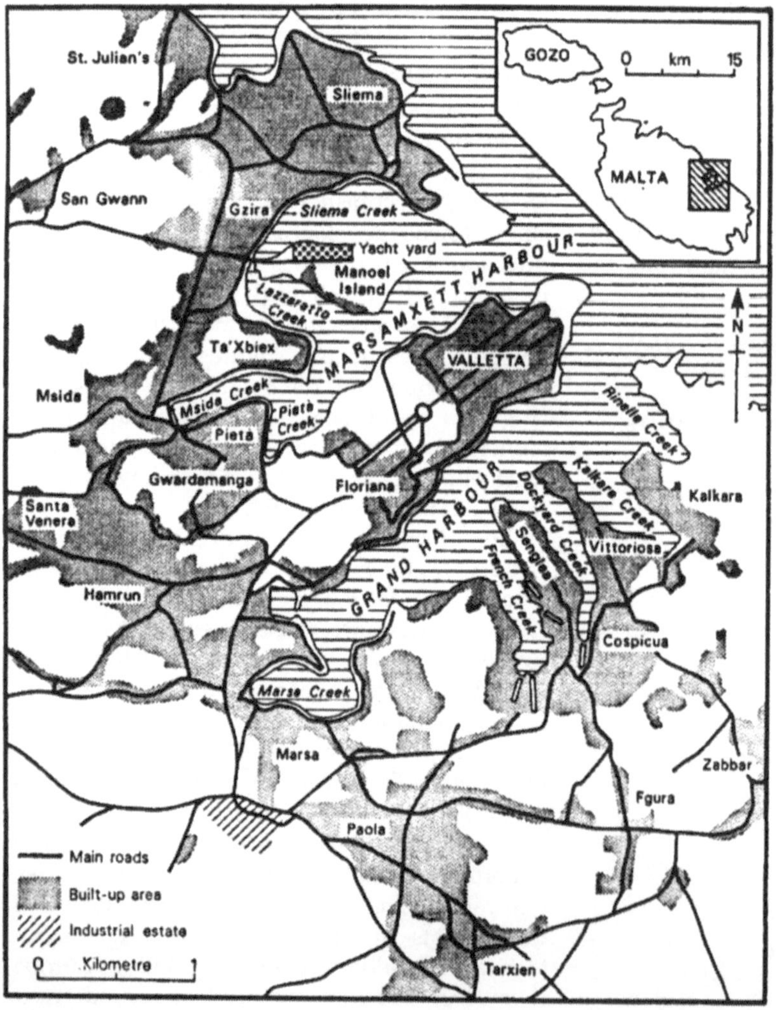

FIGURE 12.5 *Valletta and the Grand Harbour*

of the Agora in Athens and the Forum in Rome, the pre-modern urban norm was for public access and amenities to be afterthoughts, with most attention being focused on individual properties. In traditional Middle Eastern cities, for instance, the typical courtyard house was designed to be inward-looking, sometimes without even a window on to the public street.

Significance for public space was never entirely absent of course, especially in the market and places of worship, but in Renaissance Italy it was revived and promoted as a feature of the resurgent self-confidence of the state. Venice, in particular, concentrated its most prestigious and powerful institutions around the Doge's Palace and St Mark's Cathedral. The large piazza was open on one side and travellers arrived here by water amid all the architectural splendour that could be assembled by one of the major sea-borne empires of the age. 'One experienced Venice from the centre outward' (Muir and Weisman 1989).

According to Lefèbvre (1991), the Renaissance town embodied a new code that was used to create urban spaces. Façades were harmonized to create perspectives, and streets and squares were arranged in concord with public buildings. There was appearing, in effect, an interpretation of the townscape as a stage set, with buildings and street furniture as props in what amounted to an abstract space reflecting the rational, artistic and political intentions of the planners.

In Venice public space evolved as pure theatre and served a clear political purpose of establishing an hierarchy of sacred and profane spaces (*see* Figure 12.6). It sought to subvert loyalties to neighbourhood, clan and patron–client linkages through the establishment of a trusted, solid and reasonable set of centralized institutions. Regulation of the use of public spaces, such as rituals and processions, gave the authorities power over space which they later consolidated and expanded. Elaborate and frequent public ceremonies reinforced the significance of legends about Venice and established a collective awareness.

In truth Venice was never planned in the modern sense, although Cosgrove (1982) shows that attempts were made. Rather, it evolved as the reflection of a collective consciousness of the class of merchant capitalists and statesmen who were in power for a thousand years. It was initially created from the sea in a swamp where a collection of islands surrounded by a lagoon offered some protection. Its fortunes were not derived from that location or any advantage of natural resources but were gradually built from an aggressive and well-organized economy of long-distance trade in the Mediterranean and beyond. The architectural and monumental unity that we see today arose from a sense of common

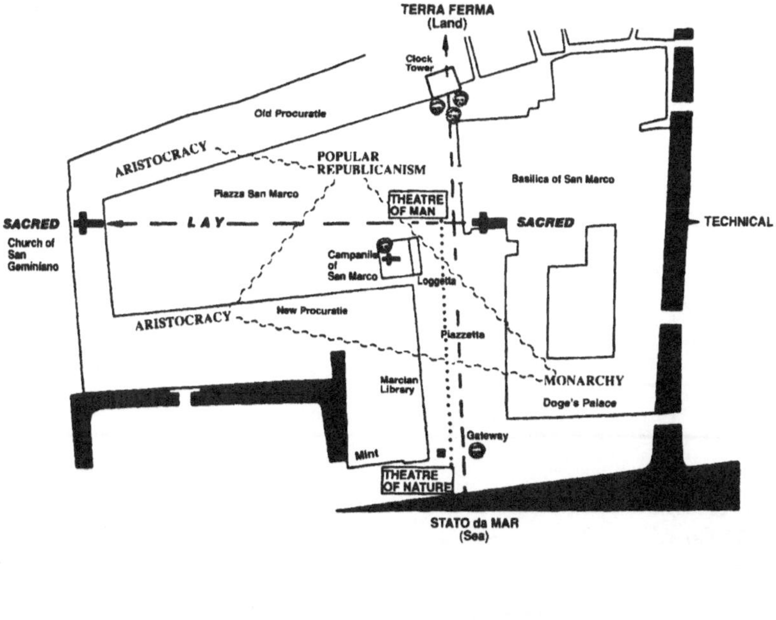

FIGURE 12.6 *Saint Mark's Square*
Source: Reprinted from Cosgrave, D. 1982: The myth and the stones of Venice: an historical geography of a symbolic landscape. *Journal of Historical Geography* **8**, 145–69

purpose that had been forged over centuries of adversity and was expressed in a common cultural language or code that was shared by that élite from the sixteenth century onwards.

SUBURBAN PRIVILEGE

Among the most important thresholds in the history of the urban landscape was the development of suburbs which were not merely accretions of additional buildings to house newly arrived migrants but rather planned estates for the wealthy. The date varies from city to city but eventually throughout the western world the city centre was abandoned to commerce and the élite decamped to the rural fringe. It is inaccurate to say that the prime force was profit on the part of entrepreneurial landlords, or indeed the greater freedom given by improvements in transport. Rather there was a range of processes which encouraged suburbanization and created an inside-out city by comparison with its predecessor social geographies.

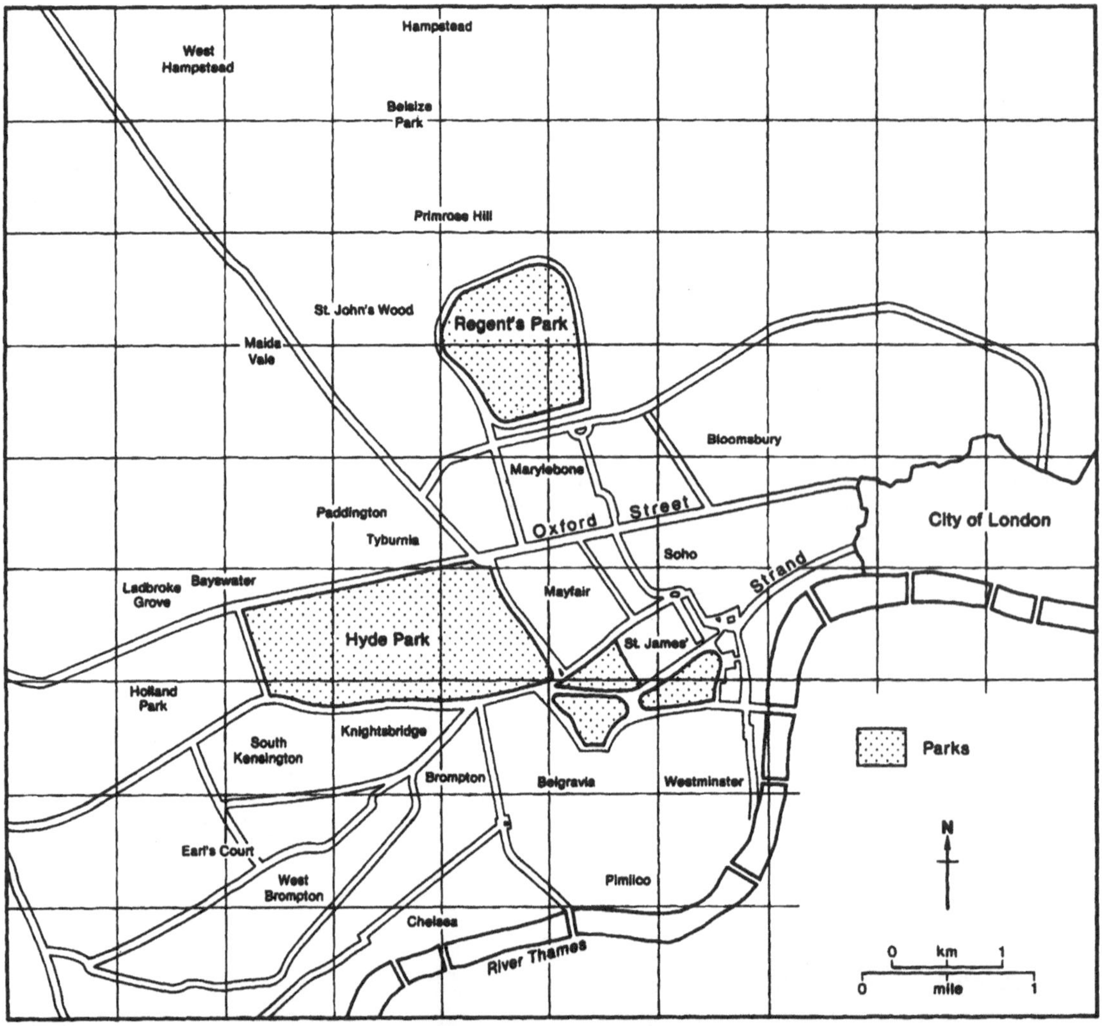

FIGURE 12.7 *The West End of London*
Source: Atkins, P.J. 1990: The spatial configuration of class solidarity in London's West End, 1792–1939. *Urban History Yearbook* **17**, 36–65

Case study: the West End of London

In Britain the capital saw aristocrats move to the western suburbs from the 1550s onwards, especially after the great fire of London had destroyed so many centrally located properties in 1666. Pockets of suburban privilege appeared at the Earl of Bedford's pioneering Covent Garden (1631), the Earl of Leicester's Leicester Fields (1635), the Earl of Southampton's Bloomsbury Square (1665), and the Earl of St Albans' Soho Square (1681). Even further west were the Earl of St Albans' St James's Square (1663), the Earl of Scarborough's Hanover Square (1717), the Earl of Burlington's scheme behind Piccadilly (1717), and Lord Berkeley's Berkeley Square (1736).

The physical morphology of these developments, so unlike the previous urban texture of London, was conducive to social change (Atkins 1990). Whereas the proximity of rich and poor, powerful and powerless, landed and trading wealth, residential and commercial land-use, was taken for granted in the hugger mugger of the pre-industrial City of London's streets, courts and alleyways, here in the planned environment of the new age there was deliberate social distancing. Each estate had a focus in a central square, with the mansion of the potentate, surrounded by the lesser houses of his clients. Some were planned as communities with a church, a market and other public facilities.

Lawrence (1993) has argued that these residential squares, because of their revolutionary introduction into the city of open space adorned with trees and gardens, were instrumental in shaping subsequent environmental ideals. They drew inspiration from *piazzi*, *plazas* and *places* in Italy, Spain and France but the subtle blend of private prestige capital and domesticated nature was new. Some squares had tree-lined gravel paths, arranged in geometric designs but others were densely planted with shrubs, hedges and

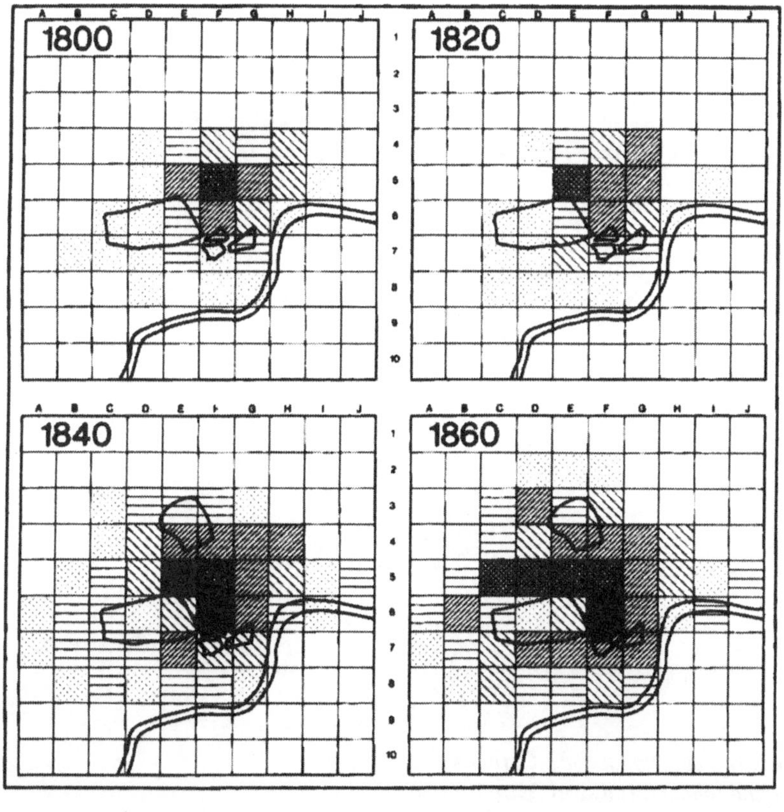

Figure 12.8 *The dark shading shows a high concentration of families of high status in 1-kilometre grid squares*
Source: Atkins, P.J. 1990: The spatial configuration of class solidarity in London's West End, 1972–1939, *Urban History Yearbook* **17**, 36–65

trees. Examples of the latter were Cavendish Square, with a garden designed by Charles Bridgeman in 1717, Grosvenor Square planted in 1725, and Bloomsbury and Russell Squares laid out by Humphry Repton in 1805. Many of the ideas were imported from the landscape gardens of wealthy rural landowners who wished to enjoy scaled-down versions of these amenities near their town houses, in gardens enclosed by iron railings with access for keyholders only. The imprint of rural authority was thus reproduced in the urban setting and the squares began to take on symbolism in line with property and social values.

By the mid-nineteenth century the West End of London had matured. It covered over 40 km^2 and was a part-time address for about 15 000 families with the best connexions, who came to town for the social season in the spring and summer (*see* Figures 12.7 and 12.8). In no other city in Britain and probably nowhere else in the world was there such an extraordinary concentration of wealth and social prestige. Some residents wished to seal the boundaries of their neighbourhoods against noisy traffic and erected gates and bars across access roads, establishing a dominance over private space which was not broken until the end of the century (Atkins 1993). In the 1990s gated roads are now reappearing, mirroring a security-conscious trend in America towards a cellular space in which public access is restricted (*see* Figure 12.8).

Conclusion

Towns and cities have come to embody so much of the essence of the economy and culture of the modern era that we may be forgiven for taking for granted a connexion between urbanization and capitalism. In this chapter we have seen that the roots of industrialization lay rather in the countryside and that it has often been true that urbanism has evolved through the amalgamation of neighbouring small cells, often

Figure 12.9 *The gate blocking the entrance to Hyde Park Gardens*
Source: Searle, M. 1930: *Turnpikes and tollbars*. London: Hutchinson

semi-rural at first, of industrial or mining activity, with later accretions of commercial functions and suburbs.

In Part 3 we will follow human impacts upon landscape and environment through into the modern era, starting in Chapter 13 with agriculture.

FURTHER READING AND REFERENCES

Atkins, P.J. 1990: The spatial configuration of class solidarity in London's West End, 1792–1939. *Urban History Yearbook* **17,** 36–65.

Atkins, P.J. 1993: How the West End was won: the struggle to remove street barriers in Victorian London. *Journal of Historical Geography* **19,** 159–71.

Blouet, B. 1987: *The story of Malta*. 3rd edn, Malta: Progress Press.

Cosgrove, D. 1982: The myth and the stones of Venice: an historical geography of a symbolic landscape. *Journal of Historical Geography* **8,** 145–69.

Langton, J. 1975: Residential patterns in pre-industrial cities: some case studies from seventeenth-century Britain. *Transactions of the Institute of British Geographers* **65,** 1–28.

Lawrence, H.W. 1993: The greening of the squares of London: transformation of urban landscapes and ideals. *Annals of the Association of American Geographers* **83,** 90–118.

Lefèbvre, H. 1991: *The production of space*. Oxford: Blackwell.

Mendels, F.F. 1972: Proto-industrialization: the first phase of industrialization. *Journal of Economic History* **32,** 241–61.

Morris, A.E.J. 1979: *History of urban form before the industrial revolutions*. 2nd edn, London: Godwin.

Muir, E. and Weissman, R.F.E. 1989: Social and symbolic places in Renaissance Venice and Florence. In Agnew, J.A. and Duncan, J.S. (eds) *The power of place: bringing together geographical and sociological imaginations*. Boston: Unwin Hyman, 81–103.

Pounds, N.J.G. 1985: *An historical geography of Europe, 1800–1914*. Cambridge: Cambridge University Press.

Sjoberg, G. 1960: *The pre-industrial city: past and present*. Glencoe, Ill.: Free Press.

Vance, J. 1977: *This scene of man: the role and structure of the city in the geography of western civilization*. New York: Harper & Row.

Wheatley, P. 1971: *The pivot of the four quarters: a preliminary enquiry into the origins and character of the ancient Chinese city*. Edinburgh: Edinburgh University Press.

Part 3

The modern era

The massive changes in the economy and ecology of the world which are labelled as industrialization came to dominance in the nineteenth century, since when they have often been regarded as the norm, against which all else is measured: *less developed* is a comparative term of that kind. The nineteenth century was not, of course, the beginning of the processes of industrialism, for its major characteristics of steam power, the substitution of machines for human labour and the spatial aggregation of production into the factory can all be seen in earlier centuries.

As we saw in the Introduction to Part 1, the human use of energy hitherto had been confined to solar power in one of its indirect forms, as in photosynthesis or meteorological elements like wind and water. The great change comes with the discovery of the efficient use of fossil fuels to power machines. This started with the use of coal to drive a water pump which made possible mining at greater depths. Steam power was then applied to railways and to shipping and was joined by the more concentrated and often more portable fuels like oil and natural gas (Table P3.1). All share one vital characteristic: unlike the products of solar energy, they are not renewable. Once the complex hydrocarbon molecules are oxidized by burning, then they are gone for ever in that form. It is theoretically possible though economically unlikely therefore that all the coal, oil and natural gas in the earth's crust could be used up by human activity.

One response to the finite nature of these fossil fuels has been the development of alternatives such as nuclear power and the hi-tech use of solar energy in modern wind-farms and via photoelectric cells, for example. Nuclear power is in theory non-renewable but the concentration of energy in the nuclei of heavy elements like uranium is so great that this is not a practical worry. In areas of plentiful falling water (natural or dammed), then large-scale hydropower is a popular addition to the repertoire.

In spite of all these developments, a large proportion of the world's population only has access to solar power, usually as biomass. Statistics for

Table P3.1 *Growth in population and energy use, 1870–1986*

Year	Industrial energy use (TW)	Per capita (watts)	Cumulative use since 1850	World population (millions)
1870	0.2	153	3	1300
1910	1.1	647	25	1700
1950	2.9	1160	100	2500
1970	7.1	1972	200	3600
1986	8.6	1720	328	5000

Note: 1 TW (Terawatt) = 1×10^{12} watts. The Table only refers to *industrial* energy use: not therefore to biofuels like wood and dung.

TABLE P3.2 *Human appropriation of net primary productivity (NPP), 1980s*

		%
World NPP	Terrestrial	132.1
	Fresh water	0.8
	Marine	91.6
	Total	224.5
NPP directly used by humans	Plants eaten directly	0.8
	Plants fed to domestic animals	2.2
	Fish eaten by both	1.2
	Wood use for paper and timber	1.2
	Fuel wood	1.0
	Total	7.2
NPP used or diverted by humans	Cropland	15.0
	Converted pastures	9.8
	Others (cities, deforestation)	17.8
	Total	42.6
NPP used, diverted or reduced	NPP used or diverted	42.6
	Reduced by conversion	17.5
	Total	60.1

Source: Diamond, J.M. 1987: Human use of world resources *Nature* **328,** 479–80

energy use usually omit these materials (wood, leaves, and dung are the commonest forms) and give only the data for the production and consumption of fossil fuels, nuclear, hydropower and new technologies for harnessing natural phenomena: *commercial energy* is the term used. In spite of all the technological developments associated with industries the world over, humans are still dependent upon the products of photosynthesis for their food and upon the solar-powered hydrological cycle for water. The extent to which control over fossil fuels has allowed the manipulation of the world's ecology is given in Table P3.2 where it can be seen that some 40 per cent of the world's terrestrial fixation of solar energy as plant tissue has been appropriated by human activity.

Compared with the explosion of cultural diversity made possible by agriculture, industry produces relatively little of it. Factories tend not to be tuned to climate like farms and all TV sets are more or less the same. The power of steam applied to railway and ship, followed by the internal combustion engine and electric telecommunications tended to spread dominant ideas quickly and they could be followed up where necessary by armies and colonial officials whose jobs it was to make them stick. So the modernism of industrialization has in general reduced the diversity of culture in the world and with it the variety of cultural landscapes: modernized cities are much the same everywhere. The expression of non-material values continued, but the cathedrals of the nineteenth century were in praise of empire and commerce; sheer size also entered music with grandiose conceptions like Gustav Mahler's Symphony of a Thousand.

The cultural landscapes of industrialism, then,

TABLE P3.3 *Population and longevity: proportion of years lived*

Date	Pop × longevity	Period covered
10,000 BC	8.6	from evolution of species
AD 1	34.2	10,000 BC–AD 1
AD 1750	28.2	AD 1–AD 1750
AD 1950	16.8	AD 1750–1950
AD 1990	12.2	AD 1950–1990

Note: The second column is a percentage of all the people who have ever lived, not an absolute number.
Source: Livi-Bacci, M. 1992: *A concise history of world population*. Cambridge, Mass.: Blackwell

include those of energy getting, the power-houses of the world; those of industrial production themselves; the cities whose increased size is only made possible by the new fuels, the communications web that binds them all together, and many other new landscapes of pleasure such as the jet-packaged holiday in the sun.

But the fruits of industrialism are spread unevenly (Tables P3.3 and P3.4), and one way of seeing the world in about 1950 was as a core of industrial nations (all consuming high per capita quantities of energy and having high levels of per capita income) with a periphery of *developing countries* whose industrialization was less complete but who nevertheless saw the core regions as desirable places to emulate. In due course we will have to question whether that is still a useful model.

Table P3.4 *Environmental impact as a function of longevity and energy use*

Date	Pop × longevity	Per cap E use (W)	Index of impact
10,000 BC	8.6	50	430
AD 1	34.2	100	3420
AD 1750	28.2	300	8460
AD 1950	16.8	1972	33 130
AD 1990	12.2	1720	20 984

Note: The Index is obtained by multiplying the two previous columns and is obviously crude; it nevertheless puts into perspective the data in Table P3.1.

13

THE IMPACT OF AGRICULTURE

In a country full of civilized inhabitants timber must not be suffered to grow. It must give way to fields and pastures, which are of more immediate use and concern to life.
(J. Morton 1712, cited in J.V. Thirgood 1989: Man's impact on the forests of Europe. *Journal of World Forest Resource Management* **4,** 127–67)

INTRODUCTION

In previous chapters we have looked at the impact of agriculture upon the landscape in the pre-industrial era, including the enclosure of the medieval open fields and commons up to the middle of the nineteenth century. Here our task will be to carry the theme forward into the modern era by investigating agricultural technologies in the twentieth century and also by looking at cases studies of how present-day rural landscapes have evolved.

THE CREATION OF AGRICULTURAL LANDSCAPES

Old settled European and Asian countries have long since carved their agricultural landscapes out of wood and marsh, achieving a relatively stable balance between humans and their environment up to a thousand years ago. This process is either recent or continuing in the Americas, Africa and Oceania (Table 13.1), and we are only too painfully aware of this as tropical rainforest yields to the chainsaw (Chapter 8). Over a third of the world's land area is now to a greater or lesser extent harnessed to agriculture, and a further 40 per cent has been disturbed by humans (Hannah *et al.* 1994).

According to Simmons (1987), the agricultural landscape was modified by the following:

- The use of fire to clear forests and grasslands for cultivation.
- The development of stone and later, metal, axes accelerated woodland clearance.
- Modifications to the soil by digging sticks, spades, and ploughs.
- The construction of terraces, mounds, ridges and furrows for agricultural use.
- Irrigation and artificial drainage works.
- The use of fences, dykes, ditches and bunds as boundaries and livestock barriers.
- The domestication, selective breeding and spread of useful animals and plants.

TABLE 13.1 *Changes in land use at the global scale (million ha)*

	1700	1850	1920	1950	1980
Forests and woodland	6215	5965	5678	5389	5053
Grasslands	6860	6837	6748	6780	6788
Croplands	265	537	913	1170	1501

Source: Richards, J.F. 1990: Land transformation. In Turner, B.L. *et al.* (eds) *The earth transformed by human action*. Cambridge: Cambridge University Press, 163–78.

Apart from the physical extent of land-use change, agriculture has involved the modification of natural ecosystems. According to Tivy (1990), the differences between agro-ecosystems and wild ecosystems are:

- There is less diversity of plant and animal species.
- There is less complex structure (spatial organization of its components).
- There is a reduction of the length of the food chain. In the wild this may be four *trophic levels* of plants, herbivores, carnivores and top carnivores. In agriculture humans replace the carnivores and in arable farming the herbivores as well.
- A larger proportion of biomass is comprised of animals, especially large ruminants.
- A much smaller proportion of the energy pool is routed through dead and decaying matter in the soil.
- Nutrient cycling is speeded up and is usually maintained by inputs of organic or inorganic fertilizers.
- Agro-ecosystems are more open in the sense that they exchange more energy and material with the outside world. Thus livestock on a farm in Britain may consume concentrated feed produced in America. Some exports from the agro-ecosystem may be unwanted and unplanned, especially the leaking of polluting chemicals into other wild or managed systems.

Modern commercial agriculture has deliberately reduced the number of crops, and varieties of those crops, upon which it relies. There are probably between 10 000 and 80 000 edible plant species in the world, of which roughly 3000 have been exploited at one time or another throughout history. Of these only 150 have become widespread and 29 species currently account for 90 per cent of food production. Table 13.2 lists the major groups of plants and animals and gives an estimate of their output by weight. It should be remembered that food products are difficult to compare, one tonne of butter hardly being equivalent to one tonne of sugar cane, unless the nutrient content of each is calculated.

FACTORS WHICH HAVE ACCELERATED CHANGE IN TWENTIETH-CENTURY AGRICULTURAL LANDSCAPES

There have been many significant economic and technological changes in modern times which have widened and deepened the environmental and landscape impacts of agriculture. We will summarize them here only in the briefest form. For greater detail, see Grigg (1987).

Von Thünen argued in 1826 that transport costs were a major differentiating factor in land use variations. He was writing before the railway age and of course he could not have predicted the easing of time schedules and unit costs that would come with better road, sea and air transport, to the extent that food is now traded globally and demand in one country may stimulate a supply elsewhere in the world. The interconnexions are now becoming so complex that one can reasonably assert that environmental change in the producing areas is driven to an extent by global factors.

The most basic elements of any farming enterprise are called the *factors of production*: land, labour, capital, entrepreneurship. In the twentieth century there has been a major shift towards the last two, especially in developed countries, with capital-intensive inputs such as machinery and chemicals being substituted for labour, and to a certain extent even for land in certain enterprises such as pigs, poultry and horticulture. As a result, farms have become larger in order to achieve economies of scale and the human contact with the soil has been greatly reduced, with an exodus of redundant farmers and farm labourers from the countryside that has undermined much of the earlier logic of landscape elements. Thus, small enclosed fields in Europe are being replaced by large, open, prairie-like fields in some areas, in order to provide easy turning circles for tractors and combine harvesters, and hedges are particularly threatened.

Marketing has also played its part. Proximity to a well-organized market of significant size undoubtedly encourages the production of perishables such as fruit and vegetables (*see* Figure 13.1), and producers' marketing cooperatives or a

TABLE 13.2 *The major plants and animals*

Group	Output 1990 (million tonnes)	Species in descending order of economic significance
Cereals	1955	Wheat, rice, maize, barley, sorghum, oats, rye, millet
Roots, tubers	597	Potatoes, cassava, sweet potatoes, yams, taro
Vegetables	442	Tomatoes, cabbage, onions, carrots, cucumbers, pumpkins, aubergines, cauliflowers, green peas, green peppers, green beans, garlic, artichokes
Fruit	342	Grapes, oranges, bananas, apples, water melons, plantains, mangoes, pears, pineapples, cantaloupes, peaches, lemons, plums, papayas, grapefruit, pomegranates, dates, strawberries, avocadoes, apricots, raisins, currants, raspberries
Beverages and stimulants	128	Hops, tobacco, coffee, tea, cocoa beans
Oil crops	74 (oil)	Soya beans, cotton seed, coconuts, oil seed rape, groundnuts, sunflower, olives, palm kernels, linseed, sesame, castor beans, safflower, hemp
Pulses	59	Beans, peas, lentils
Sugar	26	Sugar cane, sugar beet
Fibre	25	Cotton, jute, flax, sisal, hemp
Rubber	5	
Treenuts	4	Almonds, walnuts, hazel nuts, chestnuts, cashews, pistachios
Animals		
Bovine	53 (meat) 514 (milk)	Cattle, buffaloes
Poultry	36 (eggs) 40 (meat)	Chickens, ducks, turkeys
Pigs	69	
Sheep, goats	9 (meat) 17 (milk)	

Source: Food and Agriculture Organization, *Agrostat*

state scheme such as the British Milk Marketing Board (1933–94) also shift the spatial structure of production. Most recently the buying power of food processing factories and of retail supermarket chains has reduced the independence once enjoyed by individual farmers over varieties of crops and their quality.

State policies were significant in the nineteenth century, not least Britain's decision in the 1840s to abolish the Corn Laws and enter the world of free trade for its foodstuffs, but the last 50 years have seen a drift to greater intervention. Most advanced countries support their farmers financially but the European Union's Common Agricultural Policy has attracted the most comment. Its subsidies for decades encouraged the expansion of arable land through the draining of marshes, the ploughing of heath, and a general intensification of production in order to achieve the EU's aim of self-sufficiency in a range of commodities.

The breeding of new, high-yielding varieties (HYVs) of crops and animals has been remarkably successful in recent decades, with the result that many fields are now occupied by species or varieties that would not have been grown 50 years ago because of the environmental limits of climate and soil. In May much of the British landscape now turns yellow with the flowering of oilseed rape and in many poor countries the Green Revolution has enabled the spread of new HYVs of wheat, rice and other crops (*see* Figure 13.2). In the latter case these crops require substantial quantities of moisture and there has been a concerted effort by governments to introduce irrigation works into

FIGURE 13.1 *The intensification of high value fruit and vegetable crops can reach remarkable levels. In Tunisia early crops such as tomatoes, cucumbers, melons and various flowers and house plants are grown under plastic cloches to advance the growing season and thereby catch the market in Europe*
Source: P.J. Atkins

new areas, thereby modifying hydrological regimes and making different demands upon local skills and resources.

Finally, technology has made available an array of agro-chemicals. There are fertilizers to increase yields, herbicides to kill competitive weeds, and pesticides and fungicides to repel the problems of disease and attack to which the crop plants themselves are prone. World-wide about 4.4 million tonnes of pesticides are used every year and the market for agro-chemicals as a whole is worth $25 billion (Mannion 1995).

FIGURE 13.2 *A high yielding variety of cassava under cultivation in India*
Source: P.J. Atkins

NEGATIVE ASPECTS OF AGRICULTURE

Soil erosion

Unfortunately the unforeseen consequences of some of these improvements have led to land degradation, especially in the drylands (Table 13.3). Amongst the worst is soil erosion which, although it occurs naturally, has been exacerbated by human action. Estimates put the present sediment yield of the world's rivers at 2.7 to 5.0 times greater than before human disturbance of the landscape (see Chapter 4). The clearance of forest may be enough to start erosion of those fragile soils which are subject to heavy tropical downpours; elsewhere overgrazing or the lack of soil conservation measures on slopes may increase erosion by water or wind. Soil type is a crucial factor, with the easily erodible loess soils of the Yellow River basin in China losing 100 tonnes per hectare annually. Globally a loss of 10 tonnes per hectare is considered high and 20 tonnes very high. Varying estimates suggest that roughly 26–75 billion tonnes of topsoil are lost globally every year.

There are three main effects of soil erosion:

- loss of soil nutrients and organic matter, reducing the natural fertility and water-retentive capacity of the soil;
- landscape modification by gullying, which in severe cases may make farming difficult;
- increased sediment loads in streams and rivers, causing the silting of irrigation channels, dams and reservoirs, and affecting water supplies for towns and industry. There may be an increased risk of flooding downstream.

It is not surprising that scientists have emphasized climatic factors such as the amount and intensity of rainfall, and the variable resistance of the soil. We must remember, however, that one of the most important influences upon local rates of soil erosion is the attitude and customs of farmers and society at large (Figure 13.3). In particular, the adoption of intensive farming practices in the search for a short-term profit may lead to long-term ruin if the soil resource is eroded. If the costs of such environmental degradation were included in the price of any

TABLE 13.3 *Land degradation by region*

	Susceptible drylands		Other	
	hectares (millions)	% degraded	hectares (millions)	% degraded
Africa	1286.0	24.8	1679.6	10.4
Asia	1671.8	22.1	2584.2	14.5
Australasia	663.3	13.1	218.9	7.0
Europe	299.7	33.2	650.8	18.3
North America	732.4	10.8	1458.5	5.3
South America	516.0	15.3	1251.5	13.1
World	5169.2	20.0	7843.5	11.8

Source: United Nations Environment Programme 1992: *World atlas of desertification*. London: Arnold

crops produced on that land, both farmers and consumers might give pause for thought.

The Indonesian island of Java is a case in point. It has a very high population density sustained by the intensive cultivation of small farms of 0.4 ha or less. Erosion rates are highest on limestone/marl soils where losses are 19–60 tonnes/ha. The most damaging land-use is *tegal* or rain-fed cropping of maize, rice and cassava on sloping upland fields (Table 13.4). The costs of this soil erosion are very difficult to assess but best estimates suggest siltation damage to irrigation systems, reservoirs and harbours totalling £20–60 million per annum and productivity losses on site of £230 million, equivalent to 4 per cent of the value of tegal crops.

Government policies, assisted to a certain extent by overseas aid, have sought to reduce population pressure on the land by encouraging transmigration from Java to other, less densely occupied outer islands, and to give advice to farmers on soil conservation measures such as bench terracing to reduce downslope erosion. Declining oil revenues, however, have persuaded the authorities to stress the export of agricultural products such as cassava, especially since the European Community granted Indonesia the right to export about 10 per cent of its cassava output to Europe. The response has been alarming. Many producers have abandoned their traditional mixed farming systems, which have little impact on the soil, to the monocropping of cassava, sometimes even *removing* terracing in order to increase the cropped area.

The profitability of vegetable crops has also stimulated their intensive production on steeply sloping volcanic soils, which are very fertile but erodible. Vegetables and sugar cane are often

FIGURE 13.3 *Soil erosion on small-holder tea gardens in the Nilgiri Hills, Tamil Nadu, India*
Source: P.J. Atkins

TABLE 13.4 *Soil erosion in Java*

Land use	Area (million ha)	Erosion (tonnes/ha)
Sawah (wet rice)	4.6	0.5
Forest	2.4	5.8
Degraded forest	0.4	87.2
Wetlands	0.1	–
Tegal	5.3	138.3
Total	12.9	61.2*

Source: Pearce, D, Barbier, E. and Markandya, A. 1990: *Sustainable development: economics and environment in the Third World*. London: Earthscan
* Average

grown on the land of absentee land owners by share tenants who have no incentive to conserve the soil of their landlord.

We can identify two opposing types of thinking about soil erosion:

1. The classic approach. Here the problem is blamed on the farmers, who are accused of conservatism, ignorance, apathy and laziness. In addition, over-population is thought to be responsible for soil erosion through the demand for food beyond the carrying capacity of the land. Several solutions are advanced:
   (a) The authorities intervene strongly to forbid grazing in a certain area or to require labour service to build terraces and other conservation works.
   (b) Ignorance is countered by soil conservation education and the demonstration of new technologies.
   (c) It is thought that if farmers were more involved in the market economy they would have a better reason to conserve soil resources and adopt technologies that would allow a greater productivity of food.
2. Soil erosion is taken to be a socio-environmental problem. To see it only as a physical process is profoundly mistaken. The political economy approach sees soil erosion as a symptom and a result of an unjust system. Development is never in the interests of all members of society: inevitably, some groups or classes benefit more than others. In Third World countries environmental degradation has little relevance for the urban-based élite except perhaps in as much as it affects their interests as absentee owners of farm land. The peasant families who suffer the immediate consequences of soil erosion have little political power or influence on policy-making. Small-scale farming is neglected, and any government support that there is for agriculture is likely to go to the larger, commercial farmers whose exploitative attitude to the environment is often very damaging. Only when soil erosion is seen adversely to affect the process of capital accumulation by the ruling classes will attempts be made to reduce it.

Salinization

This second form of soil degradation occurs mainly in arid and semi-arid climates, where concentrations of chloride, sulphate and carbonate salts of sodium, calcium and magnesium may affect crop yields. Salts reduce the soil's ability to hold air and nutrients and they are toxic to many plants.

As with soil erosion, the process of salinization can be natural but human intervention has also played a major role. The use of irrigation without adequate drainage may raise the water table and increase the risk of salty ground water reaching the root zone and even the surface by capillary action. Australia is perhaps the worst affected country, with over 40 per cent of its soils saline or alkaline. Human responsibility is clear in the irrigated areas of South Australia and Victoria. In Syria 50 per cent of irrigated land is salinized and in Uzbekistan up to 80 per cent of some large-scale irrigated projects have been lost.

Dealing with salt accumulation is very expensive. Installing drainage systems to lower water tables and dispose of salty water is technically feasible but who should pay and what are the economic benefits? Alternatives to such physical works might be a change of land use, such as from crops to pasture, or a switch to more salt-tolerant species such as barley, sugar beet or cotton.

Desertification

Desertification, a third form of land degradation, has in recent years come into the popular vocabulary and acquired a broad meaning. It implies a reduction of biological activity by the action of humans or by climatic change, especially desiccation, to the point where desert-like conditions prevail. The popular image is of irreversible change in which sand dunes encroach upon overgrazed, semi-arid pastures. In reality the process of desertification can affect a wide range of environments through a complex set of processes.

In 1991 desertification impinged upon the lives of 850 million people (rising to 1.2 billion by the year 2000) living mainly in the arid and semi-arid areas of the world. Impact upon agricultural land has been widespread. Annual losses of crop production are estimated to be £28 billion each year, and a further 6 million hectares of land are lost each year, a process which can be prevented only by the expenditure annually of £2 billion. Corrective measures and rehabilitation of affected land would cost a further £11 billion a year.

The causes of desertification are:

- The shortage of water in desert fringe areas prompts a concentration of people and livestock around the few water sources. Overgrazing around water holes leads to the degradation of vegetation, especially in drought years when nomadic pastoralists may be driven to seek temporary settlement.
- Poor people must rely upon fuel sources for cooking and lighting such as animal dung and wood. The latter is problematic because a dense rural population will quickly clear a sparsely vegetated area of its trees and shrubs, sometimes to a distance of many kilometres from a village. The loss of the binding effect of vegetation upon the soil can lead to soil erosion. Many African countries derive over 80 per cent of their domestic energy from fuelwood, such as Mali (97 per cent), Burkina Faso (94 per cent), Tanzania (94 per cent), and Sudan (81 per cent).
- Traditional nomadic pastoralism was in balance with the harsh environment of arid areas until recently. The increase in herd numbers since the 1950s and the restriction of nomads to smaller and smaller areas beyond the settled zone, have meant a greater pressure upon the carrying capacity of the meagre pastures (Figure 13.4).
- Unfavourable climatic cycles, such as the 1968–73 drought of the Sahel belt to the south of the Sahara and the 1980s' dry period in southern Africa.
- Unsustainable agriculture, particularly the over-cultivation of soils of marginal fertility and fragile structure.

FIGURE 13.4 *Excessive grazing on marginal lands is one cause of desertification, as here in the Sahel of Sousse, Tunisia*
Source: P.J. Atkins

Reduction of biodiversity

The reduction of biodiversity is the fourth negative consequence of agriculture. Tropical forests contain 62 per cent of known plant species and their biological diversity is extraordinary. But such is the present rate of forest clearance, for agriculture and other uses, that a quarter of the world's species might become extinct within 10 to 20 years. Such wanton destruction is not only ecologically regrettable but also economically foolish. It is difficult to tell what we might lose in terms of useful genetic material for medicines and the improvement of crop varieties.

But there are also threats to biodiversity in developed countries. Many traditional crop varieties are being abandoned by farmers because they do not have desirable characteristics of high yield or marketability. Of the 9000 apple varieties grown in the UK 100 years ago, for instance, only nine are now grown on any scale. In Greece 95 per cent of native wheat varieties ceased commercial production between 1945 and 1986.

Such a simplification and concentration of genetic resources is dangerous. Many of the new HYVs produced recently by plant breeders are not adapted to pest and disease resistance in the way that the traditional varieties were after hundreds or thousands of years of careful selection by peasant farmers. Hitherto, the new HYVs have also failed to meet the varied and very specific environmental constraints of agriculture in marginal areas, away from the favoured realm of large-scale commercial enterprises. By maintaining the genetic pool of species and their varieties, future science can perhaps use desirable characteristics such as drought resistance, salt tolerance, and so on, which are presently scattered, perhaps unrecognized, throughout the fields of traditional farmers around the world. Genetic erosion, without serious attempts at preservation, is a very serious danger.

AGRICULTURAL POLLUTION

Agricultural pollution is the fifth unplanned, negative side effect. Concern here is about the use of toxic materials which may contaminate the environment and adversely affect the health of plants and animals, including humans. The use of some agricultural chemicals such as fertilizers, pesticides, fungicides and herbicides is hazardous, especially where they are persistent in the soil and may accumulate in the food chain.

World fertilizer use has increased tenfold since 1950, and pesticides by 32 times. Insufficient research has been done for us to know the full health implications for humans but the World Health Organization has estimated that 200 000 people die each year through the effects of pesticide poisoning and a further 3 million suffer acute symptoms.

Rachel Carson's *Silent spring* (1962) alerted us to the shocking effect of agro-chemicals upon the environment and was one of the stimuli of the modern conservation movement that has begun to have an impact upon national and international politics. Carson, and the many other writers who followed her example, pointed to the persistent nature of many pesticides, such as DDT, whose poisonous character became most damaging in the higher levels of the food chain, for instance, amongst birds of prey. In other words, their impact spreads far beyond the initial location of spraying, as the chemical is carried in the bodies of animals or leaches into groundwater or the drainage system.

Even natural animal waste, which normally would be welcomed in a low intensity organic farming system, can become a nuisance in excess. In the Netherlands, for instance, a surplus of manure has resulted from dairy farming and other intensive livestock rearing, to the extent that there are severe problems of storage and disposal. Some 94 million tonnes are produced each year but the land can only absorb 50 million tonnes safely as a fertilizer. Strict regulations have to be enforced about its use on the land because of a problem of nitrates and phosphates leaching into ground and surface water. In Britain the pollution of streams by accidental discharges of slurry tanks and silage towers has increased alarmingly. Slurry is a hundred times more toxic than domestic sewage and some other agricultural wastes are a thousand times stronger. Over 25 per cent of river pollution incidents in 1991 were caused by farming. Also in 1991 the government introduced a Code of Good Agricultural Practice for the Protection of Water in England and Wales.

In Britain in 1990 1.7 million people were drinking water with levels of nitrates (mainly from fertilizers) above the World Health Organization's recommended limit of 10 milligrams per litre. This is especially worrying for people living in areas of arable farming on the limestone rocks of eastern England. The government has designated several Nitrate Sensitive Areas and their number is likely to increase as the threat comes to be recognized.

CASE STUDY: HEDGE REMOVAL IN BRITAIN

The English Midlands have had a strange experience with hedges. Any that were planted in Romano-British times were grubbed out during the Dark Ages and the early medieval period, to make way for the open vistas of the common field system (Chapter 11). Here parcel boundaries were in the form of earth balks or marker stones and hedges were rare. From 1750–1850, during the period of enclosure (see Chapter 7), however, 320 000 km of hedges were again planted as edges to the new fields carved out of the open fields. This was as much as the planting in the previous 500 years put together. In the twentieth century hedges have been removed again (*see* Figure 13.5) in order to create sufficient space to turn their modern agricultural machinery. Farmers can waste two-thirds of their time turning and dealing with difficult corners. In a 40-hectare field this turning time is reduced to 20 per cent.

Between 1947 and 1993, about 380 000 km of hedges (or 47 per cent of the British total) were destroyed. The impact on wildlife, especially birds and hedgerow plants was devastating. In recent years environmentalists have campaigned

FIGURE 13.5 *Hedge grubbing, Wickham Skieth (1976)*
Source: Committee for Aerial Photography, University of Cambridge

to stop the destruction of hedgerows, attempting, for instance, to give hedges the same protection of preservation orders enjoyed by valued trees. The government responded in 1986 and 1992 with the introduction of subsidies for hedge replanting. The Countryside Commission's Hedge Stewardship Scheme has encouraged farmers to agree to a 10-year plan of replanting and hedge restoration, in return for a monetary payment of £2.50 per metre. This is more than a little ironic since previously subsidies had been paid, often to the same farmers, for hedge removal.

BIOTECHNOLOGY AND THE REFASHIONING OF THE LANDSCAPE

In the broadest sense, biotechnology has been with us since the domestication of plants and animals and their breeding to enhance useful characteristics. Recently, however, the term has acquired connotations that have moved from the conventional experimental field plots into the laboratory, with the manipulation of cells and increasingly the use of advanced microbiological techniques such as genetic engineering. There are several approaches to the use of biotechnology in plants.

Tissue culture involves chopping up a plant to make thousands of identical plants under controlled conditions. This doubles the speed of traditional plant breeding. DNA Plant Technology of New Jersey have recently used tissue culture to bring to market a tomato variety for the Campbell Soup Corporation which is high in solids. It was produced by manipulating leaf cells in an existing variety, in three years rather than the usual seven.

Genetic manipulation in the full sense is a

relatively recent phenomenon, but it holds out great promise for plants and animals with a range of desirable features such as increased productivity and resistance to pest attack. So far only a few crops have been genetically manipulated and released on to the market, not least because of the fierce opposition of environmentalists who fear that there may be effects which cannot be predicted. Two methods are employed. First, it is possible by the injection of DNA with a very fine needle into the nucleus of an individual plant cell. More common is the use of a microbe, *Agrobacterium tumefaciens*, to carry the new gene(s) into a plant's cells.

The biotechnology corporations, such as Monsanto and Calgene, seem likely to change the nature of agriculture in the long term. It is conceivable that they will develop crops tolerant of saline, very dry or cold conditions, and thus the potential of what is presently marginal land will increase. Alternatively, the enhanced productivity of agriculture in the core areas may be sufficient to satisfy world demand and there may be reduced pressure to intensify in environmentally fragile regions. The optimists even argue that there will be a reduced demand for fertilizers and biocides, and that soil erosion and fossil fuel consumption will decline (Mannion 1995). Either way, the environmental and landscape consequences are likely to be fundamental.

One might hope that biotechnology could help to solve the world's food problems by allowing poor countries to become agriculturally self-sufficient, but recent developments suggest that this chance has already been lost. The market for the new products is dominated by Western-based transnational biotechnology companies whose motivations are profit-related rather than humanitarian. Third World farmers will be charged to use the genetically improved seeds and Third World companies are unlikely to be able to replicate these seeds because of a move to allow the patenting of transgenic products.

Interestingly, Goodman *et al.* (1987) hypothesize a radical scenario that, if even partially correct, would change the face of the planet. They argue that the main applications of biotechnology may well be in the factory rather than the field, and that the manufacture of foodstuffs in future may resemble the culture of the mycoprotein which is the main constituent of artificial vegetarian foods such as Quorn™. Thus plants and animals, and agriculture as conventionally understood, would be redundant and the ten millennia of agro-ecosystem evolution would end.

CONCLUSION

The power of agriculture to transform landscapes stretches from the low-tech realm of poor countries to the high-tech of the West. But in the minds of many, the increased productive power of farmers in the developed world has been won at the expense of the environment, with a number of important negative processes set in motion. It remains to be seen whether this trend to intensification per unit of labour input can be managed sustainably, for instance, by the greater use of organic farming methods, or whether the landscapes of capitalist agriculture will spread further.

Current indications are that rural areas will become more standardized in their organization and appearance around the world, just as modern and post-modern urban landscapes have become commonplace in the cities of all continents. This is because agriculture is increasingly just another subset of the modern project (Chapter 16), exhibiting its features of simplification, capital investment in technology to boost production, regulation of working practices and quality of output, and restructuring for the extraction of maximum surplus value.

FURTHER READING AND REFERENCES

David Grigg's work is an excellent source of information about agricultural historical geography and Antoinette Mannion (1995) has written the definitive account of the relationships between agriculture and environment.

Goodman, D.E, Sorj, B. and Wilkinson, J. 1987: *From farming to biotechnology* Oxford: Blackwell.

Grigg, D.B. 1980: *Population growth and agrarian change: historical perspectives*. Cambridge: Cambridge University Press.

Grigg, D.B. 1982: *The dynamics of agricultural change*. London: Hutchinson.

Grigg, D.B. 1987: The industrial revolution and land transformation. In Wolman, M.G. and Fournier, F.G.A. (eds) *Land transformation in agriculture*. Chichester: Wiley, 79–109.

Grigg, D.B. 1989: *English agriculture: an historical perspective*. Oxford: Blackwell.

Grigg, D.B. 1992: *The transformation of agriculture in the west*. Oxford: Blackwell.

Hannah, L. *et al*. 1994: A preliminary inventory of human disturbance of world ecosystems. *Ambio* **23**, 246–50.

Mannion, A. 1995: *Agriculture and environmental change: temporal and spatial dimensions*. Chichester: Wiley.

Simmons, I.G. 1987: Transformation of land in pre-industrial times. In Wolman, M.G. and Fournier, F.G.A. (eds) *Land transformation in agriculture*. Chichester: Wiley, 45–77.

Tivy, J. 1990: *Agricultural ecology*, Harlow: Longman.

14

LANDSCAPES OF ENERGY ACQUISITION

THE GETTING OF POWER

Pylons ... however well designed and carefully routed, are too large, too alien, too unmistakably organized for a mechanical world ever to be absorbed in small-scale landscapes. They stride across our intimate countryside like files of linked giants ... I meet people who admire them, but I am not one. I abominate pylons as I do the Eiffel Tower and the old Forth Bridge and all such fidgety criss-cross Meccano-like constructions

(N. Fairbrother 1972: *New lives, new landscapes.* Harmondsworth: Penguin)

INTRODUCTION

As made clear in the Introduction to Part 3, being modern means having access to stores of energy which are not derived immediately from the sun. The main sources of such energy are coal, oil, natural gas, falling water and nuclear power and these are then applied to machines, once as steam power and now more often as electricity, though internal combustion engines rank high among the main consumers of oil and its refined products. Many places in the world, however, are still dependent on biomass energy, especially woodfuels (Table 14.1).

In order to have culturally acceptable energy at our fingertips (literally in the case of most electricity supplies) or at the gasoline pump, the materials of nature have to be extracted and transformed. As with the garnering of most resources, this entails the transformation of the pre-existing landscapes, and so new landscapes of energy acquisition and transformation are created. These are both prior to, and co-existent with, the new landscapes that the energy permits to be created. Thus the presence of a factory using steam power is not part of the landscape of energy acquisition but is made possible because of the getting of energy.

Common to all landscapes of power development is a simple piece of accounting: that over a long period of time, more energy must be gained from the tapping of the source than is invested in getting it out and making it available. Thus the energy involved in sinking a shaft, extracting coal and transporting it to a steam boiler in a nineteenth-century factory must be exceeded by the energy made available to drive the pulleys in the factory. One of the problems with some of the *alternative* energy sources of the late twentieth century, such as photovoltaic cells and passive solar collectors, is that so far more energy has been put into their manufacture than has been made available to their end-users. Eventually, of course, the balance will swing to the positive but in the early stages of a new technology, such considerations may well affect the economics of the

TABLE 14.1 *Proportion of traditional fuels in selected economies*

Region/Country	% of total consumption	High-dependence examples	% of total consumption
WORLD	6	Botswana	100
Africa	43	Mali	90
Europe	1	Zaire	91
North and Central America	1	India	41
South America	30	China	11
Asia	15	Papua New Guinea	70
Oceania	6		

Source: extracted from Data Table 12.2 of World Resources Institute *et al.* (1996): *World Resources 1996–97*. Oxford: Oxford University Press
Note: Measured in calorific content, not volume

project. Measured examples of energy input and output from modern power systems are rare but one example of open pit mining of coal in Indiana suggested that the energy in/out ratio was about 1:64, which is highly satisfactory in economic terms.

The core of the industrial revolution was coal and it is still a very important fuel in many regions of the world. It can be mined from large open pits or from galleries running off vertical shafts (*see* Figures 14.1 and 14.2). Indeed, its importance in the core developed or industrialized nations is such that special attention is devoted to it below and the emphasis here is on the other sources of industrial-scale power. Table 14.2 shows the recent position of coal production in relation to other major fuels.

FIGURE 14.1 *Easington Colliery, County Durham*
Source: P.J. Atkins

FIGURE 14.2 *Coal-burning power station, Ferrybridge, Yorkshire*
Source: P.J. Atkins

HYDRO-POWER

The nearest of these sources to solar-derived energy is the generation of electricity from falling water (Figure 14.3). The development of a plant which channelled water so as to drive a turbine which generated electric power is a nineteenth-century technology and one early instance was its use to light La Scala opera house in Milan. Roughly speaking, any flow of water can be used but the greater the head of water, the greater the amount of power to be generated. Thus the typical development is a dam across a valley which impounds water so as to provide a constant head and hence a reliable output of electricity. The turbines are usually built into the dam structure and

TABLE 14.2 *Commercial energy production 1993, Petajoules*

Total	Solid	Liquid	Gas	Geothermal and Wind	Hydro	Nuclear	
337 518	91 748	134 060	78 146	1463	8554	23 646	PJ (10)[15]
100	27	40	25	0.5	2.5	7	%

Source: extracted from Data Table 12.1 of World Resources Institute *et al.* 1996: *World Resources 1996-97*. Oxford: Oxford University Press

Notes:

1. These are data for calorific content and as such do not relate directly to bulk and hence to landscape presence. But the relationship, with *liquid* (i.e. oil) dominant, does not look unreasonable.
2. The percentages do not add to 100 due to rounding errors.
3. Only commercial energy is counted, not *traditional* fuels.

the electricity is taken to its end-users via power lines which are strung between pylons. Even though many suitable sites have now been taken up, the 1973–93 period witnessed a rise of 86 per cent in the quantity of energy generated by this method.

There is therefore a major transformation of the landscape when a new plant is installed. Inevitably, the dam impounds a large lake which drowns the pre-existing landscape. There may be displacement of agriculture, for example. In mountain and other upland areas, the valleys to be drowned may be the best land and the holdings are simply not viable without them. Along

FIGURE 14.3 *The dam on the River Zambezi at Cahora Bassa in northern Mozambique*

with the cultivated land, settlements may vanish beneath the water, only to emerge at times of drought and then form a singular tourist attraction. Communities have to be relocated and even in the late twentieth century, a few governments have proposed hydro-electric projects that have ignored the future of local people. In the case of some of the large dams in Africa, areas of savanna were flooded and many wild animals had to be rescued from islands and taken to drier areas; inundated forests yielded dead trees which choked the turbine inlets; rotting vegetation, however, added to the productivity of fish populations so there was a temporary boom in lake fisheries. There are always costs and benefits of large projects.

Secondary landscapes may of course include the use of the controlled water for irrigation, with all the subsequent changes elaborated in Chapter 13. The downstream nature of the river will also be affected since its silt content (and hence erosive power) will be radically lowered by its sojourn in the low-energy environment of the reservoir. The temperature of the watercourse may also be subject to change since thermal stratification occurs in large, still, water bodies and the draw-off into the turbines may come from a particular temperature level. This can affect the flora and fauna of the river below the dam.

The primary landscape is also changed by the technology of electricity transmission. It is rare for the end-users to be located near the generating plant itself (which obviously tend to be in relatively remote and steep places) and so high voltage lines have to be constructed from the generating plant to the towns, cities and factories that use the power. In developing countries the lines of pylons and cables are often welcomed as evidence of modernization; in densely populated industrial nations they are often despised as being ugly intrusions into cherished landscapes; where the populations of such regions are less dense, then the opposition is generally much less since the choice of routes is greater and places of acknowledged beauty can usually be avoided.

The classic example of a major dam and its effects is that of the High Dam at Aswan on the Nile, impounding Lake Nasser. It was the latest in a series of dams on the Nile and was built primarily to provide electricity to help industrialization of the Egyptian economy but also to improve irrigation water supply. The effects have been clear: the Nile valley's irrigated areas have been deprived of silt and have instead had to buy fertilizers; the silt loss has meant increased erosion of the Nile delta and decreased productivity of the fisheries of the eastern Mediterranean where the phytoplankton depended on the silt for mineral nutrition. The gross amount of water in the lower Nile has decreased because of the evaporation from the surface of Lake Nasser, which also drowned many relics of ancient Egypt, with UNESCO finding funds for the relocation of monuments like the temples of Abu Simpel. Overall, however, the creation of this landscape had a symbolic importance in proclaiming a national identity for post-monarchic Egypt which probably transcended its economic role.

OIL AND NATURAL GAS

Geology has bequeathed a situation in which these two resources are often found together, though it is not always the case that both are exploited in the same place. The sedimentary basins which contain them are, however, located both under the land surfaces and under the sea. The term *landscapes of power* must here include *seascapes* as well. The technological sequence for oil is usually that of pumping from the rock strata, transport by pipeline to a refinery and then by tanker or further pipeline to power stations, to industrial plants that use some of the products (e.g. for plastics and pharmaceuticals), or to vehicle service stations. The utility of oil products is such that world production has risen 11 per cent in the last 20 years in spite of a relatively low price during most of that period. Gas is even more popular in some contexts and so in the same period production has gone up by 72 per cent.

Part of the landscape of power extraction may therefore be hidden: pipelines are often underground except where the costs of burial would be too high, the terrain unsuitable or where terrorist damage seems unlikely. The classic case of an above-ground oil pipe is the Alaska pipeline from

FIGURE 14.4 *An offshore oil rig in the North Sea*

the North Slope of Alaska to Valdez, where the oil is transferred to supertankers. Because of the climate, the oil is heated and so the entire pipeline is built on gravel pads so that the permafrost does not melt.

Oil extraction on land can present rather different appearances in the landscape, depending mostly on the legal arrangements over landownership. Where this ownership is fragmented and every landowner or lessee wants a share of the oil revenues, then there tends to be a forest of small derricks or the reciprocating pumps known as *nodding donkeys*. If the state owns the land or the mineral rights then larger derricks tend wider areas of strata. At sea, the investment costs are so high that very large rigs, serviced by special boats and by helicopters, are scattered across the oilfield. They announce their presence, especially at night, by flaring off surplus methane, often to the detriment of migrating birds.

The most significant landscape element is probably the oil refinery. A cluster of metal tubes and emission stacks is usually adjacent to a series of storage tanks (known, curiously, as a 'tank farm'), with safety requiring that it be lit at night. The crude oil is stored in one set of tanks and the refined products in another; the pipes are basically parts of a large distillation plant, with cooling water being a necessary input. The stacks emit a number of waste gases into the atmosphere and badly maintained plants tend to leak at pipe joints with occasional fires and explosions. There is usually an unacknowledged *cordon sanitaire* around a large refinery, where housing, for instance, is discouraged.

In fact the most common unplanned landscape of the oil system is the tanker accident. There have been a series of accidents all round the world in which laden oil tankers have been in collision with other vessels or have run aground. The level of damage has been such that many millions of tonnes of oil have been shed into the sea, with consequent ecological and aesthetic effects and huge costs incurred in trying to clean up. The immediate consequences usually depend upon the grade of oil (the lighter grades may largely evaporate) as well as local conditions of wind, tide and coastal topography. The longer-term effects are not as well known because of the costs of monitoring but it seems likely that sessile plants and animals bear the marks of population change for decades. The oil thus spilled adds to the oceanic totals of the stuff which derive from smaller spills, illegal washing of tanks at sea, and seepage from natural outcrops on the sea-bed.

The great secondary landscape of oil is the ubiquity and cheapness of motor transport. In Germany, more land is devoted to the motor car (in terms of manufacturing, roads, servicing, gasoline stations and parking) than to housing. A car-oriented city like Los Angeles may devote about a third of its surface area to the car, plus multi-storey parking areas. Cheap oil underwrites a whole life-way in the West and in NICs, with a great degree of penetration now into LDCs as well.

Natural gas is simpler to chronicle, for the methane which comprises it does not need refining and is usually conveyed across land in pipelines like oil and at sea in liquid form in tankers. Accidents are less frequent and less disastrous to the surrounding areas since the gas evaporates, though explosions are possible. The secondary landscape is that of cheap electricity made by gas-powered turbines and the low-density housing with central heating to which gas can be cheaply piped.

NUCLEAR POWER

The civilian nuclear power cycle consists of the extraction of uranium and its concentration into fuel rods, the insertion of these rods into a reactor and the generation of electricity from turbines, the reprocessing of the wastes and the eventual storage of radioactive materials while they decay. Capital needs are high and the skill level needed to operate a plant is also high, not to mention the talent needed to programme the computers that control, e.g. the safety systems. These problems (and those of public acceptability) notwithstanding, the last 20 years to 1993 saw a rise of 1365 per cent in power thus generated.

The actual landscape of nuclear power generation is not spectacular; uranium mines look like most other open pits and reactors (*see* Figure 14.5) and are not much bigger than thermal power stations though they may look different if they are of the PWR type that has a domed containment structure for the reactor. From the plant, electricity is led away just as from other generating sources. People unfamiliar with the different types of reactor may sometimes mistake thermal power plants for nuclear plants: the author has heard the piles of coal at Didcot described by fellow passengers (bound for Oxford) in the train as 'Say, there's, like, gross heaps of uranium'. One other phenomenon apparent inland is the presence of a large water body since the need for cooling water per unit of

FIGURE 14.5 *Torness nuclear power station, AGR*
Source: British Nuclear Fuels

power generated is much larger for nuclear plants than for thermal generation. By the sea, this is not noticeable, but inland there is likely to be a lake or pipes from a river or lake. EdF has built many kilometres of canal to supply its plants in central France, for example, but in dry weather is still forced to reduce operating capacity.

The reprocessing plant is large and like the reactor heavily protected: even in countries like the UK it is customary to have armed guards. It emits planned releases of radioactive particles to the air and the water and stores low- and medium-level wastes while they lose significant amounts of their radioactivity. It is responsible too for holding high-level wastes (which may need to be sequestered from any form of life for 250 000 years) awaiting their final storage in an underground repository. The main characteristic of that installation is that there is no landscape at all except the headgear of the shaft; we all hope that nothing leaks from it in any shape or form.

The accidental landscapes of nuclear power are well documented. The source is nearly always the reactor and in more cases than not human error is held to blame for the accidents in which the reactor core gets overheated and begins to melt, with subsequent releases of radioactive materials. The case of Chernobyl in the Ukraine (1986) is one of the best documented, especially its effects at a distance as the cloud of radioactive materials was swept to and fro across western and northern Europe with differential scavenging out of radioactivity, and the immediate sterilization of a zone round the stricken reactor plant. The effects in terms of areas of upland Britain from which sheep may not be sold off are still with us, as is the zone where nobody can live and work, as are the many people who have and will develop cancer from the radiation.

WIND POWER

The usefulness of the windmill in pre-industrial societies has been previously mentioned (Chapter 9). In windy places such as uplands and along coasts, the technology is being re-evaluated so as to offer a *free* energy source in the face of concerns about atmospheric warming. The machines are still basically pillars with sails on them but the gearing has lower levels of friction and the blades are, like aircraft propellers, angled for maximum presentation of surface to the wind.

The usual way of generating electricity from such tower installations is to place a *wind farm* (*see* Figure 14.6) in an appropriate place and lead off the electricity from each pylon to a central collection point from which it is fed to a regional or national grid. Depending on the topography, towers are about 100 m apart and most farms have 20 plus of them. There is no doubt that they form a distinctive landscape and one which is normally an abrupt change from the pre-existing scene. If placed in uplands and along coasts they are likely at some point to conflict with people who prefer the previous landscape and controversy ensues. Apart from the visual changes, they do not produce many other transformations: if on grazing land, that use can continue. Although not customary, they could be placed in cultivated land if due care is taken with underground cabling. They do, however, produce noise: a kind of perpetual soughing accompanies the rotation of the blades. Like most noises, this is received differently by different people. Those who have fled to the hills to escape the city (either permanently or at weekends) are likely to be antagonistic to both sight and sound; their teenage offspring brought up on high decibel levels at the disco are unlikely even to notice. In Denmark, for example, large wind generation plants make a significant contribution to the national supply;

FIGURE 14.6 *Generating power on a modern wind farm, Hetton Le Hole, County Durham*
Source: P.J. Atkins

elsewhere such plants exist usually only where alternative energy sources receive a government subsidy or where indeed they are part of a demonstration designed to awake people to non-polluting energy sources. Ask enthusiasts how they are made and when the energy balance becomes positive.

CONCLUSION: CONCENTRATIONS

A last word on landscapes of power notes the uneven spread of phenomena. As noted elsewhere at several points, the developed countries are energy-rich in terms of per capita usage. They thus form islands of power consumption in a world of generally much lower values. Infra-red photography from satellites shows too that the actual sites of generation show up as islands of heat emission which are above the general level. So do areas of dense consumption (such as New York City in winter) but to a lesser degree. So the landscape of power getting is one of points of, in general, no great spatial area. These are joined by the all-important phenomena of transmission: today the pipeline and the high-voltage power line are key features not only of the visual scene but of the functioning of the economy: not for nothing are they sometimes compared with the blood vessels and the nervous system of their human creators.

FURTHER READING AND REFERENCES

Foley (1987) is an accessible text and Smil (1994) is very good on the historical aspect of the energy issue.

Debeir, J-C., Deléage, J-P. and Heméry, D. 1991: *In the servitude of power: energy and civilization through the ages*. London: Zed.

Foley, G. 1987: *The energy question*. 3rd edn, Harmondsworth: Pelican.

Hall, C.A.S., Cleveland, C.J. and Kaufman, R. 1986: *Energy and resource quality: the ecology of the economic process*. New York: Wiley.

Hill, R., O'Keefe, P. and Snape, P. 1995: *The future of energy use*. London: Earthscan.

Smil, V. 1987: *Energy, food, environment: realities, myths, options*. Oxford: Clarendon Press.

Smil, V. 1991: *General energetics: energy in the biosphere and civilization*. Chichester: Wiley.

Smil, V. 1994: *Energy in world history*. Boulder CO: Westview.

Soussan, J.G. 1988: *Primary resources and energy in the Third World*. London: Routledge.

Soussan, J.G. 1992: World energy picture. In Mannion, A.M. and Bowlby, S.R. (eds) *Environmental issues in the 1990s*. Chichester: Wiley, 131–46.

Stern, R. (ed.) 1996: *Rural energy and development*. Washington, DC: World Bank.

World Resources Institute 1996: *World resources 1996–97*. Oxford: Oxford University Press.

15

INDUSTRIAL LANDSCAPES

> All around was a lunar landscape of slag heaps, and to the north, through the passes, as it were, between the mountains of slag, you could see the factory chimneys sending out their plumes of smoke ... It seemed a world from which vegetation had been banished; nothing existed except smoke, shale, ice, mud, ashes and foul water.
>
> (G. Orwell 1937: *The road to Wigan Pier*. London: Left Book Club)

INTRODUCTION

In this chapter we will build on insights gained in Chapter 12. There we discovered that proto-industrialization had a rural emphasis in many regions of Europe, with products as varied as textiles and iron coming from small-scale manufacture in workshops and even from the domestic setting in many instances. The host landscapes were not blighted by factories, chimneys and slag heaps: these were a feature of a later phase in industrialization. Proto-industrialization was to a large extent sustainable, with the use of agricultural raw materials like wool as a major input and human muscle power driving simple machines, for instance spinning wheels and looms. Even the use of inanimate energy such as charcoal to smelt iron was sustainable because it came from managed forests (see Chapter 6).

Here we will look at the successive phases of industrial development (*see* Figure 15.1). We will divide the period from 1730 onwards into:

- the early phase of factory industry;
- the era of scientific management and Fordism;
- recent trends in post-Fordism.

This classification is basically a combination of the organization of capital investment coupled with technological changes that facilitated innovations in the manufacturing process. Alternative industrial histories of the landscape might treat sectors separately, on the basis that the environmental impact of mining or iron-making has been more visibly significant than textiles or food processing. Whichever scheme is adopted, it is worth remembering that, overall, the ability of industry to create

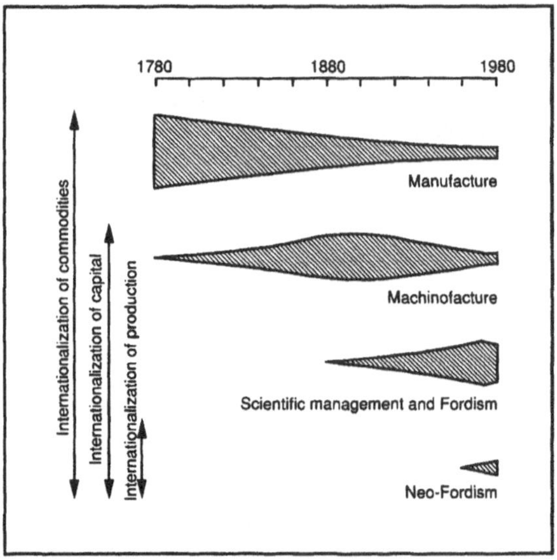

FIGURE 15.1 *The industrial process and international economic relations*
Source: Reprinted from Dunford, M. and Perrons, D. 1983: *The arena of capital*. London: Macmillan

new landscapes has accelerated exponentially during the modern era, and these are landscapes, along with those of urban areas, which are as far removed from nature as it is possible to imagine.

Palmer and Neaverson (1994) isolate six factors that help to describe landscapes of industry: sources of raw materials; type of processing plant; power sources; the nature and number of secondary industries; accommodation for the workforce; and the relationship to transport networks. The reader should also bear in mind, however, that industry has an impact that reaches beyond the immediate surroundings of its point of production. Industrial goods have been traded since prehistoric times and their reach is now global. Their consumption and eventual disposal have environmental consequences that are widespread.

THE TRANSITION TO FACTORY INDUSTRY

Proto-industrialization did not inevitably lead to large-scale capitalist-inspired factory development. The path onwards from the phase of scattered rural industry seems to have varied very considerably from region to region in Europe and, as yet, we have no fully comprehensive theoretical framework that can explain why some failed and others succeeded. One important variation lay in the way that capital was generated. In some areas it came from individual small manufacturers who were able to save, re-invest and accumulate in a kind of grassroots, incremental fashion. In others it was a more purposeful top-down investment by merchants who saw an opportunity for extending their business. To a certain extent this depended upon the industry because small entrepreneurs had less chance of succeeding in industries which required lump investments, such as in mining, metal-smelting, glass- and paper-making.

From about 1730 onwards the impact of technological innovation and the concentration of capital fundamentally changed the nature of the industrial landscape. Despite the term *revolution* often associated with economic growth from the mid-eighteenth to the mid-nineteenth century, the process was in fact never rapid, nor was it without reverses during the frequent periods of slump and depression. In fact, one can identify periods of accelerated industrial growth with stages in the economic cycles that were well established after 1800 (*see* Figure 15.2).

Five principal characteristics do set manufacturing in this period apart from earlier times. First, the scale of production was larger, often concentrated in workshops or even small factories, away from the domestic realm. This was necessary, second, because of the larger and more complex machinery becoming available and the increase of demand for consumer goods. The gathering together of workers was a third feature, requiring greater labour discipline about wages and hours of work and the divisions of tasks than had been known before. Inevitably there were clashes of interest between the workers and the factory owners on this point. Fourth, such a regulated environment allowed a greater control over the quality and standardization of the final product, bearing in mind the demands of the customer, although mechanized manufacturing excelled in coarser goods for the low end of the market. Finally, the new factory industry became increasingly urban-based for ease of labour recruitment and transport.

At first the problem of factory energy supplies was solved by the use of water power, with the result that many early textile factories were scattered in hilly country in the search for reliable

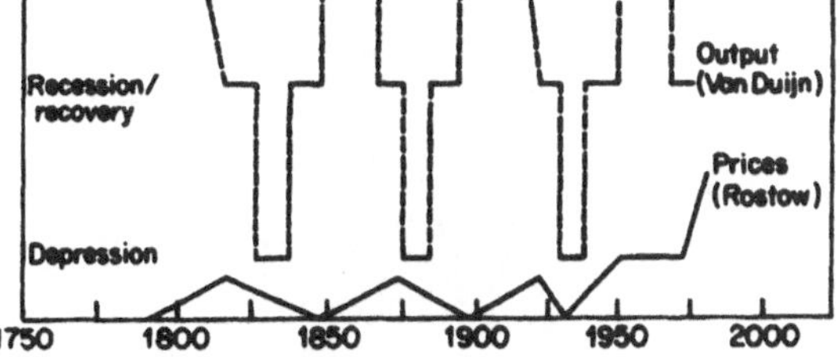

FIGURE 15.2 *Economic cycles are responsible for pulses in economic activity and therefore in landscape impact*
Source: Reprinted from Cleary, M.N. and Hobbs, G.D. 1984: The fifty year cycle: a look at the empirical evidence. In Freeman, C. (ed.) *Long waves in the world economy*. London: Pinter, 164–82

streams with a strong flow. Water continued as a major source of motive power into the middle of the nineteenth century but by then steam was well established and dominant in some sectors. Coal and coke were the main means of raising steam and fuelling furnaces. They are bulky fuels and industrial economics demanded that it be used as close as possible to its point of origin in order to minimize transport costs.

The full flowering of the Industrial Revolution was therefore concentrated, one might even say crowded, on the coalfields, facilitated by an increasingly complex network of canals, railways and turnpike roads which allowed accelerated flows of goods, people and ideas. As a result there was a rapid growth of population as migrants flocked to the job opportunities, draining the surrounding rural areas but also encouraging long-distance moves by skilled workers. Europe's economic centres of gravity and her demographic contours shifted, rewriting her human geography in the process (*see* Figure 15.3). This did not mean better conditions for the new industrial workers, as Alexis de Tocqueville found in Manchester in 1835:

> Look up and all around this place you will see the huge palaces of industry. You will hear the noise of furnaces, the whistle of steam. These vast structures keep the air and light out of the human habitations which they dominate; they envelop them in perpetual fog; here is the slave, there the master; there the wealth of some, here the poverty of most ... Here the weakness of the individual seems more feeble and helpless even than in the middle of a wilderness.

Environmental change was initiated or accelerated by industrialization. Livid scars of exploitation were common in areas of quarrying or open-cast mining (*see* Figure 15.4) where earth and rock were removed and not replaced. Much waste was also produced as a by-product of deep mining, especially of coal where there was at least

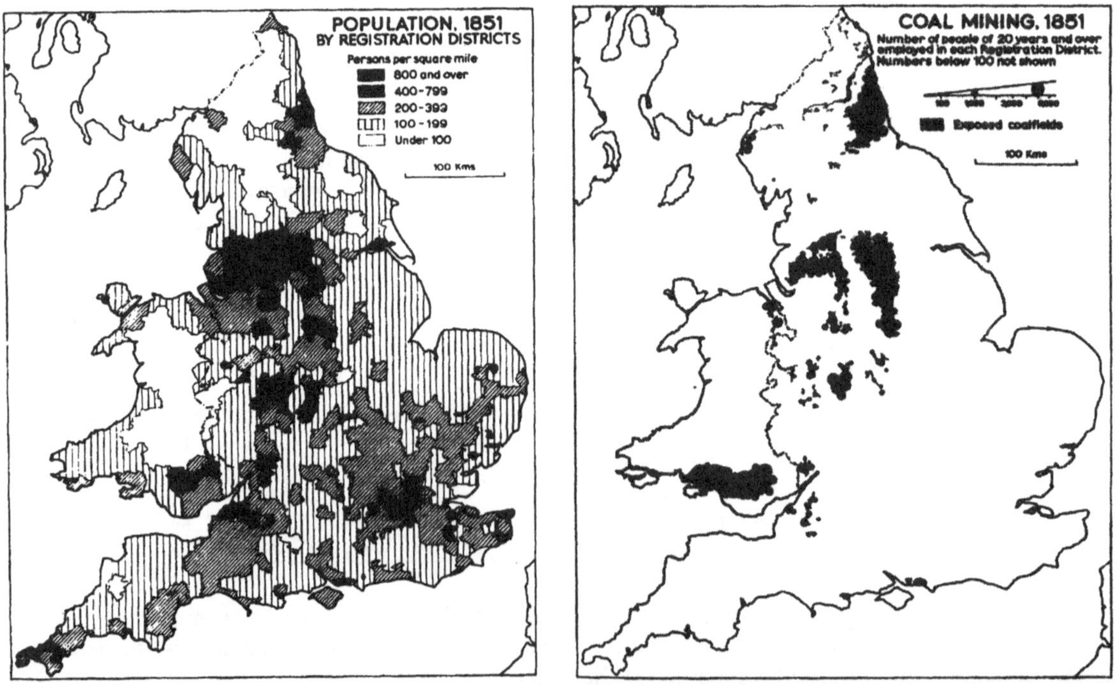

Figure 15.3 *(a) Population distribution in England, 1851, by registration districts*
(b) The location of coal mining in 1851
Source: both from Darby, H.C. (ed.) 1973: *A new historical geography of England*. Cambridge: Cambridge University Press

FIGURE 15.4 *Quarry at Mqabba, Malta. The Maltese landscape is pock-marked by quarries, many of which are now worked out and used for land-fill*
Source: P.J. Atkins

one heap produced for every pit, and of heavy industrial processes such as iron and steel making. There was subsidence caused by the collapse of underground workings, with the attendant cost of buildings damaged on the surface.

The chemical pollutants released in liquid form into waterways have had long-term implications, such as the deposition of lead and other heavy metals in the Wear river valley from the lead mining of the eighteenth and nineteenth centuries. Fumes and particulate dust from factory chimneys not only blackened buildings in industrial cities; they also had toxic effects upon vegetation over a wide area and had implications for human health. The further effect of releasing carbon dioxide into the atmosphere by the burning of fossil fuels, began the gradual process of *global warming*. A German visitor leaving the English Midlands in 1842 noted the effect of industry on the locality:

> I was delighted to have a clear view of the sky again. In Birmingham you can have no speculation on the weather. The rain is not felt till it has worked its way through the smoke, and the sun shows himself only as a yellow patch. Sunrise and sunset, stars and moonlight are things unknown.
>
> (Quoted in Trinder 1982, 8)

At both the regional and micro-scale, industry became remarkably specialized. In Britain, Lancashire was known for its cotton textiles (*see* Figure 15.5), Yorkshire for its woollens, Stoke-on-Trent for ceramics, the Midlands of England for metal trades and engineering, and so on. It was still the *workshop of the world* in the mid-nineteenth century and her sale of goods on the widening international market supported many export-orientated jobs. At Great Harwood in Lancashire, for instance, there were 23 cotton mills that made little other than turbans and loin cloths for India, while the narrow cloth of Burnley was destined for China. Such gearing is fine as long as the markets hold out, but the vicissitudes of fluctuating demand and growing competition made it very risky indeed in the early twentieth century.

The industrial map of Britain was a microcosm of Empire. There was much processing of raw materials from the colonies, including port-based food industries such as flour milling because by 1900 the country was heavily dependent upon imports to maintain its cheap food policy. Some raw material were re-exported as manufactures, such as textiles, to compliant customers throughout the imperial sphere of influence. Military conquest had also meant the conquest of markets for British exports.

The age of coal and machine-based manufacture had established new sinews and musculature in the economy of Britain by the end of the nineteenth century. In 1900 78 per cent of the population lived in towns and cities, by far the most advanced urbanization in history, and even distant rural populations felt themselves drawn into the fully integrated national economy by the pull of the market for their goods and of migration possibilities for their future employment. Prosperity had been built upon textile manufacturing, metal bashing and engineering, and other heavy industries in various sectors, such as shipbuilding and chemicals. There was also much investment abroad in expanding railway networks and generally in colonial ventures.

Competition from German and American factories, however, coupled with a rigidity of organizational structures, made the British economy vulnerable and relative decline has dogged it throughout the twentieth century. The once mighty coalfields shed much of their employment and in the 1980s coal mining itself contracted under the weight of competition from alternative fuels such as gas and political decisions to with-

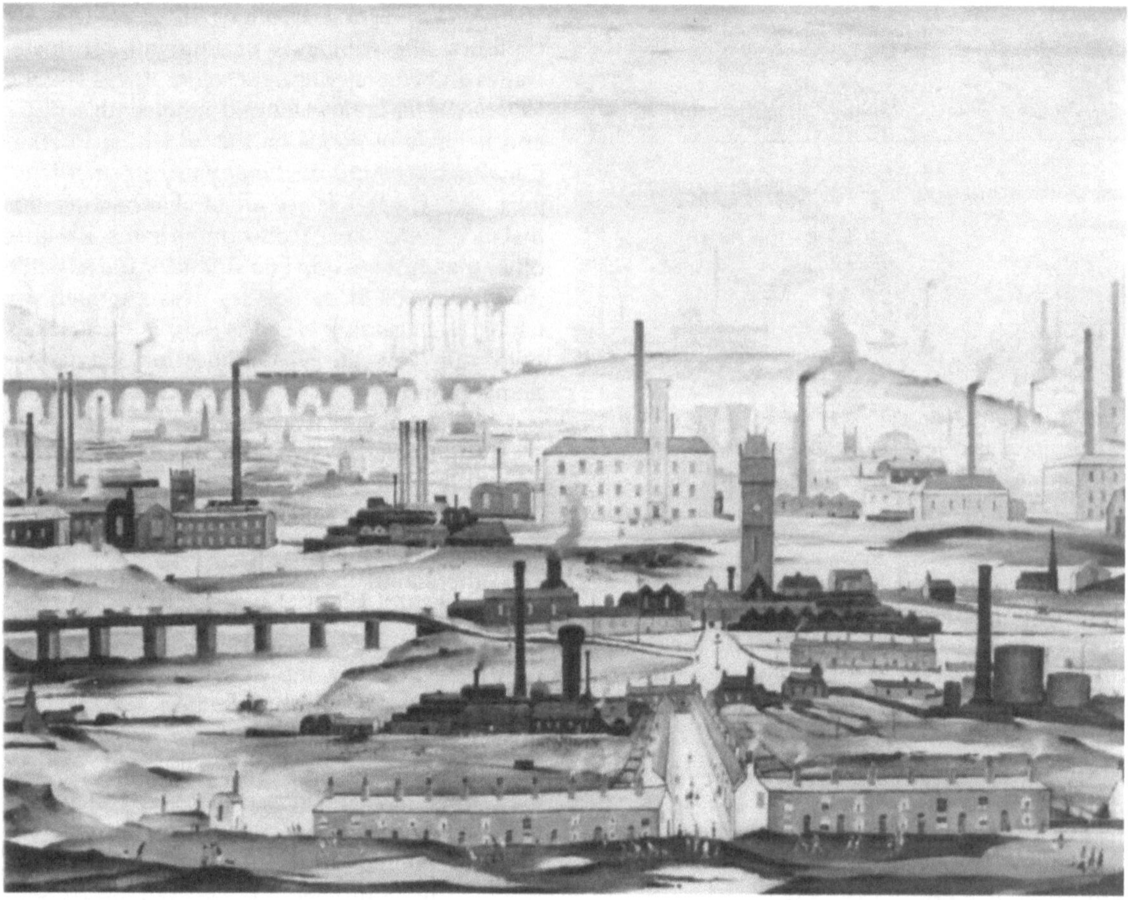

FIGURE 15.5 *Lowry: Industrial Landscape*
Source: Tate Gallery

draw subsidies and close or privatize the pits. As a result, the last 70 years have seen the emergence of landscapes of failure, where decaying factories and long dole queues have come to symbolize the visible outcome of the crises of capitalism. The mismatch between previous rounds of investment in fixed capital, such as factories, transport, houses, and services, and the present realities of economic geography has been especially painful in those areas where the Industrial Revolution flourished in the nineteenth century, including the inner cities.

Since the 1920s successive British governments have intervened to ease the transition from the older industrial realities to the new. They have scheduled regions to receive assistance to attract new industries or shore up those surviving. Along with Local Authorities, they have encouraged the establishment of suburban industrial estates and have provided funds to attract large employers into problem areas. They have also sought to clear the physical evidence of dereliction by demolishing old plant and reclaiming land such as spoil heaps (*see* Figure 15.6). There have always been industrial transitions of course, with uncompetitive technologies and organizational structures fading into what Richard Muir (1981) calls the ghosts of industry, but in the twentieth century the sheer scale of the problem has left scars that will take much longer to heal. These scars are also more than features in the landscape; they are also burned deep into the minds of many former workers by the bitterness of their rejection and idleness.

FIGURE 15.6 *Pit heap, Trimdon Grange, County Durham (1951)*
Source: Beamish Museum

It is more than a little ironic that out of the ashes of industries cremated on the funeral pyre lit by market forces there have arisen new jobs in industrial museums such as Beamish and Ironbridge. These have been created by an alliance of enthusiasts keen on the preservation of rapidly disappearing industrial artefacts and heritage entrepreneurs who see a market niche for industrial tourism. Of course, such jobs are very few in number.

CASE STUDY: DERELICTION IN THE LOWER SWANSEA VALLEY

Inevitably the extractive and heavy manufacturing industries produced much waste. The tips of coal mines and slag heaps of iron works were so common that they were taken for granted but there was also much more insidious dereliction and environmental damage that has proved very costly to deal with. A good example of all of these problems within a small compass is the Lower Swansea Valley in South Wales, a hearth of the Industrial Revolution.

The lower stretches of the Tawe valley, which passes through Swansea, saw early developments of metal smelting industries such as copper (1717) and zinc (1836) using the outcrops of coal in the valley sides. By 1860 there were 13 copper smelting works, employing 10 000, the largest concentration in the world at that time. Later tinplate (1845) and steel (1868) were added to the metallurgical tradition, along with gold, silver, arsenic and lead.

By 1960 most of the heavy industry had closed under the weight of competition from abroad. The resulting dereliction was substantial, with 7.1 million tonnes of slag and furnace waste remaining on sites between crumbling factory buildings covering 570 hectares (*see* Figures 15.7, 15.8 and 15.9). Toxic smoke and effluent over 200 years had destroyed much of the local vegetation and the top soil in certain locations, for instance on Kilvey Hill, was poisoned with heavy metals. Ayton's description in 1813 gives a flavour of the problem:

> In the neighbourhood of Swansea, there are some very extensive copper works, which are situated in a hollow, and immediately above them is not a blade of grass, a green bush, nor any form of vegetation; volumes of smoke, thick and pestilential, are seen crawling up the sides of the hills, which are as bare as a turnpike road.
>
> (Quoted in Trinder 1982, 101)

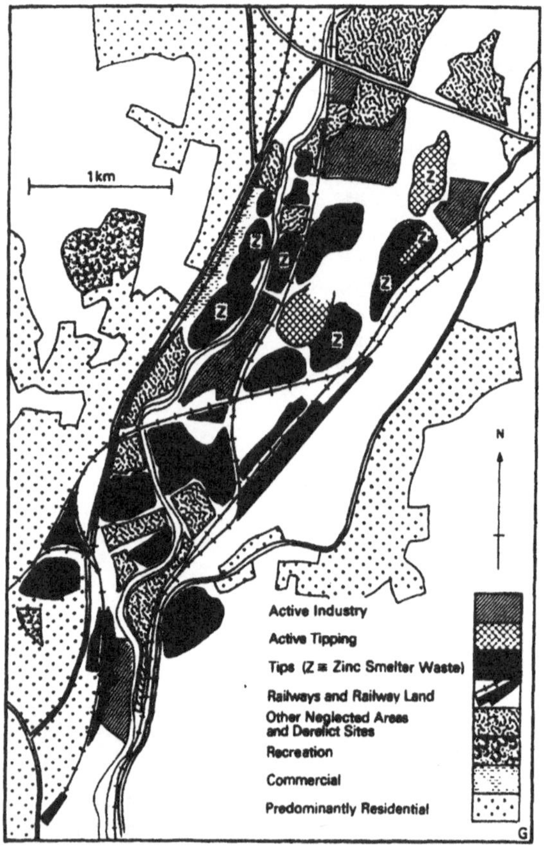

FIGURE 15.7 *Industrial pollution in the Lower Swansea Valley*
Source: Bromley, R.D.F. 1991: The Lower Swansea Valley. In Humphrys, G. (ed.) *Geographical excursions from Swansea. Volume 2: human landscapes*. Swansea: University of Wales, 33–56

In 1848 it was estimated that 92 000 tons of sulphurous acid or 65 900 cubic metres per day were released into the atmosphere by the copper works (Williams 1975).

The Lower Swansea Valley Project (1961–5) was a combination of academic study, to determine the extent of the environmental problems and their impact upon the population living nearby, and an action plan to reclaim parts of the area. Experiments in revegetation with tolerant grasses have allowed the transformation of some of the waste heaps and over 100 000 trees have been planted, giving a visual makeover. Other tips have been levelled or the waste removed. The local council acquired much of the land and, with grants from various government agencies, parts have been redeveloped for new industries, a large retail zone, a sports complex and a forest park.

CASE STUDY: IRONBRIDGE AND EARLY PERCEPTIONS OF INDUSTRIAL LANDSCAPES

Abraham Darby made iron using coke instead of charcoal for the first time at Coalbrookdale, Shropshire in 1709, a key event which made possible its mass production. In the 1770s 40 per cent of British iron smelting was located here, using minerals from close by. In the same locality the world's first iron bridge was erected in 1779 and still stands today (*see* Figure 15.10). Iron rails were another innovation of this remarkable valley and there were also important developments in steam engines and locomotives (Trinder 1981, Alfrey and Clark 1993). Today the Ironbridge Gorge is a World Heritage site, and its Conservation Area has seven Scheduled Ancient Monuments and over 200 Listed Buildings. It is truly a landscape of industry.

It was not uncommon in the early phase of the Industrial Revolution, when the experience was novel, for travellers to express their awe at the sheer power and magnitude of the experience of coming across iron works in full blast, and an Italian traveller in 1787 who found the gorge to be:

> a veritable descent to the infernal regions. A dense column of smoke arose from the earth; volumes of steam were ejected from the fire engines; a blacker cloud issued from a mountain of burning coals which burst into turbid flames. In the midst of this gloom I descended towards the [River] Severn, which runs slowly between two high mountains, and after leaving which, passes a bridge made entirely of iron. It appears as a gate of mystery, and night already falling added to the impressiveness of the scene.
>
> (Quoted in Trinder 1982, 89)

This was an aspect of the eighteenth-century sublime that was judged on the same level as romantic mountain scenery. The regimentation and drudgery of Lowry landscapes (*see* Figure 15.5)

FIGURE 15.8 *The Hafod and Middle Bank Works, Swansea, 1840*
Source: Birmingham Central Library

was more typical of the factory mode of production that predominated in the second half of the nineteenth century.

CASE STUDY: THE NORTH EAST OF ENGLAND, A 'COLONIAL' LANDSCAPE OF RAW MATERIAL EXTRACTION

> [Mining] ... thrusts into the very womb of mother earth, into infernal dark, and wrenches living rock from living rock. Smelting, forging, and casting torment the aborted foetuses with fire. Earth, air, fire and water combine in an unholy alchemical alliance... Embryo becomes artifact.
> (J.R. Stilgoe 1982: *Common landscape of America, 1580 to 1845*. New Haven: Yale University Press)

Coal has been mined commercially in the north east of England since at least the thirteenth century, though production remained limited until the sixteenth century. At first it was exploited by the Church and used locally for boiling sea water to make salt or in lime kilns. The early pits were primitive and shallow and working conditions very poor.

The demand for domestic fuel in the rapidly growing capital, made the coastwise trade to London a major economic proposition from the

FIGURE 15.9 *Industrial dereliction, Lower Swansea Valley in the 1960s*
Source: Reprinted from Hilton, K.J. (ed.) 1966: *The Lower Swansea Valley Project*. London: Longmans

FIGURE 15.10 *The original bridge is still standing at Ironbridge*
Source: P.J. Atkins

sixteenth century onwards. Over the next four centuries this was to become an umbilical cord linking the region to the economic heartland of the south east. Eventually the fleet of coal ships had a combined tonnage greater than the rest of the British merchant marine put together. By 1600 seasale exports from Newcastle amounted to 163 000 tons, rising to 5 million from the North East as a whole in 1850 and 32 million tons in 1913.

County Durham and Tyneside had the advantage of proximity to the rivers Wear and Tyne. From about 1640 much of the coal was moved on wooden railways to a *staithe* on the river bank where it was unloaded into a *keel* for transport down river to an ocean-going *collier*.

The merchant capital of the Guild of Hostmen dominated the coal trade after their incorporation in 1600. Their control of transport was such that they were able to dictate the spatial pattern of extraction and even the price of coal. Subsequently others tried to develop a stranglehold on the trade. For instance the Grand Allies, a group of mine owners bought up the mineral rights and the *way-leaves* (rights of access) over swathes of land near the River Tyne and, by this strategy of spatial monopoly, sought to prevent rivals from sinking pits or exporting their coal (*see* Figure 15.11). In the long term, however, no single interest group was able to maintain such a powerful position. There was competition from coal going via the River Wear to Sunderland and from south Durham to Middlesbrough, and other outlets were developed such as Lord Londonderry's port at Seaham Harbour.

Technology was a means extending the area of coal winning. The wooden waggonways were superseded in the nineteenth century by iron railways with steam locomotives. It is no coincidence that several of the earliest rail experiments with freight traffic were in the North East, including the famous Stockton–Darlington line (1825), and the genius of George Stephenson was of course applied here first. In addition, improvements in shaft sinking techniques allowed the opening of pits on the east Durham plateau where a thick limestone escarpment had to be punctured to give access to the coal measures. In this latter case the effect upon the landscape was dramatic, as described by an observer in 1841: 'Within the last ten or twelve years an entirely new population has been produced. Where formerly was not a single hut of a shepherd, the lofty steam-engine chimneys of a colliery now send their columns of smoke into the sky' (Mitchell 1842, 151).

Elsewhere the coal colonization also transformed the settlement pattern from one of agricultural green villages into a curious hybrid of rural and industrial, the emphasis being upon scattered exploitation by relatively small collieries with their associated pit village. There was hardly a hamlet, village or town that was not affected in some way. The visual impact was extraordinary: head gear (*see* Figure 15.12), pit heaps, smoking chimneys and the stark, mean

FIGURE 15.11 *Causey Arch, County Durham, built 1725–6 to connect with a coal waggonway which ran north to the River Tyne*
Source: P.J. Atkins

FIGURE 15.12 *The headstock at Ryhope, Tyne and Wear, 1940*
Source: Beamish Museum

streets of the typical pit village. There were no city-size industrial settlements apart from Newcastle, Sunderland and Middlesbrough, all three of which owed much of their prosperity to raw material export, and a few large towns such as Darlington. Durham City itself never achieved industrial status, relying instead on its service functions such as administration.

The extraction of coal expanded rapidly at the end of the nineteenth century and its development was increasingly intensive as larger pits were opened. It was the first coalfield to exhaust its more easily worked seams and the average output per miner fell from 370 tons per annum in the 1880s to 260 tons in the period 1909–13. Total production reached its peak in 1909–13 at 40 million tons, from a workforce of 163 000.

The North East basically had an exploitative colonial economy in which both the raw material and much of the profits of extraction were exported. Even local pit owners such as the Lambtons and the Vane-Tempests put little back into the region. The lack of re-investment in a sustainable economy and the lack of a significant multiplier from coal meant that Durham was never prosperous for ordinary people even in its heyday.

The manufacturing industry on the Great Northern coalfield was mostly of the heavy kind, needing substantial inputs of energy. In the case of Consett this meant the construction of a company town to accommodate the iron works. In the early 1840s the first iron works here was started on a greenfield site using local coal and iron ore from the coal measures nearby. The closure of the works in 1980 was a devastating blow to the local economy, with a third of the adult male workforce becoming unemployed overnight.

LANDSCAPES OF FORDISM

In the early twentieth century a new mode of production became very influential. It has been called *Fordism* after the very successful car manufacturer Henry Ford, but its reach has been far wider than the automobile industry. A principal characteristic was the tighter organization of working practices, eliminating as far as possible any wasted time in the factory and the minute subdivision of tasks into repetitive rhythms. Thus was born the production assembly line, with specialist workers who were detailed to look after one aspect of the product as it passed onward to the next stage.

In addition, industrialists sought *economies of scale*, the savings in production costs per unit of output when production plants are large (*see* Figure 15.13), and also *agglomeration economies*, which are gained when related factories are close together in space. Production units therefore increased in size, such as the mega-scale integrated iron and steel or chemical plants which gradually replaced the smaller, less efficient works. As a result, much of the Victorian investment in factories became redundant and those regions which were unable to adapt found themselves in serious economic difficulties.

Investment on such a scale required the ability to deploy financial power differently from anything seen before. The corporate framework of companies such as Imperial Chemical Industries and Ford was increasingly supported by the issue of shares and loans from large banks. The twentieth-century version of capitalism was increasingly lubricated by such financial services located in large cities such as London and New York.

Apart from the reorganization of the older heavy manufacturing industries, the twentieth century has also seen rounds of investment in technologies and products that were new. We have already mentioned the motor car, and there have been many more, including light manufacturing of various kinds (electrical, consumer goods, etc.), service industries and the information and communication sectors. Many of these operate from small units and are less drawn to sources of raw material and energy such as coal. The logic of their

FIGURE 15.13 *The Ford motor vehicle plant at Dagenham, Essex*
Source: Committee for Aerial Photography, University of Cambridge

location has been much less constrained, although market-orientation has been important.

The post-Fordist era

De-industrialization in western countries has been the result of changing technologies and new forms of capitalist organization, along with competition from the expansion of production in the NICs. Manufacturing has been restructured at the end of the twentieth century in order to cope with a new climate of investment and, although most output remains geared to the production line and mass consumption goods, some factories have crossed into a new era that has been called *post Fordist*.

In this latest stage of capitalism, factories are smaller and more flexible in their methods of production. Their location is no longer principally guided by the cost of transporting raw materials and finished products, but may take advantage of government incentives in *development areas* or of non-unionized labour in areas which have had little previous industrial history. *Greenfield sites* are a preference but derelict land is a cheap alternative (*see* Figure 15.14). The plants themselves are usually controlled from a headquarters office in a city which may not be local, and they may be opened and closed according to fluctuations in demand and of the business cycle.

In addition, research and development and high-value manufacturing on science parks, and service industries are growth areas in many countries (*see* Figure 15.15). These also tend to be more flexible than more traditional industrial sectors and their environmental impact is softer and sometimes even deliberately planned to be minimal. In landscape terms, they tend to merge with other urban and suburban forms.

Figure 15.14 *The Nissan car factory at Washington, Tyne and Wear*
Source: Courtesy Nissan

Figure 15.15 *The University of Durham's science park at Mountjoy*
Source: P.J. Atkins

Conclusion

The rounds of investment in industry have each had their own characteristic impacts. Raw material and energy-source locations replaced the rural dispersion of proto-industry, as heavy manufacturing emerged, and this gave rise to a rash of functional and somewhat spartan settlements, especially on the coalfields, as accompaniments to coal mines and textile or iron works. The degrading potential of industry in this era was insidious, both upon the environment and the employees, who worked in wretched conditions. Later, in the twentieth century, the landscapes of industry have been more carefully planned and arguably their impacts have been somewhat softer, but the overall effect has nevertheless been greater because more nations have industrialized and not all have brought pollution properly under control. In the post-industrial phase, which is just taking hold in western countries as they come to rely more upon service jobs and as manufacturers adopt a more flexible location strategy, new landscapes are emerging which bear little resemblance to the fire and brimstone of eighteenth-century representations. It is interesting to speculate how Lowry would have painted a science park or light industrial estate.

Because much of post-Fordist industry adopts a peri-urban location, and is an integral element in the birth of new 'urban' forms, it makes sense next to look at modern and postmodern urbanism. This we will do in Chapters 16 and 17.

Further reading and references

Dunford and Perrons (1983) offer a helpful account of emerging economic forms in the nineteenth century, which can be supplemented by the more traditional insights of Darby (1973). The contributions of industrial archaeologists (Alfrey and Clark, Palmer and Neaverson, Trinder) are factually informative but rather narrow in outlook.

Alfrey, J. and Clark, C. 1993: *The landscape of industry: patterns of change in the Ironbridge Gorge*. London: Routledge.

Bromley, R.D.F. 1991: The Lower Swansea Valley. In Humphrys, G. (ed.) *Geographical excursions from Swansea. Volume 2: human landscapes*. Swansea: University of Wales, 33–56.

Darby, H.C. (ed.) 1973: *A new historical geography of England*. Cambridge: Cambridge University Press.

Dunford, M. and Perrons, D. 1983: *The arena of capital*. London: Macmillan.

Mitchell, J. 1842: On the employment of children and young persons in the mines of the South Durham coalfield, Appendix to the First Report of the Royal Commission on Childrens' Employment in Mines and Manufactories. *British Parliamentary Papers* 1842, xvi.

Muir, R. 1981: *Shell guide to reading the landscape*. London: Michael Joseph.

Palmer, M. and Neaverson, P. 1994: *Industry in the landscape 1700–1900*. London: Routledge.

Trinder, B.S. 1981: *The industrial revolution in Shropshire*. 2nd edn, Chichester: Phillimore.

Trinder, B.S. 1982: *The making of the industrial landscape*. London: Dent.

Williams, M. 1975: *The making of the South Wales landscape*. London: Hodder & Stoughton.

16

MODERN URBAN LANDSCAPES

MODERN CITIES AND CITY LIFE

I well remember the effect produced on me by my earliest view of Manchester, when I . . . saw the forest of chimneys pouring forth volumes of steam and smoke, forming an inky canopy which seemed to embrace and involve the entire place. I felt I was in the presence of those two mighty and mysterious agencies, fire and water.

(W. Cooke Taylor 1842: *Notes of a tour in the manufacturing districts of Lancashire*)

INTRODUCTION

In Chapter 12 we discussed a category of urbanism that we called pre-industrial. In this chapter we will move on to investigate urban landscapes of the nineteenth and twentieth centuries and give some examples of the processes that are shaping the urban experience at the turn of the millennium. Our context is the transition from the modern to the postmodern city, and our main examples will be Paris and Los Angeles.

Nicholas Green (1990) argues that Paris in the nineteenth century developed an urban vision of itself and its surrounding hinterland that he calls the 'metropolitan gaze'. This city-centred culture and lifestyle cut across economic class divisions, drawing in businessmen and consumers from provincial towns but excluding peasant farmers and landed aristocracy. It framed the city in a flood of painted urban portraits, printed guide books, poems and novels (Prendergast 1992) and it saw nature in terms of the countryside, especially that immediately around Paris, and set about fashioning that countryside in its own image, an image that was essentially pictorial. In this view it was the visual and literary imaginations, newly and very effectively deployed, that were to create an urban identity in a way that city dwellers of the twentieth century now find unremarkable because the novelty has long since worn off.

A most effective character in Parisian novels and poetry of the day was the flâneur (wanderer), an individual who strolled around at will in the city observing both the commonplace and the unusual, without ever fully participating. Although bourgeois, and with time to spare, he was a man of the crowd and in current parlance he would be called *streetwise*. Charles Baudelaire used the *flâneur* as a narrative device in his literature, particularly his *Paris spleen* collection of 1869 and this helps us, the modern reader, to see the city through the eyes of a contemporary, albeit fictional, observer. Charles Dickens' descriptions of London at the same time, based on his own hours of wandering the streets, perform a similar function.

In the twentieth century there have also been many insights through urban literature. Robert Park, a key figure in the Chicago school of sociology in the 1920s, used the skills he acquired as a newspaper reporter to dissect the urban experience of neighbourhoods and communities.

Raymond Chandler did much the same in his detective novels based upon life in California, where he found much raw material on crime and corruption.

Such journalistic and imaginative writings have been treated with scepticism until recently by social scientists schooled in the collection and analysis of *hard* data, but current trends are very much towards accepting the legitimacy of a wide range of source materials which might help us to understand the nature of urban problems and how the identities of urban citizens are forged.

These nineteenth- and twentieth-century descriptions of the urban phenomenon conjure something so different from what had gone before that we may speak of the emergence of a new world order. This had its origins in seventeenth and eighteenth century thought and it was to become, in association with capitalism, overwhelmingly the dominant global system by the 1990s (Gregory 1994). It will help us to understand this new formation of culture, society and economy if we reflect first on the meanings of *modernity*. We will then look at the major elements as they were precipitated in the urban landscape.

Modernity

Modernity is comprised of four major elements:

1. The control of information, and its application through reason and rationality to knowledge and decision-making about the world, with a resulting growth in the influence of science and technology and the waning of religion and tradition as frameworks of behaviour. This often involves the privileging of a western and male viewpoint. *Other* views are dismissed as irrational and inferior, and are therefore fair game for aggression, domination, colonization and conversion.
2. An organization of life which owes much to an industrial market economy and to capitalism. Economic growth is not continuous, but is

Figure 16.1 *The streets of Boldon Colliery, 1968*
Source: Beamish Museum

disturbed by cycles of prosperity and depression, and occasional shocks and crises. All of those who live in modern societies are subjected to the risk of collective or individual failure, but there are in-built mechanisms in modernity that allow adjustments through a learning process of change to social institutions and to economic systems.

3. The construction of landscapes that are increasingly distant from raw nature. Cities are the chief expression of this. Their regularities of design and organization are clearly different from the more organic and informal pre-industrial urban areas, and the new spatial patterns are reflections of the forces at work in society.
4. A centralized political authority based upon the nation–state. This involves a greater freedom for the individual in thought, expression and lifestyle within certain limits set by collective rules of social discipline.

Technology

By 1900 cities were becoming increasingly dominant in articulating the economies and societies of advanced industrialized countries such as Great Britain and America. The experience of the century just ended had been quite extraordinary, with an unprecedented explosion of urban growth that created an altogether new environment. Contemporaries received this with a mixture of awe and disquiet because, although they recognized the wealth-creating opportunities innate in urbanization, it was obvious to all observers that many problems were created by uncontrolled building, pollution, poverty and social dislocation. The *shock of the new* here was the sheer scale of developments, which assailed every sense. In the Britain of 1901 78 per cent of the population was already living in towns and cities, which as a result were not only different in size to the pre-industrial urban phenomenon, but they were also palpably new in their social formation. The twentieth century has echoed to the noise of construction but its sprawling mega cities have been created on a scale that dwarfs individual human beings.

Factory technology

The mechanical technology of the Industrial Revolution facilitated a new looking urbanism. Whereas craftsmen had previously struggled to produce physical infrastructure and consumer goods at affordable prices, the mass production of the machine age ensured a steady flow of innovations such as inexpensive bricks of standard quality, sewer pipes to remove waste, iron rails for trams and trains, and steam engines for power. The steady rhythm of factory life and its artificial composition of simplified, repeatable elements, was echoed in the surrounding streets. What better way to cope with what at the time was the greatest urban expansion in history than by quickly constructing row upon row of terraced houses to a single design, on a grid pattern of streets that made any one district look like the others (*see* Figure 16.1)? Victorian architect-designed buildings may have been very varied in

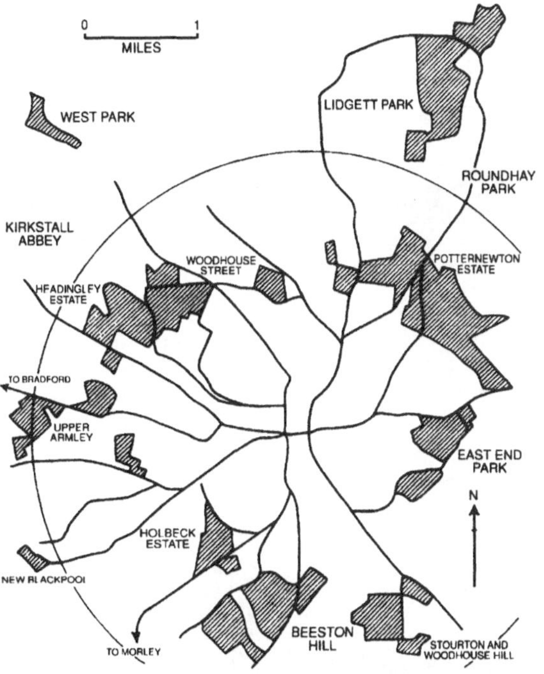

Figure 16.2 *The street-car suburbs of Leeds*
Source: Reprinted from Ward, D. 1964: A comparative historical geography of street car suburbs in Boston, Massachusetts and Leeds, England: 1850–1920. *Annals of the Association of American Geographers* **54**, 477–89

their aesthetic inspiration, but on the whole they were but embellishments on the theme of manufacturing cheap boxes for residents and businesses alike.

Charles Dickens, a very astute observer of urban life in the mid-nineteenth century, described the townscape of Coketown in his novel *Hard Times* (1854). It was based upon his personal experience of Preston:

> It contained several large streets all very like one another, and many small streets still like one another, inhabited by people equally like one another, who all went in and out at the same hours, with the same sound upon the same pavements, to do the same work, and to whom every day was the same as yesterday and tomorrow, and every year the counterpart of the last and the next.

For Lewis Mumford (1961) such Victorian cities displayed 'the most degraded urban environment the world had yet seen ... they were man-heaps, machine warrens, not agents of association for the promotion of a better life', and it is undoubtedly true that there were many associated problems of disease and social dislocation.

Transport technology

Apart from factory machinery, transport has made the most important technological contributions to city structure in the modern era. Private carriages allowed the wealthy to live away from the smells and hubbub of the pre-industrial city but public transport was almost non-existent before 1800. Horse-drawn buses started in Paris and spread to London and other cities in the 1820s, but the high ticket prices were beyond the pockets of ordinary people. It was not until the last quarter of the nineteenth century, with special workmen's fares on the railways and the electrification of trams, that working class men and women participated in a mobility transition, and street-car suburbs emerged (*see* Figure 16.2) as a new type of urban area.

FIGURE 16.3 *Gravelly Interchange, M6, otherwise known as 'Spaghetti Junction'*
Source: Committee for Aerial Photography, University of Cambridge

In the twentieth century the main force of transport restructuring has been the private motor vehicle, in four ways. First, there has been the flexibility of private trips, not dependent upon timetables and constrained only by traffic congestion and parking space. This has greatly facilitated journeys across town and fuelled ribbon developments along the major routeways. Second, the construction of urban motorways and trunk roads is space-consuming and one only has to look at a major intersection with its spiralling loops of slip roads to realize how dominating it is

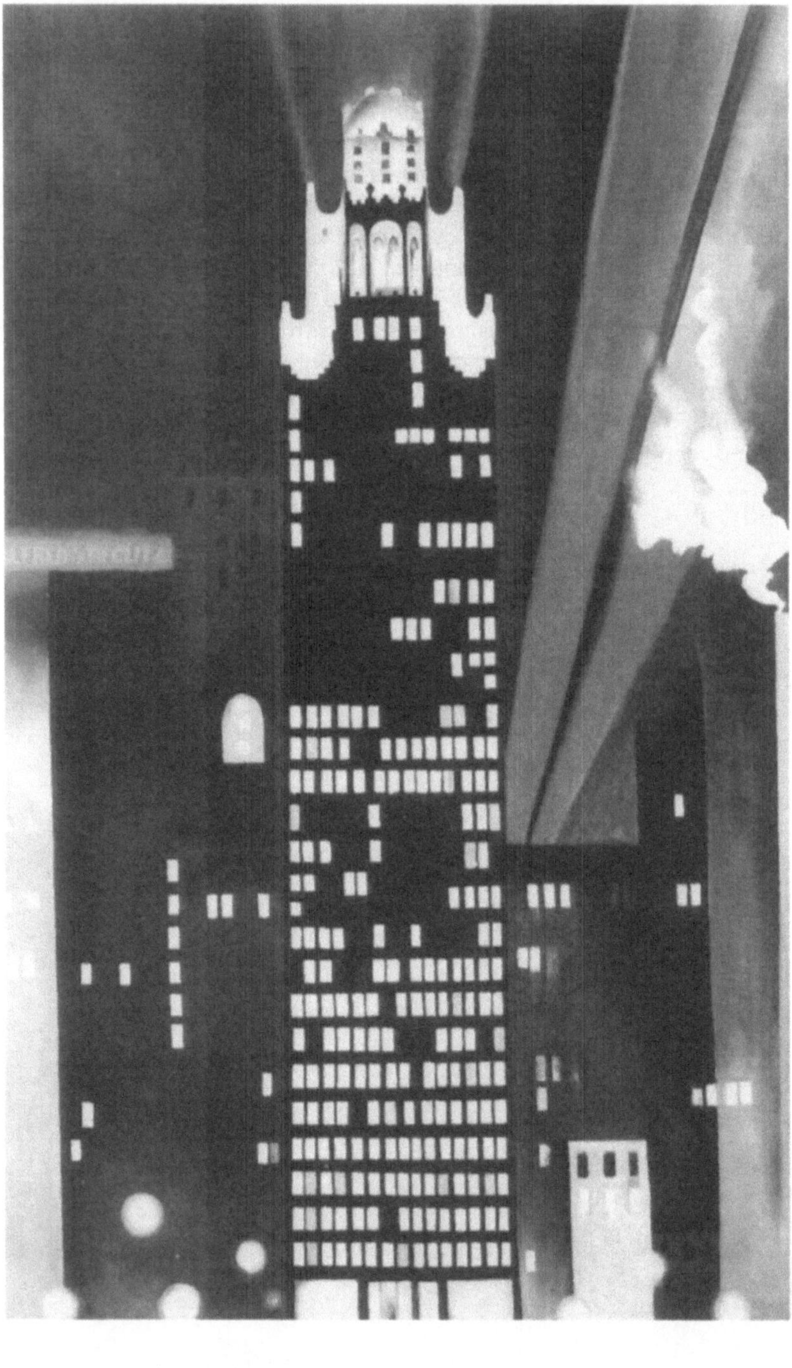

FIGURE 16.4 *Radiator Building, Night, New York, 1927, Georgia O'Keefe*
Source: The Alfred Stieglitz Collection, The Carl Van Vechten Gallery of Fine Arts, Fisk University

over adjacent land uses. Moreover, it represents one of the key visual images of the present day (*see* Figure 16.3).

Third, the car is symbolic of the individual freedoms that people in developed western societies cherish. It has been responsible for the maintenance of what Melvin Webber called 'community without propinquity', the ability of people to interact face-to-face with their family, friends and business contacts without having to live in immediate proximity to them.

Finally, the car has gradually emerged as the principal force de-centring cities. Agglomerations such as Los Angeles, which are phenomena of the motor car era, have developed with several foci, not just one city centre. Out-of-town retail parks and shopping malls attract the car-borne customer and are threatening the prosperity of the high street. Commuters are able to travel daily distances that 100 years ago would have been in realms of fantasy, and are now escaping from the noise, pollution and taxes of cities and repopulating the countryside in a process called counter-urbanization. The sprawl resulting from this expansion diffusion of urban life has, in its most extreme expression, led to the merging of neighbouring conurbations into a large urban axis known as a *megalopolis*. The example usually quoted is Boswash (Boston–New York–Philadelphia–Baltimore–Washington D.C.), but many others exist, including some dynamic super-city regions in Less Developed Countries.

Since about 1960 mass transport by aeroplane has become possible and this, coupled with the advent of efficient telecommunications and the use of the Internet, has made us global beings who can travel to or interact very quickly with people in other countries. Global cities such as London and New York draw their prosperity as international financial centres partly from this phenomenon and their landscapes of skyscraper office blocks and urban motorways are being replicated in capital cities that only a few decades ago were still extraordinarily varied in their urban experience.

FIGURE 16.5 *E.H. Suydam's drawing of 'A Group of Lefcourt Buildings', New York, 1928*
Source: Kouwenhoven, J.A. 1953: *The Columbia historical portrait of New York: an essay in graphical history*. Garden City, New York: Doubleday

Architectural technology

Changes in architectural technology and style produced some of our most visible and therefore intrusive results of modernity. The simple, clean lines of buildings by Le Corbusier and Frank Lloyd Wright were mirrored in the contemporary art of Picasso, Klee and Mondrian and the material design of the Bauhaus. But the icon of the age is undoubtedly the commercial city centre skyscraper (*see* Figures 16.4 and 16.5), with its near cousin the multi-storey block of flats. Both were made possible by the use of steel framework construction and concrete. High rise living is not new but in the twentieth century it represents one of the key modern solutions to urban problems: the intensive use of space by building upwards, and the imposition of simple, neat designs in the hope of hiding the poverty and clearing away the clutter of inner-city decay. Public reactions have always been mixed, varying from the awe of John Van Dyke's 1909 description of 'a new sublimity that lies in the majesty of mass, in aspiring lines against the upper sky' (Domosh 1987), to the disgust of tenants living in the decaying blocks of flats put up in 1960s' Britain.

ECONOMIC ORGANIZATION

A feature of industrial capitalism was the development of factory-based manufacturing, in early nineteenth-century Britain and by 1900 in other leading Western countries. Long hours were a feature of the highly disciplined regime of work, and many workers in cotton mills or steel plants could not afford the luxury of a journey to work of more than a few minutes. The bulk of the labour force therefore lived in the mean terraced streets that huddled in the shadow of every plant, with the result that the urban fabric was like a honeycomb of semi-independent modules, industrial villages within a conurbation. The common thread linking these urban villages and suburbs was the evolution of a commercial focus point which developed a variety of functions.

With the maturing of the nineteenth-century city, a greater prominence was given to the *use value* of land. This came to be judged increasingly according to its location with respect to the city centre, which had the highest values of all. Land was now a commodity like any other, to be bought and sold or rented. In the competition that followed between rivals wanting to use the same plots, the highest bidders were responsible for a sorting procedure that had not been experienced before in most cities, with commercial land uses able to outbid all others in the centre. Residences were shunted increasingly to the margin of what came to be known as the Central Business District (CBD).

As land uses became more attuned to these market forces, specialization grew and spatial patterns of functional zonation crystallized into recognizable outcomes on the ground. The separation of manufacturing industry, residential housing, and commerce into their own concentric rings or sectors had a clear logic (*see* Figure 16.6). They were described vividly by Friedrich Engels in Manchester in 1844 and were later modelled in the Chicago of the 1920s by Ernest Burgess. The patterns were not static, however, with successions of change according to competition and sectoral shifts in the economy. An example is the

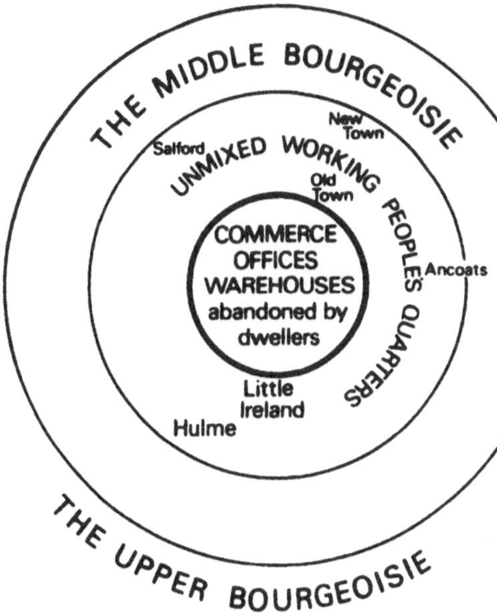

FIGURE 16.6 *A model of the land use of Manchester, as described by Friedrich Engels in 1844*
Source: Dennis, R. 1984: *English industrial cities of the nineteenth century: a social geography*. Cambridge: Cambridge University Press

flight to the suburbs of the wealthy and middle-class former residents of the city centre, whose houses were then subdivided into flats and rented to working people and immigrants.

Amongst the most interesting transformations of the nineteenth century was the emergence of fixed shops in the CBD, at first selling general goods and later specializing in particular consumer goods. Most food continued to be sold from market stalls or by doorstep delivery until the early twentieth century. Gradually small clusters of shops appeared in the suburbs but generally these fringe districts were poorly provided with services of any kind.

All of these functional transformations were of course constrained by the physical characteristics of their site and by any established fixed capital in the form of existing buildings and roads. Both affect accessibility, a crucial variable in any human interaction, and they also introduce an element of inertia into the urban fabric which influences decision-making about the use of land. In short, urban morphology (shape, texture) and its change (morphogenesis) are important considerations of townscape. The size of plots, density of buildings per hectare, and phases of accretion are signatures (a subset of what Smith (1977) calls the 'syntax of cities') of particular types of urbanism and can tell us a great deal about past societies and their economies. Perhaps the most obvious example of this is the outward growth of cities by the addition of new *fringe belts* (*see* Figure 16.7). These may be seen as equivalent to the annular rings of growth in a tree or, better, they can be related to specific economic events. The swings of booms and slumps that have characterized capitalist economies since the outset of the Industrial Revolution also have their expression in cycles of construction. When houses are in demand and house building is profitable, new residential areas compete for land on the urban fringe, but during quiet periods they give way to space-extensive uses such as golf courses, cemeteries, hospitals, and out-of-town shopping centres. Thus the residential and institutional architecture and street layouts of particular eras are frozen in space at a certain distance from the city centre in a sandwich effect.

SOCIAL FORCES

The experiences of industrialization and urbanization ultimately had profound repercussions for society. The move from the workshop to the factory meant an increased likelihood that sons

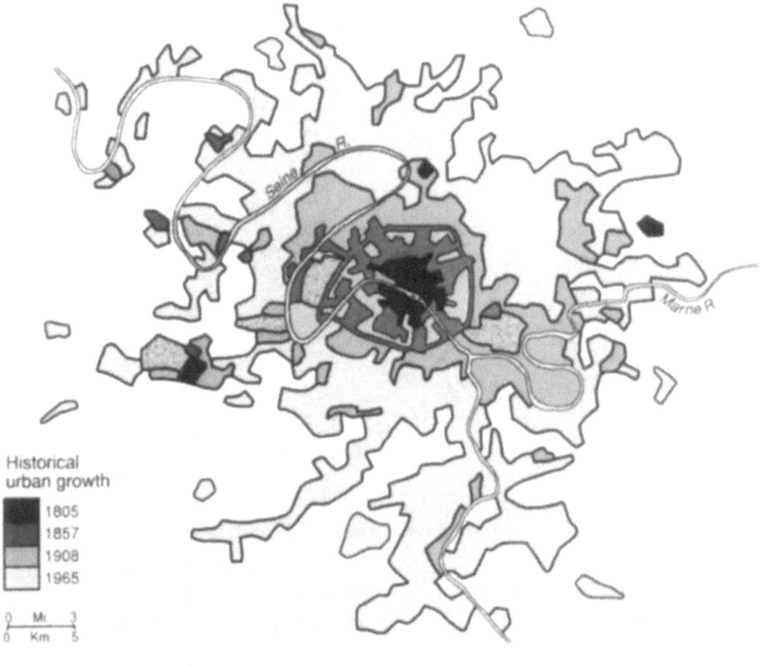

FIGURE 16.7 *The growth of Paris*
Source: Sommers, L.M. 1983: Cities of Western Europe. In Brunn, S.D. and Williams, J.F. (eds) *Cities of the world: world regional urban development*. New York: Harper & Row, 85–122

and daughters would set up their own nuclear families rather than continue to live and work in an extended family unit with their parents. This shift was reinforced in the urban environment by the greatly increased possibilities for social interaction and inevitable dilution of customs and mores that had been traditional in the countryside. The relationships that mattered to factory workers were now those with their employers and with their fellow employees, and gradually a new socio-cultural setting emerged in consequence. This provided a context in which the type of loyalties and rivalries developed that we might call a class system. The concentration of wealth in the hands of a few commercial and industrial entrepreneurs was not new but the degree of polarization between classes and the demonstration effect of power and opulence certainly were.

Modern cities, from the experience of Manchester in the early Victorian period onwards, were more complex in their organization of space and the way different groups in society interacted. Although the fragmentation of early capitalist cities into factory neighbourhoods and urban villages separated by physical barriers like canals or railway lines was to some extent mitigated by forces of integration such as public transport and the increasing availability of central services serving the whole city, in reality new forms of difference did begin to emerge. Specialized land uses had their most extreme expression in the evolution of residential districts that were segregated along class, racial, linguistic, religious or other lines. This was the result of the clustering of rural immigrants, and in the case of America of overseas immigrants, into reception areas where cheap housing was concentrated. Here there were job opportunities, at first in unskilled and often casual labour or in the informal sector of street trading, which were often arranged by relatives or friends who had arrived in the city earlier. Where the new city-dwellers for one reason or another were rejected and excluded from the opportunities of urban life by the host population, they might find themselves in slums or ghettoes where the quality of life was very low and health poor, but such an experience was by no means universal. Cohesive communities became recognizable where mutual support was common and where working class and minority cultures forged an identity that has its expression in distinctive landscapes such as Chinatown or Little Sicily.

The modern city had many social barriers that made it difficult for people to live wherever they wanted. There were many filtering processes whereby certain powerful gatekeepers were able to influence the social landscape. Owners of large estates, for instance in London and many other European cities, were able to impose their view of a desirable type of development and of the type of people who could live and work on their land. Estate agents and local authority planners and housing managers were similarly key decision-makers. In reality, any notion of land and housing markets governed only by free market forces is wide of the mark. Modernism and capitalism both have in-built biases and mechanisms that tend to ensure an outcome which is in the long-term interests of certain minority groups.

It is possible to view these barriers and biases as structural forms of power that help us to understand the evolution of neighbourhoods, but there is also evidence of this power on the micro scale. Michel Foucault (1977) pointed this out in his study of the disciplinary tendencies of modernity. He found spatial discipline in the way public institutions such as prisons and mental hospitals are designed and run, and he also argued that society maintains its self-control by an unspoken and unwritten agreement by us all that we are willing to undergo surveillance by the agents of the state such as the police, doctors and other public sector professionals. In the late eighteenth century Jeremy Bentham had a vision of a prison, the Panopticon, where every move of prisoners could be monitored from a central control point. For Foucault, modern society has moved increasingly in this direction, with the equivalent of Big Brother, from George Orwell's *1984*, watching our lives closely.

In this vein, the advent of computers has greatly improved the efficiency of keeping records on large numbers of people. This may be irritating when yet more junk mail drops through the letter box as the result of a computer mail shot, but the implications might be more sinister in the hands of a ruthless dictator wanting to monitor political opposition. The latest example of surveillance is the common use of closed circuit television systems (CCTV) to protect private and commercial property (Fyfe and Bannister 1996). Architects, of for instance shopping malls, are now designing spaces to optimize such surveillance. A 1995 survey of British city centres found that about half of councils had installed CCTV to oversee their high streets, many with government subsidies. The main reason given was to combat crime but video footage was also used to monitor the time that council employees arrived at work and there are of course many other uses to which such scanning can be put. The idea is also spreading to residential streets but, at the time of writing, few countries have legal guidelines on when, where and how CCTV can be used.

THE STATE

One of the main problems of early Victorian city life was the weakness and disorganization of the local state, which was more attuned to the medieval than the industrial age. There were several decades of chaos until the Victorians' desire for social control and their genius for civil engineering brought matters in hand. From about 1850 to 1980 there followed the golden era of state urban planning, during which existing forms of urbanism were regulated and several new forms brought into existence.

The famous civil servant Edwin Chadwick (flourished 1830s–50s) and a number of his influential successors argued for public intervention in urban areas in order to control ill-health and improve the housing conditions of the workers. The motive was less philanthropic than a utilitarian recognition that a malnourished, ill and poorly housed workforce could not perform to its maximum possible efficiency. The solutions were:

- the provision of clean water;
- the removal of human waste in specially designed tubular sewers, flushed by water;
- slum clearance in order to eliminate housing unfit for human habitation and its replacement by dwellings built according to bye-law quality standards;
- reduction of the danger of epidemic diseases spreading by reducing the density of population in overcrowded neighbourhoods.

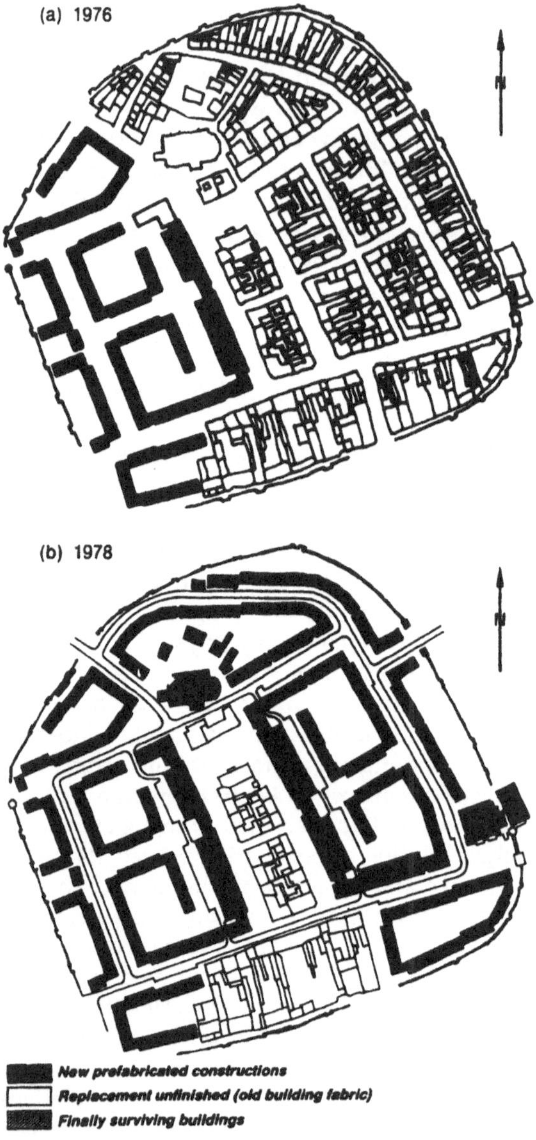

FIGURE 16.8 *The socialist remodelling of Bernau-Brandenburg*
Source: Reprinted from Denecke, D. 1992: Ideology in the planned order upon the land: the example of Germany. In Baker, A.R.H. and Biger, G. (eds) *Ideology and landscape in historical perspective*. Cambridge: Cambridge University Press, 303–29

One way to open out the urban fabric was for local authorities to provide leisure facilities, in the form of parks, bubbles of air in the urban matrix of brick and stone. These, and other elements of collective consumption such as schools, cemeteries, art galleries, libraries, and utilities like gas and water, became a yardstick of civic pride. Another solution was the construction of new suburbs and even new towns on greenfield sites away from the squalor of the inner city.

The idea of urban planning was not one which appealed to many Victorians. The values of the age were concerned more with individual freedoms and responsibilities than with any notion of imposed solutions. Yet planning there was, first in the shape of mill towns such as Titus Salt's Saltaire, sited in the countryside outside Bradford, with its unusually full provision of facilities such as a school, chapel, wash-house, institute, hospital, almshouses, bank, park and allotments. Later, Ebenezer Howard started a fashion for *garden cities* and *garden suburbs* with a street plan that included open grassed areas and roadside trees, quite unlike anything that had gone before. In the twentieth century house gardens and detached or semi-detached houses have become commonplace but they seemed revolutionary in 1900.

Immediately after the First World War it was obvious that in Britain an acceleration of house building was necessary to replace city slums and provide 'homes fit for heroes'. The private sector was incapable then of meeting the need and local councils began to provide rented public sector accommodation for their citizens. Some were built as infill on the sites of demolished slums but the majority were on *council estates* which were located on farmland beyond the urban fringe. Public facilities on these estates were sparse and transport to and from the city centre often poor but this innovation in housing provision was successful in other ways and it has become a very familiar aspect of the suburban landscape around British cities.

In the twentieth century planning has become all-embracing. Land use is zoned, with Green Belts preserved against urban developments, and planning permission is required for even the smallest construction. The British new towns of the 1950s and 1960s epitomize this new found confidence well in a capitalist context, but of course the socialist city is the most extreme example of planned modernism (*see* Figure 16.8). In many of the countries of eastern Europe where communism was imposed after the Second World War, devastated city centres and new suburbs were developed according to a model perfected in

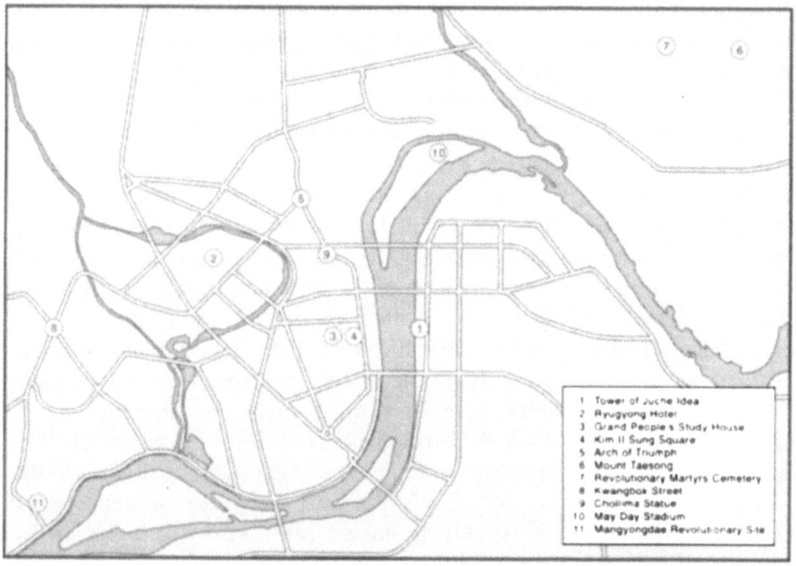

FIGURE 16.9(a) *Central Pyongyang and its monuments*
Source: Atkins, P.J. 1996: A seance with the living: the intelligibility of the North Korean landscape. In Smith, H., Rhodes, C. and Magill, K. (eds) *North Korea in the new world order.* Basingstoke: Macmillan, 196–211

FIGURE 16.9(b) *One of the major thoroughfares of Pyongyang*
Source: P.J. Atkins

FIGURE 16.10 *A shanty town in Lima, Peru*
Source: P.J. Atkins

the Soviet Union in the 1920s and 1930s. From Figure 16.9 we can see that modernist ideals of straight lines and high rise were not confined to Europe and North America. They are found even in communist North Korea where the capital Pyongyang is as close to a modernist showcase as anywhere on earth.

We said earlier that modernity and capitalism had conquered the planet, and their associated landscapes had by the 1990s become globalized. At this point we must add a note of caution with regard to the poor countries of the Third World. Asia, Latin America and Africa have experienced the impact of some aspects of capitalism, with their economies increasingly tied to the world economy by exporting raw materials and importing manufactured goods. The downtowns of their largest cities, especially in countries on the Pacific Rim, look remarkably like western cities, with office skyscrapers, apartment blocks, urban motorways and the other ingredients of modern townscape, but the scene beyond the CBD is very different. The poverty and wretched housing conditions of the marginalized shanty towns are more reminiscent of the early phase of disorganized, poorly regulated Victorian urbanization than of the late twentieth century (*see* Figure 16.10). But hundreds of millions of people live in these conditions, reminding us that modern urbanization does not have all the answers, in fact it is currently creating problems faster than it is able to solve them.

CASE STUDY: PARIS: CAPITAL OF THE NINETEENTH CENTURY

Paris was a centre of retail innovation and such a focus of cultural energy that it came to represent the spirit of the time, so much so that Walter Benjamin called it the 'capital of the nineteenth century'. Amongst the most impressive developments for contemporaries were the arcades, which first came to prominence in the 1820s and 1830s and which were so successful as a retailing and architectural experiment that they were copied all over the world (*see* Figure 16.11). They were the nineteenth-century forerunners of malls, with the shops under cover so that customers and window gazers could wander in comfort around this 'temple of commodity capitalism' in which goods were 'in window showcases like icons in niches' (Buck-Morss 1989, 83). With a concentration of fashionable luxury goods in glittering surroundings of marble and glass, the arcades were a vision of the pleasures and possibilities of the age. Walter Benjamin, an historian and philosopher of modernism, regarded the arcades as emblematic of the nineteenth-century experience, a modern allegory of the labyrinth myth. Further developments of this 'phantasmagoria of display' were the world expositions (held in Paris in 1855, 1867, 1889 and 1900) and the department store, which started here with the Bon Marché about 1865.

FIGURE 16.11 *The Central Arcade, Newcastle-upon-Tyne*
Source: P.J. Atkins

Paris was also the capital of urban planning, as a result principally of the efforts of Georges-Eugène Haussmann (1809–91), who was appointed Prefect of the Departement of the Seine in 1853. By that date Louis-Napoléon had already sketched out a plan for public works and now, as newly declared Emperor, he put considerable pressure on the civil authorities to start improvement schemes immediately. Haussmann, in his period of office which lasted until 1870, was not so much the designer as the driving force behind the changing face of the capital. The Haussmannization of Paris principally meant the demolition of insanitary and disease-ridden slums and their replacement by the wide boulevards which are so characteristic of the city centre today (Chapman 1953). With extensive compulsory purchase powers, Haussmann was able to create a new street pattern that could better accommodate the growing volumes of traffic and in his view go some way to fulfil Napoléon III's ambition of making Paris 'the most beautiful city in the world' (Sutcliffe 1970, 32). The intention of his Paris cross, made up of new north–south and east–west through routes, was to revitalize the commercial centre by giving it better links with the burgeoning suburbs and also to reduce the threat of public disorder by allowing the rapid disposition of troops. Figure 16.12 shows the Avenue de l 'Opéra, which was finally completed in 1876–77, after Haussmann's time.

Haussmann's efforts in central Paris have been called 'the biggest urban renewal project the world has ever seen' (Sutcliffe 1993). Certainly a great deal of capital was invested and the result was a classical Second Empire style of architecture that impose a formulaic unity that had not existed previously. This harmony was false, however, as was soon to be revealed in the Commune of 1871.

The arbitrary nature of power exercised by Napoléon III caused unrest. The acceleration of capitalist accumulation evidenced in the property speculation that accompanied Haussmann's rebuilding caused the gap between rich and poor to open up. Many poor people were swept out of the refurbished centre and repression of working-class political opposition was resented. Strikes multiplied, there were many public meetings and agitation in the late 1860s led to riots. But it was defeat in the war with Prussia in 1871 which precipitated the Commune. An army of worker-revolutionaries took over and held Paris until eventually defeated in May of that year (Harvey 1985; Ferguson 1994). The barricades they erected blocked Haussmann's boulevards and created a temporary landscape of resistance that had resonance for decades to come.

The contradiction of Paris as commercial and cultural mecca, on the one hand, and revolutionary volcano, on the other, was not lost on Benjamin:

> As a social formation, Paris is a counterimage to that which Vesuvius is as a geographical one: a threatening dangerous mass, an ever active June of revolution. But just as the slopes of Vesuvius, thanks to the layers of lava covering them, have become a paradisiacal orchard, so here, out of the lava of Revolution, there bloom art, fashion, and festive existence as nowhere else.
>
> (Quoted in Buck-Morss 1989, 66)

The Basilica of Sacré Cœur strategically and symbolically dominates the northern skyline of Paris. Its foundation stone was laid in 1875 but it

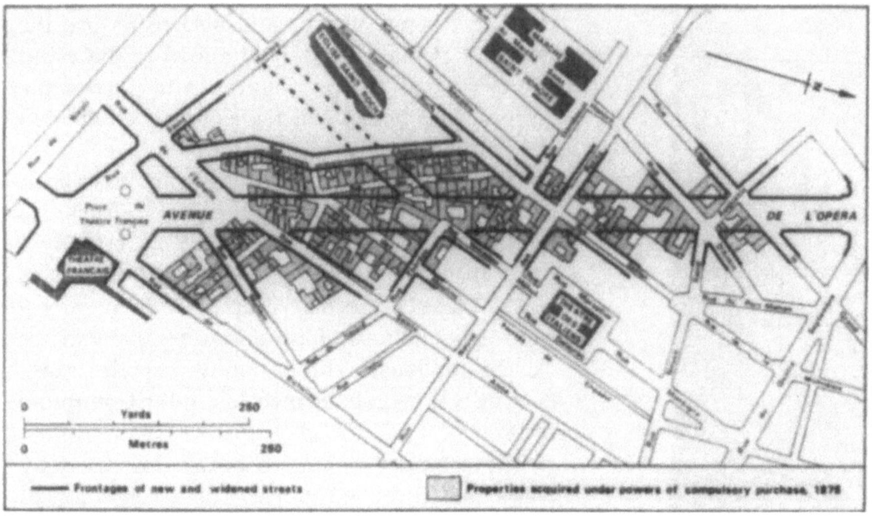

FIGURE 16.12 *Avenue de l'Opéra, Paris*
Source: Sutcliffe, A. 1970: *The autumn of central Paris: the defeat of town planning 1850–1970*. London: Arnold

had been a reaction to the Franco-Prussian War (1870–1) and especially to the Paris Commune of 1871. A group of ultra-conservative Catholics, enthusiasts for the cult of the Sacred Heart, allied with those advocating restoration of the monarchy, managed to secure legislation to acquire a site on the heights of Montmartre. This was the location of key events in the Commune and the whole project inflamed the passions of the mainly republican Parisians who saw it as a means of subjugation. Harvey's (1979) account of Sacré Cœur implies that it remains a political symbol to the present day.

CONCLUSION

The advent of modernism was a social and intellectual event of the most fundamental significance. It ushered in a different way of looking at the world, the consequences of which are still with us today. The rational ordering of life was introduced to all aspects of human endeavour, from cultural manifestations such as architecture to the planning of economic structures like assembly-line manufacturing. The principal landscape impacts were concentrated in urban areas and it is theorists such as Benjamin who help us to appreciate the long-term implications of this. He realized that it is the act of consumption which symbolizes the modern way of life, and therefore that landscapes of retailing and service provision essentialize the shift from previous eras. Other writers have looked at a variety of issues concerned with the consumption of meaning in landscapes, and it is to this that we now turn in Chapters 17 and 18.

FURTHER READING AND REFERENCES

The works of David Harvey and Derek Gregory are inspirational on the theme of modernism and post-modernism. More traditional urban histories are provided by Mumford, Dennis and Sutcliffe.

Buck-Morss, S. 1989: *The dialectics of seeing: Walter Benjamin and the Arcades project*. Cambridge, Mass.: MIT Press.

Chapman, B. 1953: Baron Haussmann and the planning of Paris. *Town Planning Review* **24,** 177–92.

Dennis, R. 1984: *English industrial cities of the nineteenth century: a social geography*. Cambridge: Cambridge University Press.

Domosh, M. 1987: Imagining New York's first skyscrapers. *Journal of Historical Geography* **13,** 233–48.

Ferguson, P.P. 1994: *Paris as revolution: writing the nineteenth century city*. Berkeley: University of California Press.

Foucault, M. 1977: *Discipline and punish: the birth of the prison*. London: Allen Lane.

Frank, W. *et al.* 1979: *America and Alfred Stieglitz: a collective portrait*. Revised edition. New York: Aperture.

Fyfe, N.R. and Bannister, J. 1996: City watching: closed circuit television surveillance in public spaces. *Area* **28,** 37–46.

Green, N. 1990: *The spectacle of nature: landscape and bourgeois culture in nineteenth century France.* Manchester: Manchester UP.

Gregory, D. 1994: *Geographical imaginations.* Oxford: Blackwell.

Harvey, D.W. 1979: Monument and myth. *Annals of the Association of American Geographers* **69,** 362–81.

Harvey, D.W. 1985: *Consciousness and the urban experience: studies in the history and theory of capitalist urbanization.* Baltimore: Johns Hopkins University Press.

Mumford, L. 1961: *The city in history.* London: Secker & Warburg.

Prendergast, C. 1992: *Paris and the nineteenth century.* Oxford: Blackwell.

Smith, P.F. 1977: *The syntax of cities.* London: Hutchinson.

Sutcliffe, A. 1970: *The autumn of central Paris: the defeat of town planning 1850–1970.* London: Arnold.

Sutcliffe, A. 1993: *Paris: an architectural history.* New Haven: Yale University Press.

17

POSTMODERN LANDSCAPES

> Los Angeles ... [is] built on the power of dreamscape, collective fantasy, and façade.
>
> (Zukin 1991, 219)

INTRODUCTION

The philosophical currents of social science have shifted in recent years. We will look at the methodological consequences of this for landscape studies in Chapter 18, but here we will carry forward our study of urban landscapes by looking at one major intellectual drift, *postmodernism*. In effect this chapter announces the rupture of the standardized, ordered lines of *modern* thinking and investigates a sample of the effects in the physical environment of rich countries at the end of the century.

POSTMODERNISM

Since 1980 a great deal has been written on postmodernity and postmodern cities. The literature is so large and diffuse that it is far from easy to encapsulate but in this chapter we will build on our discussion of modern cityscapes by giving an outline of the postmodern condition and its urban forms at the turn of the millennium.

It may be helpful to start with a distinction between postmodernism and postmodernity. Postmodernism is usually thought of as a style, particularly of architecture but also in art, design, literature and philosophy that has come to prominence since the 1970s. Its roots lie firmly in modernist experimentation with collage and montage in the 1920s, and the style is therefore an eclectic and fragmentary one which no longer adheres to the rules of modernist utility, simplicity and standardization, and which often juxtaposes incompatible designs in neighbouring buildings or districts (sometimes called *heterotopias*). Postmodern architecture is very concerned with the façade of a building, and for that reason it has sometimes been accused of superficiality, but the visual impression is often very interesting and entertaining due to the variety of shapes and bright colours that are used to enliven even the dullest of office blocks (Harvey 1989, 66–98; Gottdiener 1995, 119–37). The borrowing of ideas and visual tricks has provoked another common criticism, that postmodern architecture is pastiche. It is occasionally said to be *hyper-real* (Eco 1986), having meanings that stretch beyond the familiar realm of modernist symbolism. The Lloyd's building in London (*see* Figure 17.1) has its lifts and ducting visible to passers-by on the street, giving the impression that it has been turned inside-out, while the Gateshead Metro-Centre has shops arranged around a Roman forum to amuse and attract customers. Neither seeks to merge into an established context but rather they create their own blend of fiction and reality.

Postmodernity is the playing out of a new set of conditions in society and economy. These are complex but for our present purpose we can begin with a brief summary. The first point is that there has been a breakdown in the authority of the major theoretical and ideological underpinnings of much twentieth-century life. The monolithic voices of modernist socialism and

FIGURE 17.1 *Lloyds building, City of London*
Source: P.J. Atkins

FIGURE 17.2 *Docklands, London*
Source: P.J. Atkins

capitalism are crumbling, along with the world-view of objective science. Knowledge is now said to be more fragmented into separate units that have no overarching theories (*metatheories*) drawing them together, and which can no longer claim superiority over each other in their search for some elusive, absolute truth. This new relativism most encouragingly means that voices can be heard other than that of the white, male, Western, middle-class, English-speaking Protestant of the traditional stereotype of power. Feminism, multiculturalism and a post-colonial view of the world are all struggling to be heard.

Second, postmodern consumers can choose the place and the time they wish to experience. Television is the main vehicle for delivering a vast increase in the range of culture and entertainment into the living room but the design of buildings and other urban spaces has recently performed the same function by simulating exotic and historical styles, what has been called *elsewhereness*.

Third, there are significant shifts in what were once the secure social underpinnings of modernity. The glue that bound the family together for centuries, for instance, is now dissolving as a majority of marriages break up and traditional gender roles are no longer taken for granted. Lifestyles have greater flexibility and variety as advances in technology allow rapid communications and a range of consumption alternatives that could only have been dreamt of 50 years ago. Even individual identities, classes, sexual and cultural, are negotiable as never before in the modern era. Ironically, such is the speed of change in the postmodern era that the future is now less predictable than it has ever been: this is a further destabilizing element in what has become a disturbing and confusing experience for many people.

Fourth, what appeared to be the immutable foundation of the economy, a manufacturing sector based in large assembly line factories and producing standardized products for a mass market, has been questioned. Cracks have begun to appear in the highly successful but somewhat rigid Fordist (named after Henry Ford the car manufacturer) version of capitalism, and a more flexible mode of production has been born in agglomerations of smaller plants where vertical integration and scale economies, traditional concerns of Fordism, have been replaced by new financial arrangements, new types of sourcing of components, the recruitment of labour in new industrial regions, and a release of innovative energy to create new products. The resulting industrial landscapes (Chapter 15) are filled with relatively small factories and science parks rather than large plants that were common up to the 1970s.

Fifth, the innate fragmentation of post-modernity has geographical implications in a tessellation of cultural and political space. The nation–state, so characteristic of the centralizing tendencies of modernity, is losing credibility and not a little power to local formations of identity. The Soviet Union has split into its component parts, and even the unity of Federal Russia is threatened by secessionist tendencies in Chechnya. The bloody sundering of Yugoslavia and the inter-communal fighting in Bosnia-Herzegovenia are further testimonies to the struggle of submerged nationalities for self-determination.

This desire for the forging of identities through localization is the counterpoint of the parallel trend of globalization. A compression of time and space has taken place during the twentieth century through international transport links, computer-based global financial transactions, and world-wide telecommunications facilitated by satellites and cable links. A global culture, advertised and marketed by transnational corporations such as Coca Cola and McDonalds, is at a nascent stage of development (see Chapter 21).

Sixth, Jameson (1991) has argued that post-modernism is the 'cultural logic of late capitalism'. By this he means that culture has become like any other commodity that is packaged, sold and consumed. It is increasingly bound up with the money economy and no longer a separate dimension of life. A good example of this is the heritage industry which has mushroomed in Europe and North America in recent years, with historical experiences such as Yorvik, a reconstruction of Viking York, complete with smells of pigsties and recordings of children's songs in ancient Norse. In a sense this might be said to be fake history (Hewison 1987) but it appeals to the postmodern eclectic demand for small bites of cultural experience from any time or space. Tourists encourage the provision of such recreations and distillations of history by their wish to see and move on, but they are merely mirroring the habits of the television viewer who switches channels or the Internet surfer who moves from one web site to another. All are post-modern consumers whose desire for variety and stimulation is greater than their attention span.

If there is fake history, then can there also be fake spaces: landscapes without depth, context or meaning? California is probably the best place on earth to look.

Case study: Los Angeles and Disney

> There is nothing to match flying over Los Angeles by night ... You will never have encountered anything that stretches as far as this before ... The irregular, scattered flickering of European cities does not produce the same parallel lines, the same vanishing points, the same aerial perspectives either. They are medieval cities. This one condenses by night the entire future geometry of the networks of human relations, gleaming in their abstraction, luminous in their extension, astral in their reproduction of infinity.
>
> (J. Baudrillard 1988: *America*. London: Verso, 51–2)

Everyone knows that Los Angeles represents something different. For decades Hollywood has been manufacturing dreams for the rest of the planet and it is therefore a second spiritual home for us all. Perhaps fewer of us are aware that LA is the quintessence of car-orientated urbanism, a vast sprawl of a city where private motorized movement has become the way of maintaining an economic and social existence. It has been called 'a hundred suburbs in search of city' because of its decentralized structure, and there is a fascination in this fragmentation which Ed Soja (1989; 1996) picks up in his work.

The structure of LA is essentially threefold. There is a small downtown where much corporate and even global (for instance, Japanese) capital has been invested recently in providing financial services, and there are also administrative functions. The next zone out is a cluster of immigrant communities, of Latinos, Asians and other groups, who live in segregated and self-sufficient neighbourhoods. These ethnic enclaves are small reproductions of Third World cities near the heart of one of America's largest and most prestigious metropolises. Beyond them, in a series of satellite cities that Soja calls 'exopolis', there lies a belt of urban creations that in a sense resemble the Hollywood film sets across town. They are the packaged landscapes of post-suburban America (Knox 1992).

Exopolis is the true home of hyper-reality. Here the housing schemes and shopping malls are manufactured self-consciously as sanitized copies of some historical or exotic experience and for the residents this image becomes more important than the reality they might experience outside. Umberto Eco (1986) called these theme parks 'real fakes' because they are more reassuring and comfortable than the originals from which they are simulated. Here space is also fragmented into gated residential estates such as Mission Viejo, a 250 000 ha development sponsored by the tobacco corporation Philip Morris. Here there are 30 000 homes, most built in a Spanish architectural style, with golf courses, parks, lakes, shopping centres and businesses that provide a balanced community of employment, services and leisure. Mission Viejo is self-contained and has a pleasant, crime-free environment. All tenants have to sign a contract that restricts them from cleaning their car in the street and from painting their house a colour that does not please their neighbours. These restrictions are thought to be a small price to pay by most people.

Walt Disney was a film entrepreneur who made the animated cartoon one of the most successful Hollywood products, from the 1920s, using characters such as Mickey Mouse. Disney was more than just a film maker, however. He had a vision of American cultural values that he wished to share with the cinema audience and also visitors to his theme parks: Disneyland® Resort, California (opened 1955) and Walt Disney World® Resort at Orlando, Florida (1971). Disneyland was an extraordinary reconstruction of elements of the vernacular landscape, carefully selected to purvey an image of homely and folksy small-town middle-America (Sack 1992, 162–68; Kunstler 1993; Gottdiener 1995, 99–118). Main Street USA is a central feature joining a number of amusement parks. It was said to be 'what the real Main Street should have been like', and the park is 'Disney realism, sort of Utopian in nature, where we carefully programme out all the negative, unwanted elements and programme in the positive elements' (Zukin 1991, 222) (*see* Figure 17.3).

Apart from a set of rides and amusements, Walt Disney World includes EPCOT® (The Experimental Prototype Community of Tomorrow, partially opened in 1982) which, according to Disney:

> will be a planned, controlled community, a showcase ... There will be no landowners and therefore no voting control. No slum areas because we will not let them develop. People will rent houses instead of buying them ... There will be no retirees. Everyone must be employed.
>
> (Zukin 1991, 224)

This was to be authoritarian paternalism writ large on the American landscape but in fact it was

FIGURE 17.3 *Main Street, USA, Magic Kingdom, Walt Disney World, Florida © Disney*

not completed in the way originally intended. Disney foresaw an entirely enclosed city but the town actually started in 1994, called Celebration, was more in the mould of his earlier vision of Main Street. It will be on a human scale, with grocery stores rather than supermarkets, and a full range of community services provided within walking distance. The school will give each pupil a personal learning programme and the hospital has the aim of encouraging the citizens' *wellness* rather than just treating disease. The whole development, when it is completed early in the next millennium, will be an experiment in the application of traditional community values to a postmodern landscape. The idea is so popular that lotteries have to be held to decide who the lucky tenants will be.

The consumption of fantasy spills over from the amusement park to the manipulated space of Celebration. Disney, who was criticized before his developments of Disneyland and Walt Disney World began, has been proved remarkably perceptive in his dual foresight of a trend towards themed postmodern spaces and his idea that the global corporations might come to replace the state as the provider of amenities for a section of the population.

FIGURE 17.4 *Roman Forum, MetroCentre, Gateshead*
Source: P.J. Atkins

CASE STUDY: SHOPPING MALLS AND MALLEABLE CUSTOMERS

> One person's 'text' is another person's shopping centre or office building.
>
> (Zukin 1996, 43)

Shopping has become a leisure activity, in fact the second most important one after watching television (Goss 1993). Despite the convenience of supermarkets and malls and advertisements claiming that they save time, in fact we spend more time shopping now than ever before. The design and ambience of shops and shopping centres seek to make the experience a pleasurable and guilt-free one so that we will revel in conspicuous consumption and be seduced into spending more time and money on shopping (Sack 1992, 144). This urban form, then, affects all of our lives and is of increasing economic

FIGURE 17.5 *MetroCentre, Gateshead*
Source: P.J. Atkins

significance, but it is also important for the social and cultural messages that it conveys (Gottdiener 1995, 81–98).

Shopping centres as single planned units have their prehistory in the America of the 1920s but they did not proliferate until after the Second World War. The first mall to be fully under cover opened in 1956 at Edina, Minnesota and the idea caught on in the 1970s in North America and the 1980s in Europe (Judd 1995). This type of shopping is still a novel experience for most Britons, for whom the predictably warm, dry environment, away from the weather outside, is of course one of the principal attractions.

All malls are controlled spaces. Congestion is controlled by the manipulation of pedestrian flows, and crime minimized through the use of security guards and CCTV surveillance (Chapter 16). Also there are usually rules about the exclusion of undesirables and the prevention of anti-social behaviour such as drinking and dropping litter. Some malls in the US have banned anyone under the age of 18 and others require teenagers to be accompanied by an adult. Making speeches or distributing political leaflets is generally frowned on and collectors for charity may be discouraged if they are thought to embarrass or annoy shoppers.

The larger malls offer a large range of services in addition to the retailing, from leisure facilities to hotel accommodation. They are *cities within cities*, although a residential element is not common. For Judd (1995) 'typically they reproduce a stylized, romanticized, even fairy-tale interpretation of city architecture and culture. They attempt to make perfect what is flawed.' For Goss (1993, 33) they are 'artificial fantasy worlds hermetically sealed against the unsanitary and unsafe outside world'.

Shopping malls are a form of institutionalized spectacle that ultimately have a similar objective to the Disney experiences that we discussed above. In the malls they are selling material goods instead of pure entertainment but the methodologies have many common features. Both are ideologically charged landscapes that are complex to read but which deserve attention for their considerable cultural energy.

There were 30 000 malls in North America by the mid-1980s. They have become familiar and trusted but that has not protected them from the economic fluctuations that beset all capitalist investments. The largest ones may be threatened if their image becomes at all tarnished and this innate risk has meant a new trend in the last ten years towards smaller, more flexible and more specialized malls, which are more responsive to local conditions and their local clientèle.

Conclusion

Shopping malls are the postmodern equivalent of the nineteenth-century Parisian arcades described by Walter Benjamin (Chapter 16). But they are carefully planned and controlled spaces, not the anarchic extravaganza of cultural expression without any underlying purpose that is sometimes attributed to postmodernism. One might very reasonably argue that the theme parks, malls and gated residential areas of California are merely the latest manifestation of corporate capitalism and, as such, nothing fundamentally new. What is novel rather is the extraordinary explosion of culturally based work in the social sciences which has been brought to bear upon the investigation of contemporary landscapes. This exciting *cultural turn* is yielding insights that are shifting the centre of gravity of disciplines such as human geography and Chapter 18 will introduce some of the ideas for landscape study that have emerged in recent years.

Further reading and references

Harvey's book is the best single source but Sharon Zukin's work is also helpful.

Eco, U. 1986: *Travels in hyperreality*. New York: Harcourt Brace Jovanovich.

Goss, J. 1993: The 'magic of the mall': an analysis of form, function, and meaning in the contemporary retail built environment. *Annals of the Association of American Geographers* **83**, 18–47.

Gottdiener, M. 1995: *Postmodern semiotics: material culture and the forms of postmodern life*. Oxford: Blackwell.

Harvey, D. 1989: *The condition of postmodernity: an enquiry into the analysis of cultural change*. Oxford: Blackwell.

Hewison, R. 1987: *The heritage industry: Britain in a climate of decline*. London: Methuen.

Jameson, F. 1991: *Postmodernism, or, the cultural logic of late capitalism*. Durham: Duke University Press.

Judd, D.R. 1995: The rise of the new walled cities. In Liggett, H. and Perry, D.C. (eds) *Spatial practices: critical explorations in social/spatial theory*. Thousand Oaks, Calif.: Sage, 144–66.

Knox, P.L. 1992: The packaged landscapes of post-suburban America. In Whitehand, J.W.R. and Larkham, P.J. (eds) *Urban landscapes: international perspectives*. London: Routledge, 207–26.

Kunstler, J.H. 1993: *The geography of nowhere: the rise and decline of America's man-made landscape*. New York: Touchstone.

Sack, R.D. 1992: *Place, modernity, and the consumer's world: a relational framework for geographical analysis*. Baltimore: Johns Hopkins University Press.

Soja, E. 1989: *Postmodern geographies: the reassertion of space in critical social theory*. London: Verso.

Soja, E. 1996: *Thirdspace: journey to Los Angeles and other real-and-imagined places*. Oxford: Blackwell.

Zukin, S. 1991: *Landscapes of power: from Detroit to Disney World*. Berkeley: University of California Press.

Zukin, S. 1995: *The cultures of cities*. Cambridge, Mass.: Blackwell.

Zukin, S. 1996: Space and symbols in an age of decline. In King, A.D. (ed.) *Ethnicity, capital and culture in the twenty-first century metropolis*. Basingstoke: Macmillan, 43–59.

18

METAPHORS AND MEANINGS IN MODERN LANDSCAPES

READING THE LANDSCAPE

There is no stone in the street and no brick in the wall that is not actually a deliberate symbol – a message . . . as much as if it were a telegram or a post-card.
(G.K. Chesterton, quoted in D. Schmid 1995: Imagining safe urban space: the contribution of detective fiction to radical geography. *Antipode* **27**, 242–69)

INTRODUCTION: METAPHOR AND MEANING

At one level, it is possible to *read* the landscape as if it were a text or a painting. Even the most humble, innocuous-looking landscape carries with it cultural meaning, sometimes clearly visible but often in a sort of code that is absorbed by local inhabitants at a subconscious level. This is why exotic areas, outside our normal, everyday experience are difficult for us to understand and appreciate, and sometimes even unsettling and threatening. Yet we cannot say why our own normal surroundings are so comfortable by comparison.

In ancient and complex societies in particular, the landscape is a composition of hundreds or thousands of years of small additions and modifications, and sometimes large-scale changes, overlaid on top of each other. It is therefore a cultural *artefact* or construction and acts as a record, not only of people's physical efforts in clearing the wood or draining the marsh, but also of their ways of thinking, the technology they used, and even of their social and political structures.

In studying a long-lived landscape many of the insights may have to be by indirect inference, unless we can date archaeological remains or find other concrete historical evidence. But to those who are trained in its interpretation, the landscape itself is a kind of document, rather like a medieval *palimpsest* (*see* Introduction) on which the handwriting may be difficult to decipher due to erasures and superimpositions.

To continue this metaphor, the *text* of the landscape may have been composed in a vocabulary and with a grammar that may be archaic, so that reading it is like rediscovering a dead language. This is because much of the original landscape may have disappeared, leaving only isolated relict features. Thus most lowland Bronze Age and Iron Age landscapes have been ploughed out by farmers from later periods, leaving only evidence such as burial mounds and hill-forts. These are no longer functional but nevertheless they tell us a great deal about prehistoric society and polity.

At the subconscious level, landscape influences us on a day-to-day basis. Like an electric fire it radiates comfort at us, and we bask in the familiarity almost without noticing it. This is because of a subtle inculcation of the ideas and feelings that go together to make up a culture (Duncan and Duncan 1988, 121). The recipients of these subliminal messages may be unaware of this process of *naturalization*, by which familiar objects and their relationships become second nature, because of many years of repetitions which come to be accepted unquestioningly as normal.

Complex and large-scale societies are only stable and prosperous when the majority of their citizens are willing to adhere to the rules and norms of behaviour that evolve over time. These structures of life are consciously and unconsciously negotiated through political, cultural and social institutions such as religion and education, and also by means of the written word, verbal messages and by visual communication through television, art, and even static objects in the landscape. As far as the landscape is concerned, there are many signs and symbols which inform and persuade people to behave in a certain fashion. To give a few examples, we know, usually without having to be told, that we must keep off the grass in a formal park; a learned reverence for religion in most societies forbids people to be noisy or disrespectful in the grounds of a temple or church; a barbed wire fence implies ownership and a prohibition to trespassers; and golden arches may make us salivate whether or not we can see a burger restaurant. All of these landscape elements trigger an automatic response and in a sense our lives are directed and constrained by the meanings of the landscapes that we live in and move through. Such meanings in turn are decided by cultural discourse (Box 18.1).

Landscapes may also 'speak to notions of how the world should be, or more accurately how it should appear to be' (Cosgrove 1989, 104). Thus we are heavily influenced by conservative media images of unspoilt countryside, and may choose to holiday there or perhaps move to live there, but at the very least we seek to preserve it and protect it from development. The moral element here is deeply ingrained yet it is also extraordinarily complex because the countryside is in fact not unspoilt but rather the product of intrusive actions by former generations. Anyway, development is not always bad because country people need jobs.

Box 18.1 *Some definitions*

Discourses are the complexes of social and cultural meanings that help us to interpret the world. They are the ideas, understandings, rules, practices, and ideologies that are shared by different groups of people. They are constantly evolving by the processes such as negotiation, debate and coercion. The material objects which precipitate out of discourses are loaded with meaning, but that meaning may be interpreted differently according to context.

Iconography is the study of images with a view to uncovering their inner message through a qualitative analysis. It started as the interpretation of paintings where the artists had deliberately included certain qualities to convey meanings.

Semiology or semiotics deals with signs and symbols, and thereby helps us to understand landscapes, which can be seen as systems of signs.

Deconstruction considers texts (such as landscapes) from the standpoint of the viewer, considering the different views of the sexes, classes, races, religions and cultures. This helps us to appreciate that no two individuals have exactly the same view of what they see. This is important because it shows that the concept of landscape is slippery and unstable.

Ordinary landscapes are created or refreshed on an everyday basis by the individual actions of ordinary people. Élite groups may appropriate much of the space produced, thereby marginalizing others, and they often lead with new ideas about desirable change but, in the final analysis, it is society as a whole which will approve or reject the propositions put to it in the form of new landscape elements. If acceptance is granted, then, in the words of Rotenberg (1995, 6): 'landscape forms [will] constitute a system of socially recognized signs that enter into people's conversations about their understanding of nature, the city, social power, and their experiences with these realities'.

CASE STUDY: THE KANDYAN KINGDOM

The island of Sri Lanka became the object of interest for Western powers for its resources, especially spices such as cinnamon. In the seventeenth and eighteenth centuries the Sinhalese retained a semblance of independence from European rule in its interior kingdom of Kandy. The Portuguese, Dutch and then British controlled the coastal margin (*see* Figure 18.1) and made occasional forays against Kandy. Eventually the British conquered the whole island in 1815.

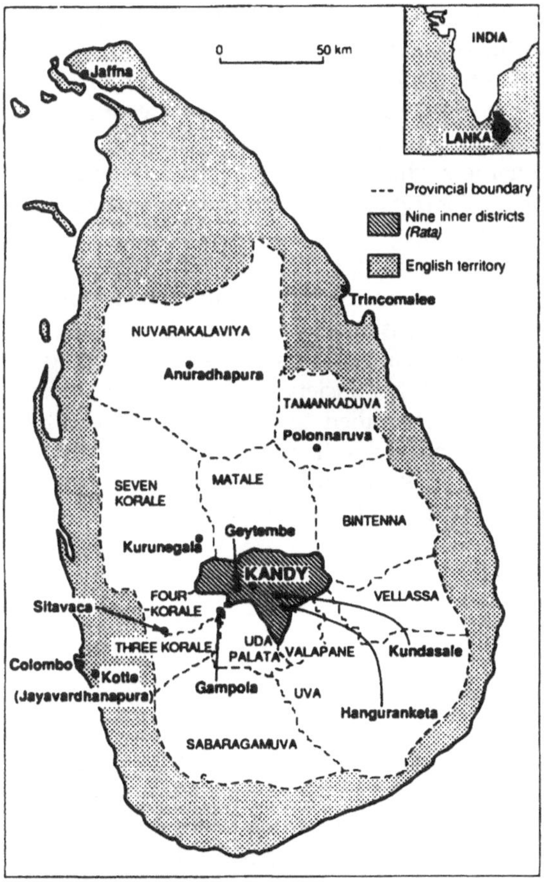

FIGURE 18.1 *Lanka in 1815*
Source: Duncan, J.S. 1990: *The city as text: the politics of landscape interpretation in the Kandyan kingdom*. Cambridge: Cambridge University Press

At the time of its fall Kandy had achieved a maturity of townscape that was based upon the symbolism of kingship (Duncan 1990). There was a Sinhalese consensus that texts held an important authority for their society and thus it was that two narratives had an important bearing upon the layout of the city.

It was in the king's interests to create a landscape that would impress his subjects by its magnificence and also convince them of his legitimacy by a faithful rendering of his religious responsibilities. The latter involved building suitable sacred structures and spaces and the city was therefore planned to resemble the Buddhist cosmos in miniature form (*see* Figure 18.2). The city's morphology was a mirror of the celestial city of Sakra on the sacred Mount Meru, divided

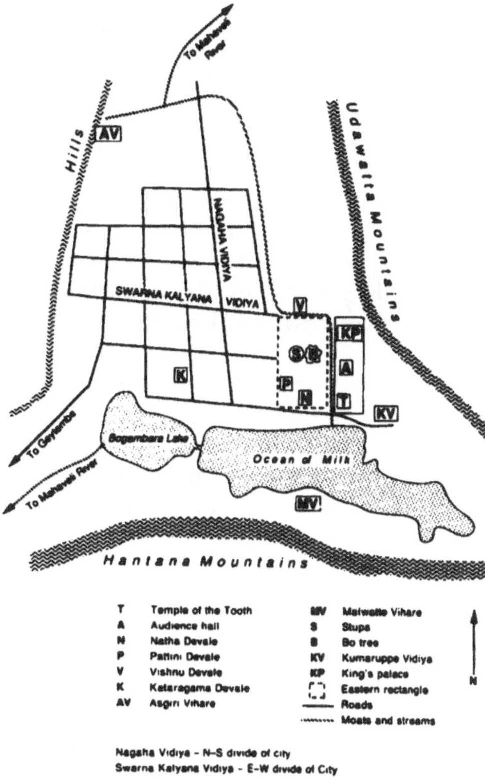

FIGURE 18.2 *Kandy in 1815, after L.J.B. Turner (1918)*
Source: Duncan, J.S. 1990: *The city as text: the politics of landscape interpretation in the Kandyan kingdom*. Cambridge: Cambridge University Press

into four quarters, with four shrines and four gates, and with the corners of its rectangular shape pointing in the four cardinal directions. The cosmic axis located the city at the centre of the world and gave it a direct connexion with heaven. The lake also had a parallel in the story of the Ocean of Milk, an allegory of the creation and renewal of fertility.

King Sri Vikrama defeated the British in battle in 1803 but was unable to impose his authority over his whole kingdom and its nobles. He therefore embarked on a building programme at Kandy which he no doubt hoped would embellish his reputation and establish the charisma of a god-king. This involved several years of forced labour by his subjects to dam a stream and thus enlarge the lake, to increase the number of streets in the city, and to extend the royal palace. Unfortunately for him, Sri Vikrama's calculations went awry because his people did not read his new landscape in the way intended. For many of them the cosmic balance of the city was upset and the kingly interpretation went beyond his rights and duties. The readings of the nobles and the peasants were just as political as that from the throne and internecine conflict seemed inevitable. The British invasion of 1815 was made possible by the defection of a number of nobles disaffected by Sri Vikrama's rule.

The analysis of landscape in these terms is rarely so clear as for historic Lanka because of the many complexities that are overlaid in the modern and postmodern worlds. Today we are confronted by many possible alternative *readings* of our surroundings and one purpose of cultural geographers and other social scientists is to provide insights into what these analyses might be.

IDENTITIES OF TOGETHERNESS/OTHERNESS

Nations may be bound together by a collective view of landscape. This is often mythological but no less powerful for that. For Australians the *outback* is a shared experience that has forged their self-view of a nation of rugged individualists who have battled against inhospitable nature in the dry heart of their vast country. Most visit it only occasionally but know what to expect when they do because they live with it constantly in the representations of literature, painting, films and television.

Landscapes may also, however, represent *otherness* (*see also* Chapter 20). The most penetrating example of this is the representations of the countries that they had conquered that were made by the European imperialists, especially the French and the British. These became dominant perceptions and were reproduced in school textbooks and literature to such an extent that they were taken for granted in the Western world. Edward Said (1978) has written about the power of Western views of the colonies in his unpacking of the concept of orientalism. Not only were the subject peoples oppressed by the set of stereotypes imposed on them but they were also put into an unwelcome category of difference, of otherness which has stuck, even after the process of decolonization was completed after the Second World War.

Other landscapes or indeed familiar landscapes are in a sense created out of the prejudices, preconceptions and simplifications that we carry in our heads. In turn these representations may be naturalized by society and accepted without question. They then become the basis on which subsequent generations forge their identities. This culture–landscape–culture feedback loop is a principal means by which we see others and see ourselves.

CASE STUDY: THE HILL STATION IN BRITISH INDIA

During the British raj in India an extraordinary set of landscapes were established which are symbolic both of the imperialists' view of their realm and of their own part in it (Kenny 1995). The 80 or so hill stations (*see* Figure 18.3) were established ostensibly as summer retreats for the colonial masters to get away from the worst of the tropical heat and the disease associated with the monsoon season. The most important were Ooty, Simla, Poona, Darjeeling, hill stations for Madras,

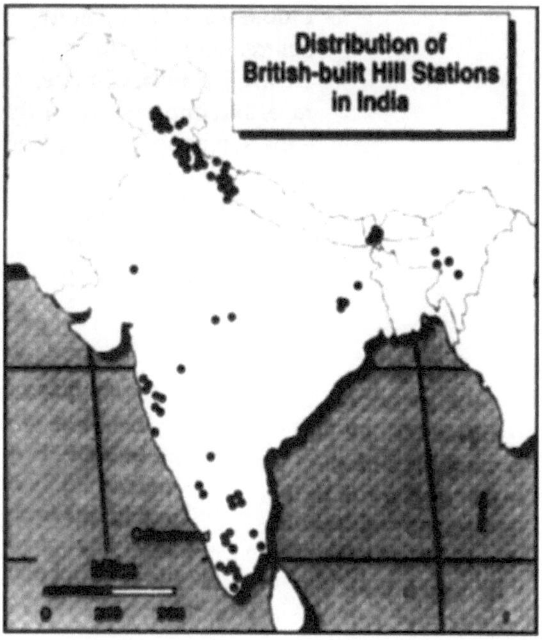

Figure 18.3 *The main hill stations of India*
Source: Reprinted from Kenny, J.T. 1995: Climate, race, and imperial authority – the symbolic landscape of the British hill station in India. *Annals of the Association of American Geographers* 85, 694–714

Figure 18.4 *The Ooty Club*
Source: P.J. Atkins

Agra, Bombay and Calcutta, respectively: the litany of names is enough to recall a life of pampered leisure away from the responsibilities of power. Some of the Britishness survives in these places, with English-medium public schools and exclusive clubs (*see* Figure 18.4) still reproducing the élitist experience, but for the last 50 years with a different clientèle.

We may argue that the hill stations provided a number of functions other than a cool climate. They were sparsely settled by indigenous people and could therefore be made into micro-scale British-style landscapes without disturbing local sensibilities. But perhaps more important they established a physical separation between the rulers and the ruled. This was born out of nineteenth-century views of racial difference, which emphasized the greater efficiency of Europeans in cooler conditions, but it also helped the British define themselves in terms of their own choosing.

In the case of Ootacamund (Ooty), the official seasonal residence of the government of the Presidency of Madras from 1870, these terms were established by John Sullivan in the early nineteenth century. His vision was of a pastoral Utopia, with European species of trees, flowers, fruits and vegetables, a serpentine lake similar to those found in English landscape gardens, and the full range of facilities from churches and schools to parks and clubs (*see* Figure 18.5). In appearance it must have seemed like a spa, although the health-giving properties were not always guaranteed, as a cholera epidemic in 1877 showed.

By 1901 the permanent population of Ooty was 19 000, doubling during the summer. The town reached its zenith in the early twentieth century but after the First World War it attracted the attention of Indian nationalists, who emphasized its exclusion of Indians from hill station society, even Maharajahs. In the 1930s the official summer exodus to the hills was stopped and Ooty retained only its function as a private resort.

The hill stations of the British empire were encoded with power. Within the settlement itself status was related to elevation, with the Governor having the most splendid residence on a high slope and the native population living huddled on low ground where drainage was poor. In the regional context the modest infrastructure was mobilized at great expense for the benefit of a relatively small number of sahibs. This was a

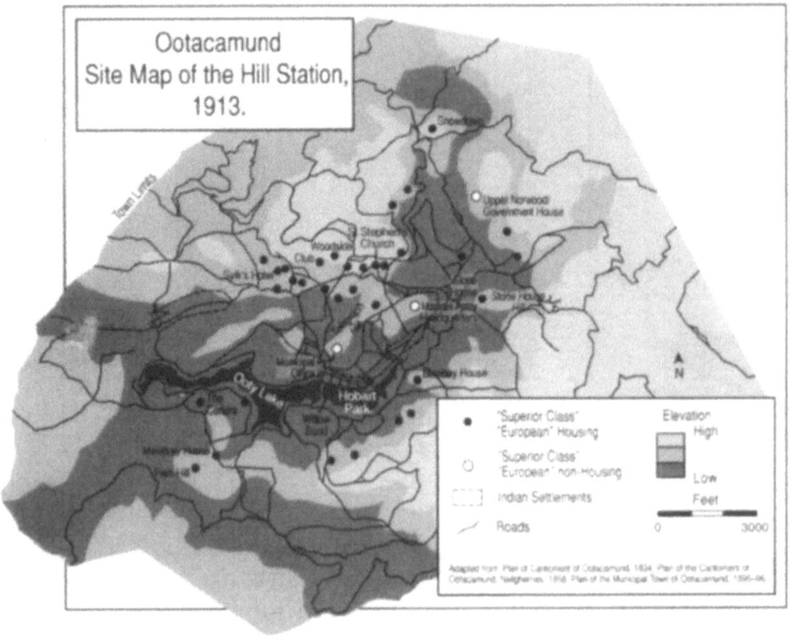

FIGURE 18.5 *Ootacamund*
Source: Reprinted from Kenny, J.T. 1995: Climate, race, and imperial authority – the symbolic landscape of the British hill station in India. *Annals of the Association of American Geographers* **85**, 694–714

constant reminder to Indian people that their rulers were both separate and sufficiently powerful to command resources in a display of conspicuous consumption. Little Britain simulated was therefore India subdued.

THE POWER IN LANDSCAPES

It is also possible to argue that landscapes are not just passive objects that lie silently waiting for successive human imprints. In a sense they are active participants in channelling socio-economic evolution because they set the physical and psychological constraints within which people must act. A crude example of this might be the very distinctive landscapes (buildings, monuments, transport systems) of the Soviet Union or Nazi Germany, which were dripping with the symbolism of state power to such an extent that ordinary citizens were oppressed by their physical surroundings. The suggestion of this approach is that all landscapes transmit messages of power, subtly in most cases, that reinforce and reproduce existing cultural norms and class structures.

CASE STUDY: IDEAS OF LANDSCAPE IN EIGHTEENTH- AND NINETEENTH-CENTURY BRITAIN

In the eighteenth century Britain participated in the Age of Enlightenment. Harmony and rational values were prized and much of the lead in society was given by the ideas of philosophers, writers, artists and poets. European influence was strong and young male aristocrats made the *Grand Tour* to Italy to breathe in the ancient culture and returned enthused by classical architecture and literature. Education at this time was dominated by Latin and Greek, so many were predisposed to think of the ancient Mediterranean as having experienced a golden age of civilization, and it became fashionable for the wealthy to express their taste by commissioning neo-classical buildings (for instance in the style of Palladio), geometrically designed formal gardens, and painting with classical allusions (Poussin, Claude and imitators). One source of inspiration was Virgil's *Georgics*, a model of how the wealthy and powerful should behave. Virgil suggested a combination of beauty and use, pleasure and profit, within a strict but benevolent social hierarchy.

A romantic school of pastoral painting flourished, with bucolic scenes of cows grazing on peaceful meadows matched later by Constable's chocolate box icons, the Hay Wain (1820–1), Salisbury Cathedral (1822–3) and the Cornfield (1826) (Bermingham 1990). Previously patrons had preferred paintings of country houses or horses (by artists such as Stubbs). It was an unthreatening world that was depicted, with all signs of rural poverty eliminated and even the farm labourers reduced to the background as miniatures.

This rural idyll has become a peculiarly British dream which has survived down to today (Williams 1973). Britons are obsessed by the countryside even though few actually live there, and many have strong views on how it should look, no doubt much to the annoyance of farmers and others who may be refused planning permission for unsightly but functional developments.

A related cultural focus was the more intellectually aesthetic philosophy of the picturesque and the sublime. Sir Uvedale Price (1794), Richard Payne Knight (1794) and others promoted landscapes in which the beauty perceived was that of roughness and variety, as in wild and rugged mountain landscapes. Some were not satisfied with what nature provided in the flat and to them boring areas and they set out to discover the picturesque. Thus began tourism in Britain, particularly with the discovery of the Lake District, North Wales, and the Scottish Highlands, and these areas have been popular ever since (Chapter 20).

The need for the romantic and the picturesque was a direct counterbalance to the horrors of the Industrial Revolution which had so shocked the nation from 1750 onwards. There was a revulsion at the noise, pollution, disease and poverty of the new and rapidly growing urban industrial areas. The Victorian city in particular was commonly seen as a *threat* to civilization, with its social problems of crime and vice, and the countryside was an important psychological counterbalance with its pleasant and wholesome life. This perception was erroneous because rural slums

FIGURE 18.6 *Mr and Mrs Robert Andrews, Thomas Gainsborough (1748–50)*
Source: reproduced by permission of the Trustees of the National Gallery

were just as bad as those in the towns and life was very hard indeed for agricultural labourers and their families.

The counterbalance to views of the beauty and harmony of both wild nature and of the pastoral was the eighteenth-century notion of rational improvement for profit and pleasure. This Whig idea was the basis on which élite landscapes were designed for the wealthy and powerful in the form of country parks and gardens. Landscape parks were at their peak of numbers, size and prestige in the eighteenth and early nineteenth centuries. They were symbolic of the power of the ruling class in taming the landscape to suit their taste, and it is not accidental that many landowners chose to have their portraits painted standing in the grounds of their estate (*see* Figure 18.6) (Berger 1972, 108; Daniels 1993, 80).

Eighteenth-century garden designers mostly rejected straight lines or the geometric shapes of the Tudor knot garden. Equally, they disliked the formal gardens of Italy or France, such as Louis XIV's Versailles. But classical allusions were *de rigueur*, such as temples, statues and groves, and it was not uncommon to create romantic ruins and even employ a hermit to live in them. The magnificence of such gardens is still breathtaking today, not least when one considers the cost and effort of design and construction. In the case of Stourhead (*see* Figures 18.7 and 18.8) this was funded from the sizeable fortune (made from banking and slave trading) of Henry Hoare (1705–85). He based the symbolism of the garden's design on Virgil's epic poem the *Aeneid* and the influence of the landscape painter Claude is clear. Lord Cobham, at Stowe in north Buckinghamshire, was more explicit in his patriotic symbols, building a Temple of Ancient Virtues, a Temple of British Worthies, a gothic temple dedicated 'to the liberty of our ancestors', and a Temple of Friendship.

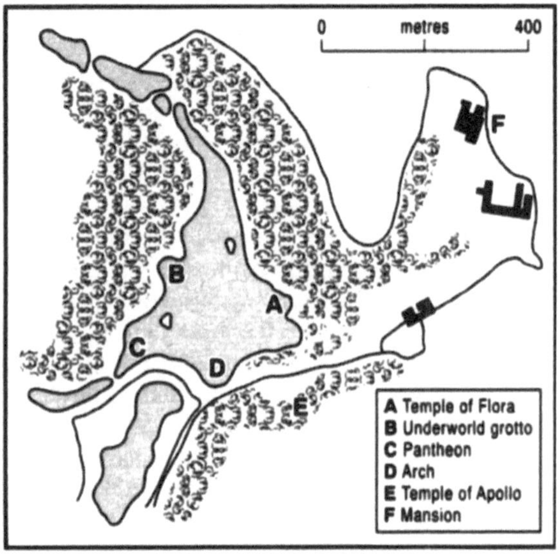

Figure 18.7 *Stourhead gardens, Wiltshire*

Figure 18.8 *The Temple of Apollo, Stourhead*
Source: P.J. Atkins

Garden design was a highly valued art and some of its practitioners became famous and influential, for instance Sir John Vanbrugh (1664–1726), Charles Bridgeman (d.1738), Stephen Switzer (d.1745), William Kent (1685–1784), Lancelot 'Capability' Brown (1715–83) and Humphrey Repton (1752–1818). Their skill was partly the practical ability to organize the civil engineering of earth moving and water control, but there was also a need for botanical knowledge of the species of plants and trees that would flourish, and a painterly ability to visualize a transformation of the landscape. On this last point, Horace Walpole declared that 'an open country is but a canvas on which a landscape might be designed' and Alexander Pope thought that 'all gardening is landscape painting'.

Box 18.2 *Techniques of changing and improving nature*

In addition to the generic grass, clumps of trees and expanses of water there were:

- curving rather than straight lines for roads and canals;
- temples, colonnades and obelisks to terminate a vista;
- serpentine walks, lakes, plantations and clumps of exotic trees;
- gothic follies, grottoes and temples;
- chinese artefacts (chinoiserie) such as pagodas, lattice work bridges;
- a ha-ha or sunken fence to abolish the visual separation of house and garden/park;
- a *ferme ornée*, a combination of woodland, pasture and tillage; and
- many new species of ornamental plants such as magnolia, rhododendron, hydrangea, buddleia, fuschia.

William Kent started as a painter, had spent several years in Italy, and was himself a collector of Italian landscape paintings. His gardens were still partly formal. Capability Brown designed 188 major landscape parks, including many of the great houses, such as Blenheim and Chatsworth. Humphrey Repton had 220 major commissions, including Regent's Park. These designers were working mainly for rich landed magnates, although Repton towards the end of his career was reduced to designing small gardens.

The landscapes created were exclusive and élitist. They made a variety of social and even political statements. For instance, there was an element of naked power in moulding the landscape to the taste of the landowner. It made a clear statement to all of the tenants and workers that he was firmly in charge, and was backed up by laws such as the Black Act of 1722 which made it a hanging offence to 'cut down or otherwise destroy any trees in any avenue, or growing in any garden, orchard or plantation, for ornament, shelter or profit' (Daniels and Seymour 1990).

In order to create the extended vistas of a large landscape park there was inevitably a sweeping away of existing fields and farms, and sometimes even inconveniently sited villages. Vanbrugh demolished Henderskelfe church, castle and village in his making of vistas at Castle Howard and three villages perished for Stowe gardens.

The trees planted in some parks were an important economic resource. Several parks in the Dukeries (Nottinghamshire), for instance, yielded considerable income from the sale of timber, pit props for the nearby coal mines, hop poles, bark and charcoal (Seymour 1989). But trees also had a symbolic function. Growing oaks was seen as patriotic because it helped provide timber for the navy's ships and they were also a potent symbol of landed authority. They were considered to be venerable, patriarchal, stately, guardian and quintessentially English, like the landed families themselves (Chapter 6). The quicker growing conifers were seen as vulgar, grown only on the estates of the new breed of greedy, money-making, *parvenu* industrialists.

By the early nineteenth century some parts of the country had the very firm imprint of these designed landscapes, but after that few new landscape parks were commissioned other than the municipal parks that were created in many Victorian cities. However, the prestige of garden design and plantsmanship was by now well established amongst the burgeoning middle classes and the twentieth century has seen several designers such as Vita Sackville-West and Gertrude Gekyll who have influenced generations of ordinary gardeners. Even the smallest suburban garden now has its share of bedding plants and herbaceous perennials, and many have scaled-down features of historic landscape gardens in the form of water features, clumps of shrubs and back lawns.

To conclude this case study, we have argued that a particular *view* of landscape, called by some writers *the gaze*, conjured by the combination of intellectual, cultural and other circumstances of a particular period, helps people to re-evaluate their surroundings and, for the powerful and wealthy at least, to impose new limits of taste that have a profound environmental impact.

Ideological landscapes

The power innate in most landscapes is subtle and often invisible behind coded signs. It may be a world of the powerful spirits of ancestors or perhaps a newly settled region interpreted through imported lenses of religion or culture. Throughout history there have also been examples of deliberate expressions of the control of nature in the ideological frameworks of political movements. We may then have concrete expressions in the world of a manifesto which seeks to impose its will or celebrate a particular perception of improvement. This is well put by J.B. Jackson (quoted in Meinig 1979) 'Landscapes are expressions of a persistent desire to make the earth over in the image of some heaven.' In the twentieth century the Nazis in Germany and the Communists in the Soviet Union and other countries have had clear visions of desirable landscape change coupled with a ruthless will to succeed, brooking no opposition.

Case study: the North Korean landscape

The attitude of the ruling Communists in North Korea towards their environment must be seen in the context of twentieth-century Korean history, which has been a series of momentous struggles. The Japanese colonial phase (1905–45) was a particularly difficult time, with widespread and systematic cruelty, exploitation and repression, and the Korean War (1950–53) was appallingly destructive. Ever since, North Korea has seen itself as an embattled state, with few friends. A hostile natural environment, with a mountainous topography that restricts the amount of arable land, and climatic extremes swinging from severe floods one year to drought the next, is but another enemy to be defeated by socialist ingenuity. For Kim il Sung, the leader from 1945 to 1994, 'it is the duty of communists to master and remake nature'.

From the 1950s Kim developed his *juche* philosophy, a curious blend of Marxism–Leninism, Stalinism, and Korean nationalism, which stressed the self-reliance of Korea and her need to solve her own problems. A central tenet is that humans are not constrained in their ability to modify nature, other than the limits of the technology they are currently employing. Juche Korea has boundless optimism about its powers of intervention and control and relatively less concern about the negative environmental consequences of its actions, unless production is thereby curtailed, for instance where excessive tree clearance and hillslope terracing are reported to have caused soil erosion.

It is not possible to approach North Korea with an innocent eye. The ideology of juche and communism permeates every city, every village and every field. The country may not have achieved a high standard of material wealth but the impact of its polity upon the landscape has been extraordinary. Perhaps their biggest achievement in remaking nature has been the North Koreans' outstanding record in irrigation (*see* Figure 18.9). The irrigation of 1.4 million hectares of cultivable land has been completed, about 70 per cent of the total and one of the highest proportions anywhere in the world. This has been achieved by the planning of a complex system of 1700 reservoirs fed by 25 800 pumping

Box 18.3 *Problems with coastal reclamation in the DPRK*

1. The cost of the construction work has been a drain on the meagre national budget.
2. The salty tideland soil has required more fresh water for flushing than the nation can reasonably afford.
3. Labour has been in short supply because the DPRK keeps a standing army of over one million, the fifth largest in the world.
4. There will inevitably be major environmental disturbance as a result of these reclamation works. These will be ecological and geomorphological.
5. There have been open political tensions about the priority which the tideland projects should receive in the overall economic planning process.
6. The yields achieved on reclaimed sites have been disappointing, only 4–5 tonnes of rice per hectare compared with the national average of 7.6 tonnes.

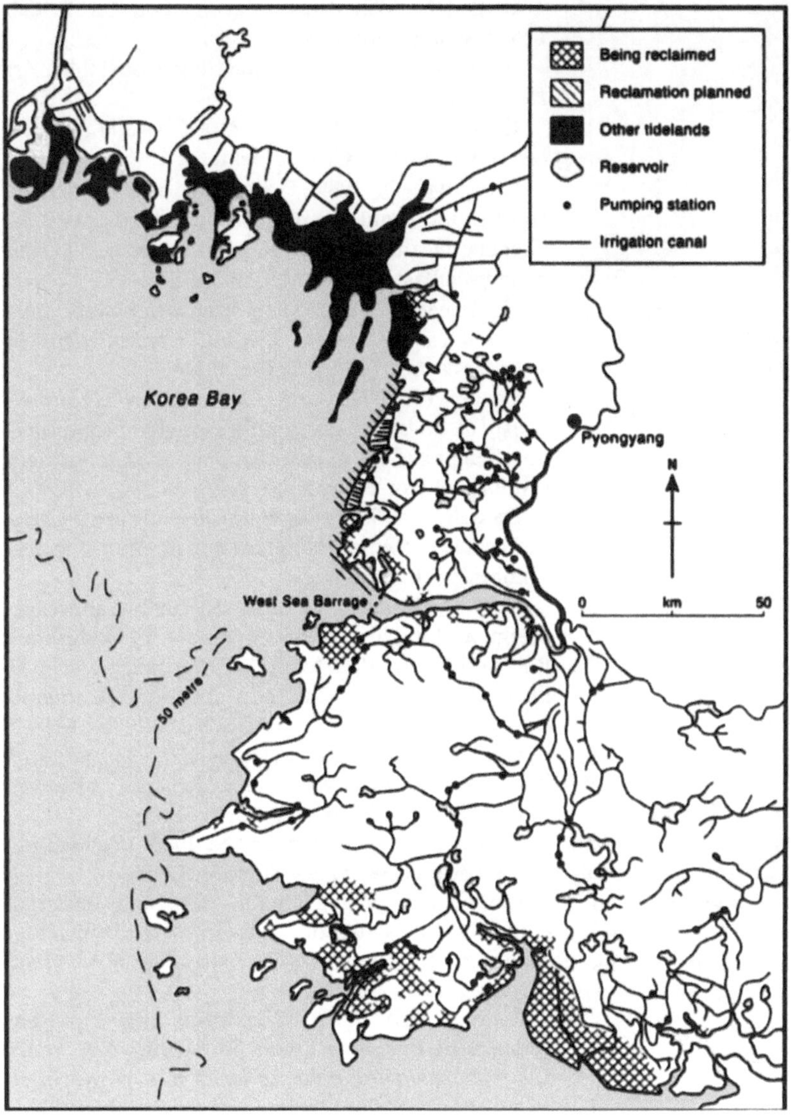

FIGURE 18.9 *The irrigation pumping network and coastal reclamation in North Korea*
Source: Atkins, P.J. 1993: The dialectics of environment and culture: Kimilsungism and the North Korean landscape. In Mukherjee, A. and Agnihotri, V.K. (eds) *Environment and development: views from the East and the West*. New Delhi: Concept, 309–31

stations, and 40 000 km of major irrigation canals. Especially important in the recent expansion of the irrigated area has been the construction of the West Sea Barrage near Nampo (*see* Figure 18.10). This stretches 8 km across the mouth of the Taedong River, and ponds back the fresh water of the river to create a large lake. Apart from irrigation, this water is also used for domestic and industrial purposes. The temperature increase of 1°C in the river is reported to have encouraged fish and changed the micro-climate of riverain farms, but other, negative ecological implications seem to have been ignored.

What Cosgrove (1992) calls the *rhetoric of reclamation* is amply demonstrated in the DPRK's (Democratic People's Republic of Korea) desire to create new arable land by reclamation of the tidal mudflats and saltmarshes of the low-lying west coast. During the Korean War a team of geographers from Kim Il Sung University was given the task of surveying the west coast for potentially reclaimable land. They reported that

FIGURE 18.10 *The West Sea Barrage, which ponds back the River Taedong*
Source: P.J. Atkins

150 000 ha were available but this figure was later increased to 500 000 ha by the politicians, an extraordinarily ambitious target when one considers that the Netherlands, the most successful nation in history at creating dry land from the sea, drained only 165 000 hectares of land in the Zuyderzee between 1927 and 1968.

In conclusion, North Korea has developed what Unwin (1992, 191) calls a 'human meaning of landscape'. The interpretation has been single-minded, from the single mind of Kim il Sung. The resulting landscape of the DPRK is an outcome or a by-product of socialism but also a key medium through which the society has been transformed. The confrontation with nature has in turn transformed the Korean version of socialism.

CONCLUSION

Some of the concepts introduced in this chapter have been complex, but we have in fact only skimmed the surface of the theory and practice of landscape analysis current in the 1990s. Chapters 19 and 20 provide elaboration for the role of power, pleasure and otherness in landscape creation.

FURTHER READING AND REFERENCES

Dennis Cosgrove and Stephen Daniels are among the most important and prolific authors in this area of landscape study.

Berger, J. 1972: *Ways of seeing*. London: British Broadcasting Corporation.

Bermingham, A. 1990: Reading Constable, *Art History* **10,** 38–58.

Cosgrove, D. 1984: *Social formation and symbolic landscape*. London: Croom Helm.

Cosgrove, D. 1989: Geography is everywhere: culture and symbolism in human landscapes. In Gregory, D. and Walford, R. (eds) *Horizons in human geography*. Basingstoke: Macmillan, 118–35.

Cosgrove, D. 1992: *The Palladian landscape: geographical change and its cultural representations in sixteenth century Italy*. London: Pinter.

Cosgrove, D. and Daniels, S. (eds) 1988: *The iconography of landscape: essays on the symbolic representation, design and use of past environments*. Cambridge: Cambridge University Press.

Daniels, S. 1993: *Fields of vision: landscape imagery and national identity in England and the United States*. Cambridge: Polity Press.

Daniels, S. and Seymour, S. 1990: Landscape design and the idea of improvement. In Dodgshon, R.A. and Butlin, R.A. (eds) *An historical geography of England and Wales*. 2nd edn. London: Academic Press, 487–520.

Duncan, J.S. 1990: *The city as text: the politics of landscape interpretation in the Kandyan kingdom*. Cambridge: Cambridge University Press.

Duncan, J.S. and Duncan, N. 1988: (Re)reading the landscape. *Society and Space* **6,** 117–26.

Kenny, J.T. 1995: Climate, race, and imperial authority – the symbolic landscape of the British hill station in India. *Annals of the Association of American Geographers* **85,** 694–714.

Meinig, D.W. 1979: The beholding eye: ten versions of the same scene. In Meinig, P.W. (ed.) *The interpretation of ordinary landscapes: geographical essays*. New York: Oxford University Press, 33–48.

Rotenberg, R. 1995: *Landscape and power in Vienna*. Baltimore: Johns Hopkins University Press.

Said, E. 1978: *Orientalism*. New York: Random House.

Seymour, S. 1989: The 'spirit of planting': eighteenth century parkland 'improvement' on the Duke of Newcastle's north Nottinghamshire estates. *East Midland Geographer* **12,** 5–13.

Unwin, P.T.H. 1992: *The place of geography*. Harlow: Longman.

Williams, R. 1973: *The country and the city*. London: Chatto & Windus.

19

LANDSCAPES OF POWER AND PLEASURE

Humans are preceded by forest, followed by desert.

(Parisian graffito)

Introduction

Other than the bitter wars in the former Yugoslavia, Europe has not experienced a major conflict for fifty years, the longest period in its history. It is understandable, now that even the Cold War is fading, that many people feel complacent about peace and have begun to take it for granted. The same cannot be said of other continents where tank battles, guerilla infiltration, armed stand offs and even genocide with spears and machetes are common currency. The norm for the world is still instability and preparedness for war. It has been ever thus and in consequence there have been a myriad of modifications to the environment, both before and after battle.

Ritualized forms of violence channelled into sport go under the umbrella term of hunting. This may be the relatively sedate élitist pastime of fox hunters riding to hounds through the hedged pastures of the English Midlands, but more typically on the continent and in North America it is a pedestrian and macho male preserve, the whiff of cordite blending with that of testosterone. In Malta this has a physical outcome because the island is littered with the hides, shooting butts and trapping grounds of hunters (*see* Figure 19.1). This one small nation has a gun culture which surpasses almost any in the world and as a result no migrating bird is safe. Up to 500 000 turtle doves, and many rare birds of prey are shot each year as they cross to or from North Africa. In some villages over 40 per cent of the male population have gun licences (Fenech 1992).

Leisure time is an increasing part of the lives of many in Western countries and rising disposable

Figure 19.1 *A hunter's hide, Marfa Ridge, Malta*
Source: P.J. Atkins

incomes have enabled people to indulge their desire to experience other cultures and environments. A visa in the passport and a suntan are relatively recent proofs required of holiday makers that they have achieved something worthwhile. They are phenomena of the jet age, since mass tourism in the 1960s discovered the delights of sun, sea, sand, sex and sangria. Before that, vacations, involving any more than Sunday and the occasional Saint's day, were rare for any outside the wealthy class.

LANDSCAPES OF CONFLICT AND WAR

Pre-industrial warfare, despite its relatively primitive technologies, has left enduring marks on the landscape. Iron Age hill forts such as Maiden Castle in Dorset are still impressive today and the concentric castles of Edward I in North Wales even more so (Chapter 11). Both were principally defensive but they were also laden with a symbolism of imposed power that could scarcely be missed by the local populace and any potential enemies.

The principal impact of these ancient relics of war on our landscape has been their authorship of place. They created nuclei around which people wished to congregate, at first perhaps only in times of strife but in many cases permanently in the form of a town. In the Old World the defensive factor has been a key to much urbanism.

Second, castles and battlefields have an enduring fascination for a public hungry for historical interest and leisure diversions. They have become important elements in the heritage landscape, with guided tours and battle re-enactments (*see* Figure 19.2).

In the age of large-scale industrial warfare, beginning with the first globalized conflict between 1914 and 1918, there has been a shift from what we might call the picturesque of war towards devastation of landscapes over wide areas. The Western Front of the First World War is the example of this *par excellence* because of the reliance upon fighting from networks of interlocking trenches that were bombarded by heavy guns. The horrific results of shelling (*see* Figure 19.3) were visually dramatic because of the de-

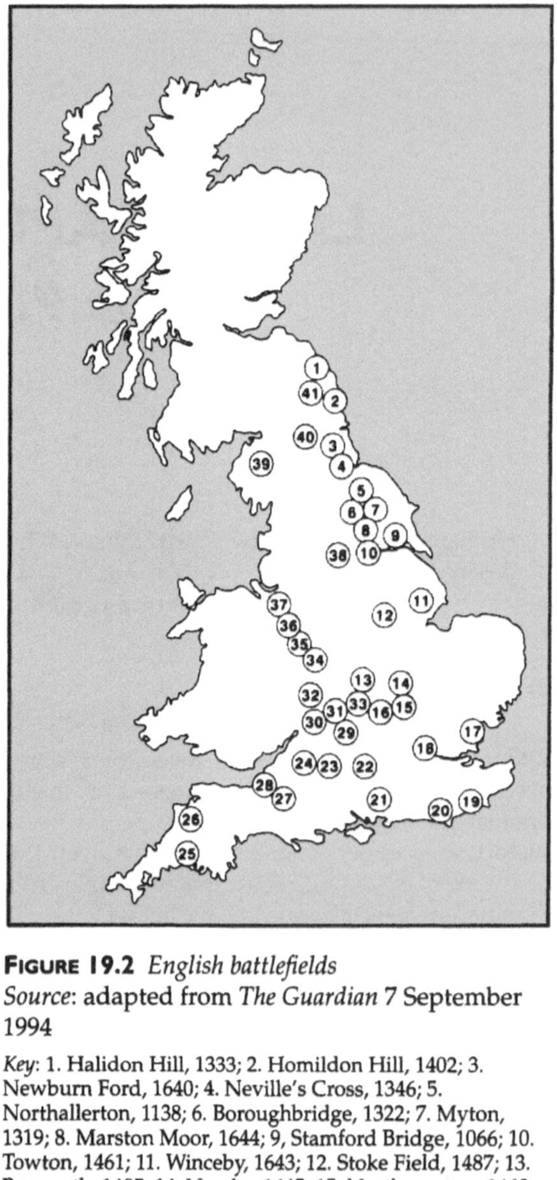

FIGURE 19.2 *English battlefields*
Source: adapted from *The Guardian* 7 September 1994

Key: 1. Halidon Hill, 1333; 2. Homildon Hill, 1402; 3. Newburn Ford, 1640; 4. Neville's Cross, 1346; 5. Northallerton, 1138; 6. Boroughbridge, 1322; 7. Myton, 1319; 8. Marston Moor, 1644; 9, Stamford Bridge, 1066; 10. Towton, 1461; 11. Winceby, 1643; 12. Stoke Field, 1487; 13. Bosworth, 1485; 14. Naseby, 1645; 15. Northampton, 1460; 16. Cropredy Bridge, 1644; 17. Maldon, 991; 18. Barnet, 1471; 19. Hastings, 1066; 20. Lewes, 1264; 21. Cheriton, 1644; 22. Newbury, 1643; 23. Roundway Down, 1643; 24. Lansdown Hill, 1643; 25. Bradock Down, 1642; 26. Stratton, 1643; 27. Langport, 1645; 28. Sedgemoor, 1685; 29. Stow-on-the-Wold, 1646; 30. Tewkesbury, 1471; 31. Evesham, 1265; 32. Worcester, 1651; 33. Edgehill, 1642; 34. Hopton Heath, 1643; 35. Blore Heath, 1459; 36. Nantwich, 1644; 37. Rowton Heath, 1645; 38. Adwalton Moor, 1643; 39. Solway Moss, 1542; 40. Otterburn, 1388; 41. Flodden, 1513

FIGURE 19.3 *Zonnebeke Church, near Paschendale, 1917*
Source: War Office 1924: *Notes on the interpretation of air photographs* London: War Office

natured appearance of the land in certain localities, but there were also more insidious long-term dangers for farmers from minefields and other unexploded munitions. The officially declared *régions dévastées* in France covered 3 337 000 ha (*see* Figure 19.4) and included 620 settlements completely destroyed and another 1334 severely damaged (Clout 1996a; 1996b).

North East France and Western Belgium are crowded with memorials, perhaps the most concentrated impact of warfare on the planet. They range from the massive Menin Gate at Ypres, intended explicitly by the architect Blomfield as a symbol of 'the enduring power and indomitable spirit of the British empire', to the many cemeteries and individual graves established and still tended by the Imperial (later Commonwealth) War Graves Commission. These landscapes of remembrance were carefully designed, with immaculately kept lawns, tree plantings and beds of native English flowers, to be a peaceful, even sublime haven amongst the destruction. By 1930 there were 891 such cemeteries for the British and colonial dead (*see* Figure 19.5), with 540 000 headstones.

Heffernan (1996) argues that such memorial landscapes are more than just records of the folly of war, or even sacred spaces. Their awesome grandeur overwhelms the senses and dulls the critical faculties of any visitor. It reifies and mythologizes the experience of war, a calculated decision by a government which refused to allow bodies to be repatriated and buried privately. 'Thus memorialized, Britain's war dead imposed a powerful block on the development of popular moral or political criticisms of the Great War, the political and military leaders who organized it, and the post-war world which these same leaders constructed' (Heffernan 1996, 313).

Unfortunately the First World War was not 'the war to end all wars' but a curtain raiser to the most conflict-ridden century that history has seen. Increasingly powerful and efficient weapons have had a widespread impact, not just explosive devices but also chemicals. In the latter category defoliants were used by the Americans

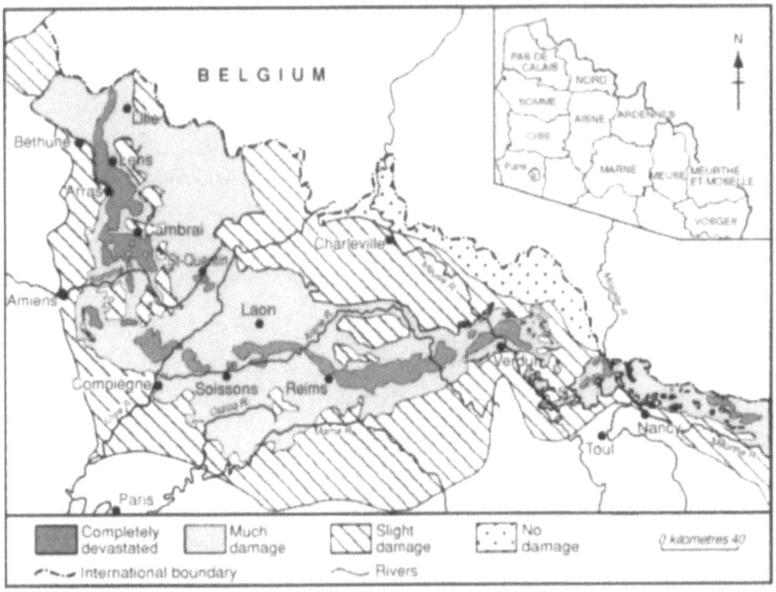

FIGURE 19.4 *Areas devastated in the First World War*
Source: Clout, H. 1996: Restoring the ruins: the social context of reconstruction in the countrysides of northern France in the aftermath of the Great War. *Landscape Research* **21**, 213–30

in Vietnam to devastating effect upon its ecosystems and people, and nerve gases and biological agents are also available. The Gulf War (United Nations versus Iraq) in 1991 illustrated the possibilities of precision bombing but the main impact of modern warfare remains indiscriminate, such as the long-term effect of landmines.

Weapons need to be tested and troops trained and the manifestations of war therefore also include the preparation areas reserved for the armed forces. Although these occupy vast swathes of the countryside and restrict access by the public, interestingly their environmental consequences are sometimes positive because ecosystems may be preserved that would otherwise have been destroyed by intensive farming, house building or other land uses.

What might be termed post-industrial warfare depends upon nuclear devices. Obviously, the impact of the use of such weapons might be terminal for global civilization and would be destructive of environments and ecologies well beyond the immediate impact zone. So far, the visible effects have been restricted to Hiroshima and Nagasaki in Japan and to the various desert testing ranges around the world.

The impression of warfare given so far in this Chapter has been of landscapes exploded by shot and shell. If we modify the vocabulary slightly to *conflict*, then the outcome is different but equally significant. Conflicts may last for generations or even centuries without resolution, the slow drip of hatred and low-level violence rather than hot war, nevertheless still having profound implications for the landscape. Over the last 30 years

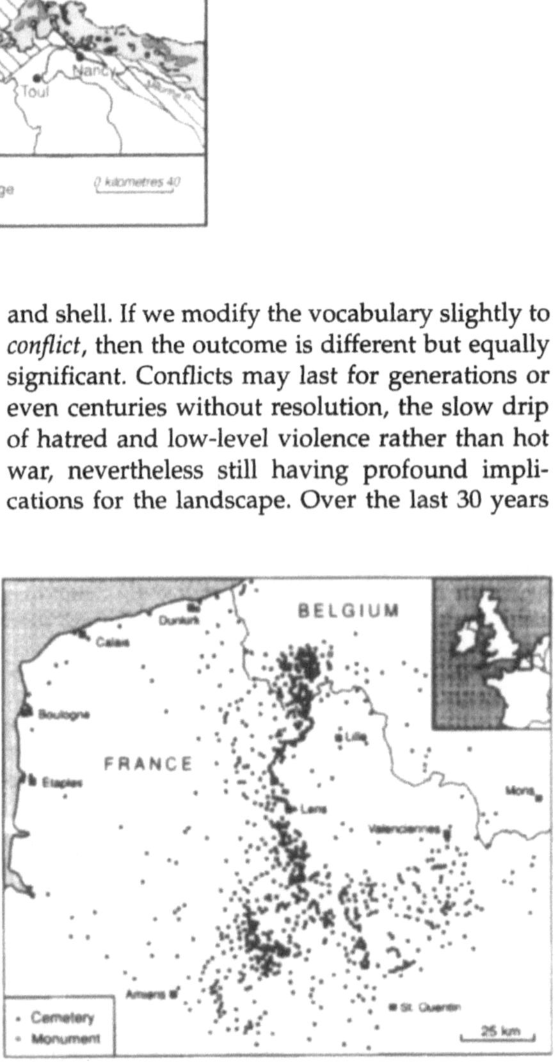

FIGURE 19.5 *First World War cemeteries*
Source: Heffernan, M. 1996: For ever England: the western front and the politics of remembrance in Britain. *Ecumene* **2**, 293–323

Belfast has been blighted by bombs but the most lasting impact of the Troubles has been its segregation into Protestant/Loyalist and Catholic/Nationalist districts. Rather than the class war which Marxists argue has divided modern cities into rich and poor landscapes, here we have a religious polarization with its roots in history and its manifestations on the ground in the form of blocked streets and high walls between the sectarian communities. This is a city physically divided and with a symbolic separation enhanced by street art.

Equally divided, but with the centre of gravity of power clearly biased in one direction or another we have landscapes of domination. In the South Africa of the apartheid era both the countryside and cities were divided into racially segregated zones where access to resources was very clearly arranged in favour of whites. In the West Bank the siting of Israeli settlements over the last 30 years has been calculated to occupy as much land as possible and to perform a surveillance role over Palestinian villages and towns (Chapter 20). Here even the environment has been politicized by conflicts over water for drinking and irrigation, and the visual appearance of the landscape has been dramatically altered to suit the goal of domination.

Landscapes of leisure

As we have seen in Chapter 18, the eighteenth century was an important threshold in the calculated investment of capital for the purposes of pleasure. Most landscape gardens created then had a subtext of status and power, but a few were dedicated to the profit motive by providing leisure facilities in semi-rural surroundings. These were the so-called pleasure gardens which formed a necklace adorning the suburbs of European cities.

The London pleasure gardens are probably the best known. Names such as Vauxhall, Ranelagh and Sadler's Wells were famous for such low-key delights as bowling greens, skittle grounds, chalybeate springs for taking the waters, cream teas, open air concerts, fireworks and zoos. They were varied in the number and status of their clientèle but the largest were singularly impressive to contemporary observers. For concerts and assemblies, Ranelagh Gardens had a rotunda which had an internal diameter of 50 metres. Miss Lydia Melford in Smollett's *Humphrey Clinker* likened it to 'the enchanted palace of a genio, adorned with the most exquisite performances of painting, carving, and gilding, enlightened with a thousand golden lamps that emulate the noonday sun' (Wroth 1896, 203).

A more exclusive part of the eighteenth-century leisure landscape was the spa. Bath, Cheltenham and Tunbridge Wells were as much part of the social season as they were health resorts and they were forerunners of the sea-bathing which became popular in the nineteenth century at Brighton. The coastal vacation experience of the twentieth century, with variations from Blackpool to Benidorm, has its roots solidly in the past.

Our case studies will take a broad view of leisure. We will look at two British examples that are rarely considered by the British themselves and will be unknown to most foreign readers. The first looks at the search by middle-class and working-class people for a rural retreat in the form of a small plot of land and a cottage. The nearest American equivalent would be a log cabin in the woods or mountains. The second concerns the creation of collective spaces for growing vegetables.

Case study: the spontaneous landscapes of the British plotlands

All countries have their vernacular landscapes, authored anonymously by the spontaneous decisions of countless ordinary people. These vary infinitely in their modes of origin and their form, and most lack the regularity and geometric symmetry that are characteristic of planning from above. But such spontaneity has been squeezed in the twentieth century by the bureaucratic framework of modern town and country planning and by the construction sausage machine of property developers and builders. Any new development which escapes these strictures seems like a popular triumph.

One example of the makeshift landscapes which emerge under loose planning controls is the *plotlands* that were settled in Britain between the world wars (Hardy and Ward 1984). These were on cheap marginal land or were acquired by squatting. They were often areas liable to flood or on steep slopes that were not attractive to conventional development. They were occupied by people at the low end of the income scale and the accommodation was in wooden huts, small bungalows, recycled railway carriages or redundant buses. Their physical infrastructure of roads, sewers and main utilities was poor, and other community facilities, such as schools and shops, were minimal.

The plotlands were of essentially two types. In Essex and south of the Thames there were plotlands established by people migrating out of London for a quieter life, usually individually or in small groups. Some had small holdings attached for keeping horses or growing vegetables but most were tucked away in patches of woodland or in inaccessible locations. On the south coast the motivation was creating a second home for holidays, and here the clusters were larger and more concentrated. Contemporary commentators, who were mostly middle class, at best, deplored their untidy appearance and, at worst, called for their demolition as rural slums.

The plotlands are a modern manifestation of the desire for a piece of Arcadia. This has long been part of the British psyche: from the seventeenth-century Diggers to the 1970s' hippy teepees of central Wales, there has been a tradition of seeking one's own salvation away from city-centred mainstream civilization. The newly established roots have never been deep, however, and within a generation or two such experiments dissolve and leave little trace. Thus it was with the plotlands, many of which have been abandoned or upgraded into more substantial properties.

Peacehaven on the Sussex coast may be taken as one example of a *people's landscape*. It was not typical of the plotlands perhaps, because it had a greater appearance of order and was not hidden discretely down an unmade track. It was conceived by the developer, Charles Neville, in 1915 as a garden city by the sea. He bought a large tract of land and divided it into a grid. Individual plots were sold somewhat haphazardly to holiday-makers, ex-servicemen, those seeking the health-giving properties of fresh air, retired couples and a miscellany of other incomers. The result over the next two decades was not a normal housing estate but a random collection of low-cost bungalows and chalets that were called 'a colony of shacks, a long ungainly street of houses that all seem ashamed of themselves' (S.P.B. Mais 1938, quoted in Hardy and Ward 1984, 84).

Canvey Island in Essex, had a similar history. Large tracts of land were bought up by Frederick Hester from 1899 to 1905 and divided into plots for the development of a holiday resort to be known as Canvey-on-Sea. In the event, Hester was declared bankrupt but the construction continued without him as plot owners multiplied, particularly from the 1930s when London's East Enders enjoyed day trips to Southend and many decided to build second homes on Canvey. In 1951 three-quarters of the permanent dwellings on the island were 'light structures of the bungalow or chalet type' but many were destroyed in the 1953 floods which affected the low-lying island particularly badly. Relatively little of the plotland landscape survives today.

CASE STUDY: ALLOTMENT GARDENS

Allotments have varied origins. In eighteenth-century England, during the process of parliamentary enclosure (Chapter 7), cottagers

FIGURE 19.6 *The 'boathouses' of Armier Bay, Malta: eyesore or popular landscape?*
Source: P.J. Atkins

were often allotted small areas of land to compensate for their loss of access to common resources, and in the nineteenth century it was customary in many areas for farmers to allow their labourers to cultivate patches on the edge of the fields. Communally grouped vegetable garden plots seem to have been provided by some estate owners and the Poor Law authorities in a few parishes from about 1800 onwards. It was popular as a device to reduce the Poor Rates and by 1833 allotments were provided in 42 per cent of parishes, especially those in the south of England where agricultural wages were low (*Report* 1969, 5).

Birmingham, Nottingham and a few other cities made vegetable gardens available for rent but it was not until 1907 that this became a statutory duty. These urban allotment sites, still visible today in the suburbs and alongside railway lines, were dominated by industrial workers. The plots are usually standardized in units such as ten rods (250 square metres), but the regularity of the boundary baulks is overwhelmed by the informality of planting and improvisation of recycled materials into potting sheds which has made the allotment an expression of the ordinary male arcadia (*see* Figure 19.7). The use of the gendered adjective here is deliberate because allotments have traditionally, especially in working-class areas, been misogenistic landscapes, the alternate of female domesticity.

For miners in particular, the allotment seems to have provided not only an element of subsistence during periods of short time working but also a compensating pride in self-help away from the alienating conditions of labour at the pit. Now the mines have largely gone and the welfare state has reduced the need for voluntary community initiatives but the allotment leisure culture remains strong. In the north east of England there is a continued loyalty to the cultivation of prize vegetables, especially leeks, and the rearing of pigeons in *crees*, together making a popular stereotype of Geordie life.

FIGURE 19.7 *The allotment garden landscape of North East England*
Source: P.J. Atkins

The use of allotments for purposes other than vegetable and fruit growing is common on the continent where *chalet gardens* on small plots are popular as summer retreats from city life. In Britain the weather does not invite this use, nor would most councils grant planning permission for second homes, even if they were only huts.

CONCLUSION

In this chapter we have discussed landscapes that are often overlooked in the literature. The reasons for this are plain. Warfare creates landscapes of destruction which are too negative in connotation for the aesthetes who have written much of the published appreciation of scenery, and popular leisure landscapes have similarly been lost to view in the somewhat élitist vogue for landscape gardens. Fortunately the new wave of cultural studies is now redressing this imbalance.

FURTHER READING AND REFERENCES

There are few substantial texts yet published on the cultural interpretation of the landscapes of warfare and leisure, but the following references contain much interesting and relevant information. In particular see also Simmons (1996, 335–48) and Urry (1990).

Clout, H. 1996a: Restoring the ruins: the social context of reconstruction in the countrysides of northern France in the aftermath of the Great War. *Landscape Research* **21**, 213–30.

Clout, H. 1996b: *After the ruins: restoring the countryside of northern France after the Great War*. Exeter: Exeter University Press.

Fenech, N. 1992: *Fatal flight: the Maltese obsession with killing birds*. London: Quiller Press.

Hardy, D. and Ward, C. 1984: *Arcadia for all: the legacy of a makeshift landscape*. London: Mansell.

Heffernan, M. 1996: For ever England: the western front and the politics of remembrance in Britain. *Ecumene* **2**, 293–323.

Report, Departmental Committee of Inquiry into Allotments. 1969: [Chairman: H. Thorpe] Cmnd 4166. London: HMSO.

Simmons, I.G. 1996: *Changing the face of the earth: culture, environment, history*. 2nd edn, Oxford: Blackwell.

Urry, J. 1990: *The tourist gaze: leisure and travel in contemporary societies*. London: Sage.

Wroth, W. 1896: *The London pleasure gardens of the eighteenth century*. London: Macmillan.

20

'OTHER' LANDSCAPES

Our ancestors have turned a savage wilderness into a glorious empire.

(Edmund Burke 1775: *Speech on conciliation with America*)

INTRODUCTION

Hunters and gatherers in jungles and deserts live in surroundings which are unimproved by the standards of modern civilization. They have no concept of *wilderness* because to them the wild resources are familiar and a source of more or less reliable sustenance. This chapter will look at how such perceptions changed as the advent of agriculture and later, industry, transformed the economy and drew people into new ways of living and new ways of experiencing nature. In particular we will show that the *othering* of landscape can create and facilitate new modes of behaviour that may change the world.

A WALK ON THE WILD SIDE

From the earliest stages of agriculture, the beneficiaries of the harvests learned to fear the envy of others. At Jericho they built a massive walled enclosure to protect themselves from raiders, possibly even before they settled to permanent cultivation (see Chapter 2). The cliché of enmity between 'the desert and the sown' seems to have had substance in the early Middle Eastern neolithic and, throughout history, at least until the last few hundred years, the balance between village-based farmers and migratory resource users (hunters, gatherers, pastoralists) was delicate and a source of friction in marginal areas.

Fear of the world beyond the civilized realm was sometimes less specific than the threat of human violence. At the religious level animists are more comfortable with their whole environmental context than worshippers of sky gods, who have abandoned sacred associations with trees, rocks and rivers (Short 1991). For the latter group the wilderness was a desolate place full of wild beasts and bandits; it was beyond the accepted order of society and was a symbol of evil. On the other hand, the wilderness could also be a place of renewal and deliverance. In the Christian story, Jesus withdrew there for 40 days and nights; he was tempted by the Devil but resisted and was cleansed and reinvigorated by the experience.

Taming wild nature and thereby reclaiming it for God was considered an act of religious piety throughout the civilized world until the eighteenth century. Very much the same notion of the mastery of nature for human betterment has survived in socialist dogma until very recently in the former Soviet Union and even today in North Korea. As we saw in Chapter 8, much of the clearing of the wood in Europe was inspired by this concept of duty, which during the Enlightenment of the seventeenth and eighteenth centuries was rationalized through ideas of the superiority of science and technology. Educated landowners and entrepreneurs were encouraged to improve upon nature for pleasure and profit, whether in a landscape park (Chapter 18) or the reclamation of moorland (Chapter 10) or drainage of a marsh (Chapter 9) for cultivation.

FIGURE 20.1 *The Wilderness, 1860, Sanford Gifford*
Source: Toledo Museum of Art

This is not to say that wilderness was considered entirely barren and useless. Pre-industrial sustainable economies were at least partly based upon the full use of natural resources beyond the shadow of the village or town, and their preservation was energetically pursued by the owners of common rights at times of threat to their livelihood. Rather than a sharp dichotomy between civilization and wilderness, we had better think of an overlap whose boundaries always have been, and still are, bitterly contested according to the relative strengths of the interested groups.

FROM THE ROMANTIC TO THE GREEN MOVEMENTS

John Berger, a famous interpreter of images, has argued that all art which is based on a close observation of nature eventually changes the way nature itself is seen. Thus it was that an eighteenth-century literary movement that favoured a romantic view of the world eventually opened the eyes of the public to the glories of wilderness (Figure 20.1). English poets such as Wordsworth and a small but influential school of painters found spiritual currency beyond the usual subject matter of art and their ideas gradually seeped into the consciousness of other nations. Novelists and film makers have taken up the theme, especially in societies which see themselves pitched against nature in their efforts at nation building. America and Australia are examples that spring to mind of countries that have been carved out of wilderness in a modern era when fear and loathing of mountains, forests and deserts was replaced by respect and even a love of wild places.

The changing regional concentration of landscape paintings shown in Figure 20.2 is one way of plotting the increasing popularity of the rugged sublime. The convenience of owning an image of Snowdonia or the Lake District was that one could enjoy the aesthetic experience without having to travel and suffer the rigours of poor roads and inadequate accommodation. It is these artistic visions we seek to preserve when we attach the label heritage to portions of the countryside such as Constable Country (north Essex), Hardy's Wessex, or the more recent

(a)

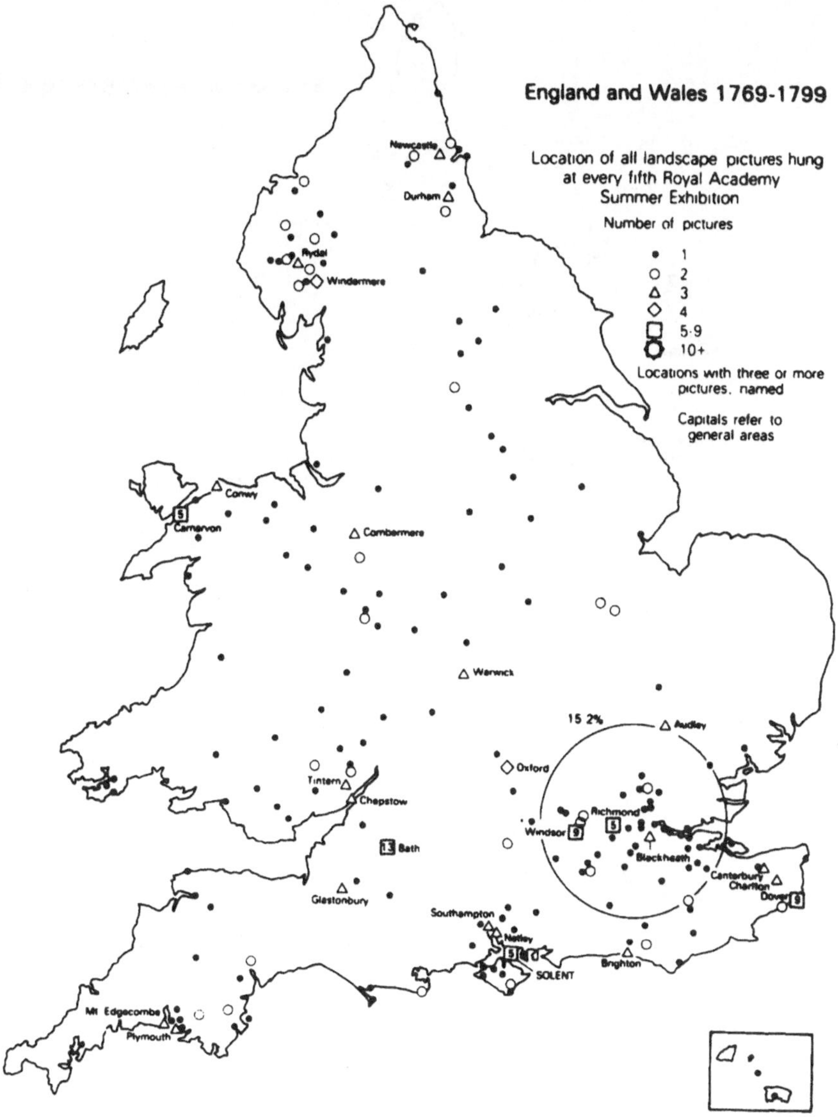

Herriott Country (Vale of York and Wensleydale) (*see* Figure 20.3). Thus artists have reinterpreted our landscape and, by presenting us with selected powerful images, have helped to create in our mind's eye a model set of valued landscapes.

The picturesque view of the Lake District was very different from an earlier view that deserted landscapes were barren, infertile and hostile, and art, in all its forms, has helped forge our view of it as a precious resource. Wordsworth's guide book (1st edition 1810) was in effect an outstanding piece of early regional geography and his poetry was as effective in its day as a modern Tourist Board advertising campaign. It is possible to identify in all this the origins of what would now be called *green* ideas of the appreciation of nature and its conservation in as wild a form as possible by resisting human interference.

Representational art is potentially even more powerful than this suggests. Daniels (1993) has argued that it may be one of the vehicles by which a national self-consciousness is created. The manufacture of symbolism and mythology as reference points in the moulding of the oneness of a collective identification is possible through landscape art.

(b)

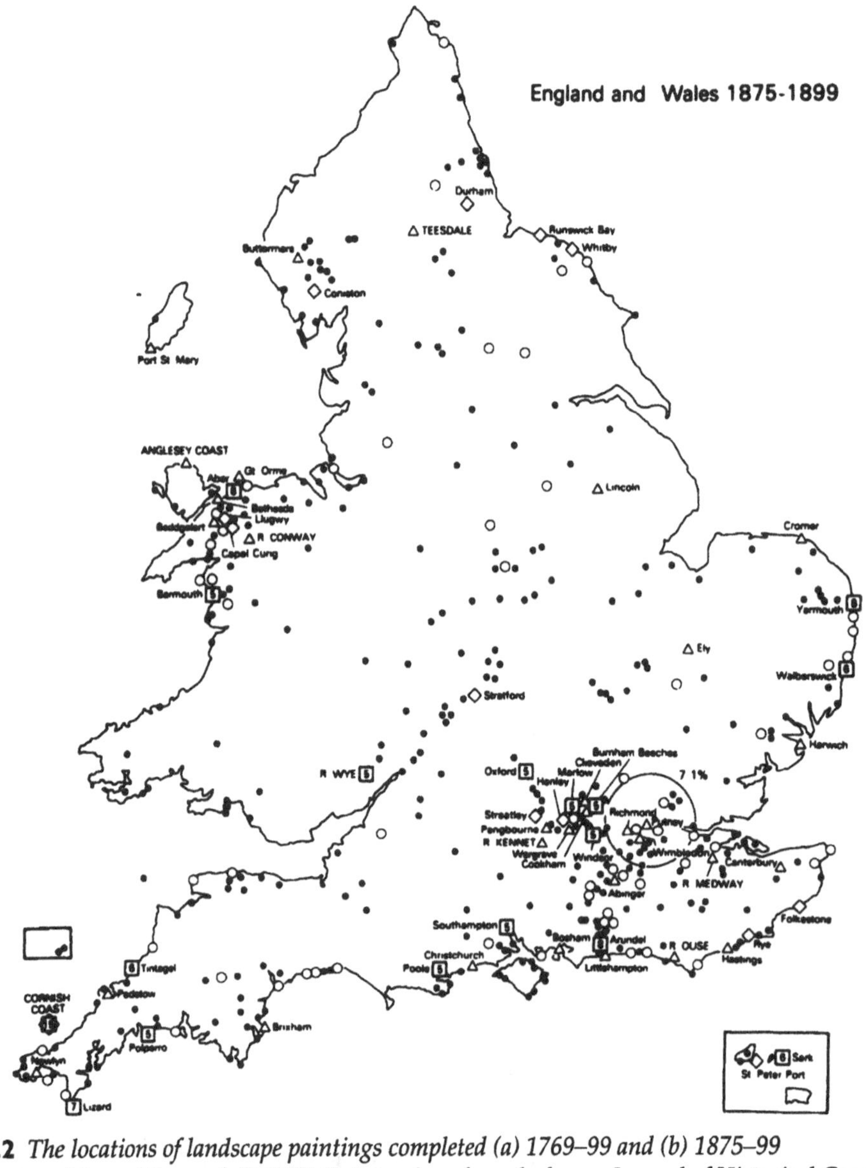

FIGURE 20.2 *The locations of landscape paintings completed (a) 1769–99 and (b) 1875–99*
Source: Reprinted from Howard, P. 1985: Painters' preferred places. *Journal of Historical Geography* **11**, 138–54

THE FRONTIER AND THE NATION

> Human beings are expressions of their landscape.
> (L. Durrell, 1969: *Spirit of place: letters and essays on travel*. London: Faber)

Frederick Jackson Turner (1861–1932) is most closely associated with the view that the collision between people and raw nature can be a positive, national-character-forming experience. According to this idea, after independence from the British (1776) it was the challenge of expanding America's boundaries by pushing the edge of settlement further to the west and south that was the principal medium of transformation from a European to a new way of thinking (*see* Figure 20.4). There was also a 'purity and pristine beauty that gave it a special moral power superior to Europe' (Cosgrove 1984, 184).

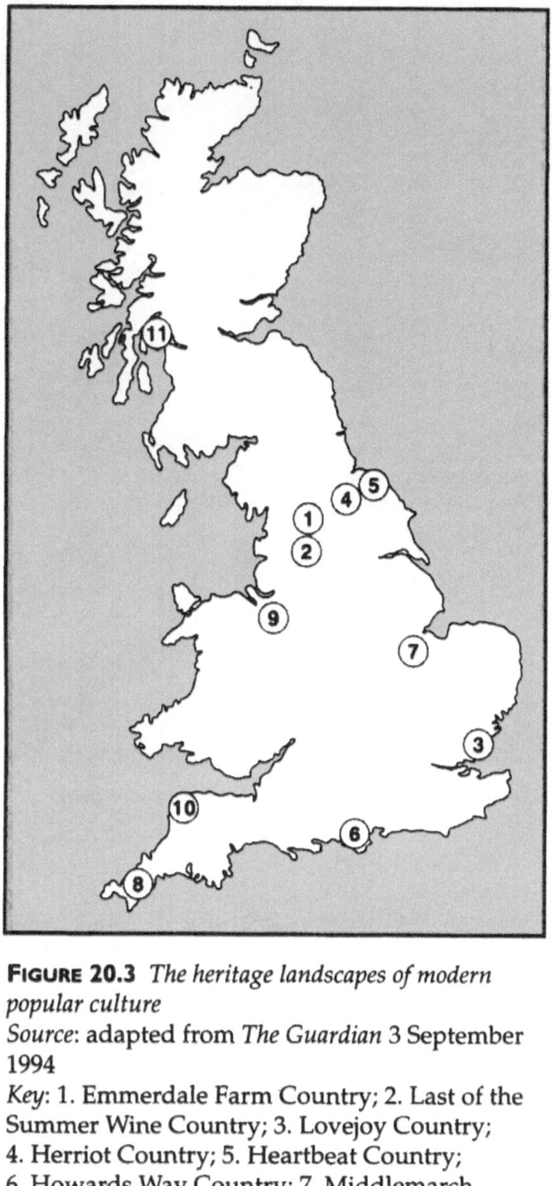

FIGURE 20.3 *The heritage landscapes of modern popular culture*
Source: adapted from *The Guardian* 3 September 1994
Key: 1. Emmerdale Farm Country; 2. Last of the Summer Wine Country; 3. Lovejoy Country; 4. Herriot Country; 5. Heartbeat Country; 6. Howards Way Country; 7. Middlemarch Country; 8. Daphne Du Maurier Country; 9. Brother Cadfael Country; 10. Tarka Country; 11. Para Handy Country

Turner argued that it was at the frontier, with its individualism and optimistic view of social and economic change, and not on the eastern seaboard where much of the wealth was still created, that the new American democratic identity and self-belief definitively emerged. In reality this was no more than an extension of the eighteenth-century European passion for the *pastoral*, which in turn was a revival of certain themes in classical literature: the American wilderness therefore presented a wonderful opportunity to experience and build upon an ideal that could not be achieved fully in Europe's lived-in landscape. A new American interpretation of wilderness was possible through an awareness of raw nature that was denied earlier pioneers in Britain, France or Germany, and this was brought to a peak in the nineteenth century by a distinguished group of painters, poets, travel writers, novelists and philosophers, including Thomas Cole, Washington Irving, Ralph Waldo Emerson and Henry Thoreau (Nash 1973). The Catskill Mountains were especially valued for their wild landscape and the Hudson River School of painters forged a style depicting the primeval grandeur of untamed nature.

Although scholars would no longer agree with much of Turner's frontier thesis, there is no doubt that the wilderness theme has played a deep and fundamental role in shaping the American consciousness. Those of us in the rest of the world share this enthusiasm for the American frontier and wilderness through reading Walt Whitman's poetry, Norah Ingalls Wilder's *Little House on the Prairie* or our consumption of images of the *wild west*, especially those of pioneers in their covered waggons or cowboys and indians riding through the spectacular scenery of Monument Valley (Utah). Our perceptions are moulded by such fictional stereotypes and the powerful medium of television has embedded them still further by constant repetition. Reinterpretations of American identity through the conquest of new frontiers in even more exotic and alien surroundings are now available through television series such as *Star Trek* where Captain Kirk and his crew explore *the final frontier.*

CASE STUDY: THE MORMON CULTURE REGION

In 1847 Brigham Young founded the Salt Lake City community of the Church of Jesus Christ of the Latter-day Saints in the Wasatch Oasis. He and his followers felt that they had found their

FIGURE 20.4 *Across the continent: 'westward the course of empire takes its way', Frances Palmer for print published by Currier and Ives (1868)*
Source: Museum of the City of New York. The Harry T. Peters Collection

promised land after wanderings through the American wilderness that were analogous to the sufferings of the biblical Children of Israel in their exodus from Egypt. They very quickly identified with what to them was a metaphorical landscape.

The founder of the Mormons, Joseph Smith, had resolved to take his people to the Salt Lake Valley in 1845 and later a pioneer group had scouted the region and reported soil of 'most excellent quality' and 'very luxuriant' vegetation. There were risks of course in their trek west but these were later played up in order to magnify the achievement of the early settlers. By 1852 the Mormon journey had become 'one of the greatest miracles since Moses had crossed over the Red Sea', and the they had found 'a barren desert, as barren as the desert of the Sahara' (Jackson 1992). Here environmental perceptions became myth and were harnessed as theological ammunition, but Bowden (1992) has found more generally that such invented traditions were characteristic of the American frontiersperson's idea of wilderness. There were many myths about primeval forest, infertile virgin prairies, and the Great American Desert.

The Mormons laid out their New Jerusalem in a gridiron plan and they were equally systematic in their determination to colonize as wide an area for themselves as possible. They sent groups to occupy outlying stations in what became the state of Utah, and beyond (*see* Figure 20.5). This planned spread from the core was a deliberate strategy to provide a broad ecological base for a self-sufficient community, along with controlled routes to the outside world for converts to use when migrating from Europe and the Atlantic seaboard. Meinig (1965) calls the forging of this *culture region* a 'bold assertion' and sees the expansion on all sides as a battle with nature and with the surrounding cattle and mining interests.

As with other American frontier settlements, there were problems. Some outposts were

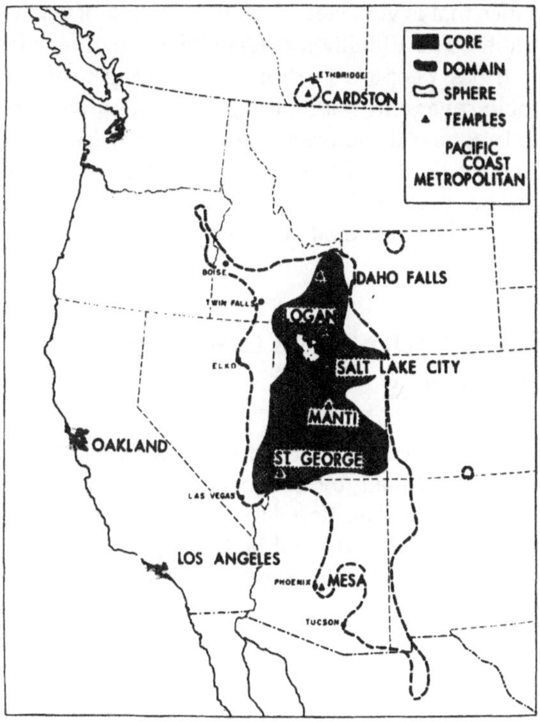

FIGURE 20.5 *The Mormon culture region*
Source: Reprinted from Meinig, D.W. 1965: The Mormon culture region. *Annals of the Association of American Geographers* **55**, 191–220

abandoned due to drought or remained isolated, especially in the push south where desert and mountain eventually constrained development. To the north, opportunities were lost because of a mistaken idea about the risk of frost. Yet the regions marked core, domain and sphere on the map were at least as distinctive and probably more homogeneous in their Mormon character than other regions of the country dominated by one ethnic or religious group. The core and domain together, for instance, are about 90 per cent Mormon in affiliation. The identification of cultural regions is notoriously difficult and fraught with theoretical pitfalls, but in the case of the Mormons their encounter with nature was a metaphor for their struggle with the wider world of the gentiles and their self-conscious 'otherness' and solidarity has been responsible for a unique signature upon the landscape.

'OTHER' LANDSCAPES

> When we discover that there are several cultures, instead of just one ... Suddenly it becomes possible that there are just *Others*, that we ourselves are an 'other' among Others.
> (P. Ricoeur 1965: *History and truth*. Evanston, Ill.: North-Western University Press)

One important departure in cultural studies has examined how individual people, communities and whole societies relate to their surroundings and identify with certain elements of the environment. They often derive important parts of their identity from the landscape and in turn modify those landscapes to take on the forms which most readily mesh with already established identities. The interaction may be voluntary or it may sometimes be imposed, as in the case of colonial powers or authoritarian regimes.

Edward Said is a key writer in this area, specializing in how oriental cultures have been viewed from the west and how both western and oriental identities have been forged as a result of the stereotypes of the east established during the colonial era. Travel writers, colonists, administrators, novelists, painters, playwrights and a motley army of others felt it was their duty and their right to portray the colonized realms in Asia (and Africa) as objects of interest and interpretation. The view was always of *there* not *here*, inhabited by people who were strange, unknown, exotic and fundamentally different, a 'people without history' (Gregory 1994, 170). From this otherness of the oriental it was a short step to a denigration of many of its characteristics, such as the irrationality of the culture, the supposed laziness and ignorance of the natives, their wiliness or inscrutability, and their need for education in the ways of western civilization. Such a logic then found a ready justification for conquest, dispossession, exploitation, and sometimes even genocide, while on the other hand it boosted the self-esteem and self-confidence of the conquerors and colonial masters.

As we have seen in Chapters 2 and 18, the European nations took many elements of their agro-ecosystems and their cultural landscapes with them to their imperial outposts. This was legitimized by the environmentally alien character of the tropics and by the fact that most of the territory invaded was still at a pre-industrial stage of economic development. In the words of Carter (1987) the inhabitants were still 'children of nature' and were there to be tamed along with their environmental adaptations.

Such images of otherness are an important and often unrecognized part of our everyday world. Famine and Third World poverty are objectified on our television screens and this encourages us to think of poor people as hopeless, helpless, and in the minds of some as feckless and degraded. Lutz and Collins (1993) have shown that other media are complicit in this, their example being the famous American popular magazine the *National Geographic*. They argue that the *National Geographic* has over the century of its existence shaped the American view of the world outside, and the editorial line has had an underlying ideological stance that is very clear from its selection of photographs. The 10 million readers have a regular diet of tribes people's portraits, with images of the noble savage in traditional garb, topless women, and women in domestic situations also featuring high on the list. Obviously any choice of photographs could be thought to be biased, but this one tends to be traditional, sexist and patronizing.

Case study: the landscape of Palestine

Said's discussion (1978) of orientalism has been partly informed by his own experience as a refugee from Palestine. His family left for Egypt during the 1947 war and he has since lived as an exile from his homeland. We will use this

Table 20.1 *Security landscapes*

Landscape scale	Elements	Artefacts
National	Military	Airfields, ports, army camps, training areas and firing ranges, military industries, fortification systems along borders.
	Security and defence	Transportation and settlement networks, public buildings, civil defence, national planning for dispersion of population, emergency economic structures.
	Results of wars	Cemeteries and monuments, remembrance days, media events, education, literature, folklore.
Regional	Military	Army camps, training areas and firing ranges, military industries, fortification systems along borders, human-made forests, direction signs, industries.
	Security and defence	Regional transportation network, physical planning of settlements and other infrastructure elements.
	Results of wars	Memorial monuments and parks, old forts and fortifications, ruined or abandoned villages.
Local	Military	Army camps, industries (in cities).
	Security and defence	Bomb shelters, guard towers, fences around settlement wired for lighting, bomb disposal holes, special architecture planning, check points on roads.
	Results of wars	Monuments, cemeteries, streets and parks named after fallen, folklore, literature, guard duty at public buildings, behaviour on streets, ruins, old fortifications.

Source: modified from Soffer, A. and Minghi, J.V. 1986: Israel's security landscapes: the impact of military considerations on land uses. *Professional Geographer* **38**, 28–41

tormented country as an example of the implications of the exclusionary geographies that have been routinely practised in occupied territories as a means of domination and suppression.

The foundation of the state of Israel has never been fully accepted by either her Arab neighbours or by the indigenous Arabs of Palestine. The latter group lost their land and homes in the west in 1948 and migrated to the West Bank, the Gaza Strip and beyond. In 1967 a further war led to the Israeli occupation of all the land west of the River Jordan and ever since, regional politics have been dominated by the need for the occupied territories in Lebanon, Syria and the West Bank to be disgorged and returned to some form of constitutional normality.

Both sides in this conflict see the opposition as the *other*. There is relatively little cultural or social contact, and the regular closure of check points on the Green Line has minimized economic links in the form of flows of goods and labour. The open violence by both parties to the conflict has legitimized extreme perceptions of the other point of view and a war mentality is never far from the surface. In short, there are two antagonistic communities living side by side who are apparently incapable of compromise. The landscape reflects this, with its overt security component (Table 20.1).

Part of the intractable problem in Palestine is the Zionist aim of controlling all of the traditional biblical lands, including the West Bank (Judaea and Samaria). This has meant a very strong urge for certain groups in Israeli society to create settlements among the predominantly Arab farm lands and villages and thereby to harness nature in the cause of nationalism. By a combination of legal purchase and armed occupation, a large number of settlements have been established and more are planned. Approximately 145 000 people now live in 128 new foundations (*see* Figures 20.6, 20.7 and 20.8), many sited deliberately close to Arab areas of population and most having strategic hill-top locations which are both defensible and have a surveillance function. Eventually the aim is to have such a physical and numerical presence that it will be impossible for the Palestinians to reclaim their land.

According to Falah (1996), the activities of the settlers (Gush Emunim) are a logical outcome of Israeli policies since 1948 that have seen over 400 Palestinian villages razed to the ground and all traces of Arab history removed (*see* Figure 20.9). This amounts to a form of ethnic cultural cleansing of the landscape that has allowed new cultural imprints to be established.

Our landscape and *theirs* jostle for position in the West Bank. The juxtaposition of the traditional Arab villages and the modern Jewish settlements is very noticeable. The latter are constructed of highly visible materials and are surrounded by high security fences and armed

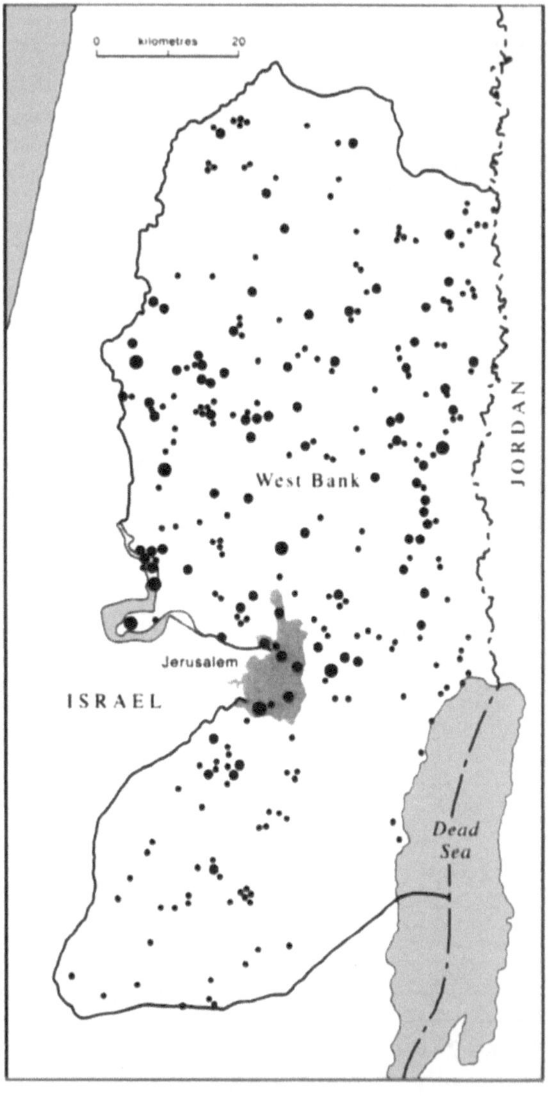

FIGURE 20.6 *Israeli settlements on the West Bank*

FIGURE 20.7 *Israeli settlement in the West Bank*
Source: G. Rowley

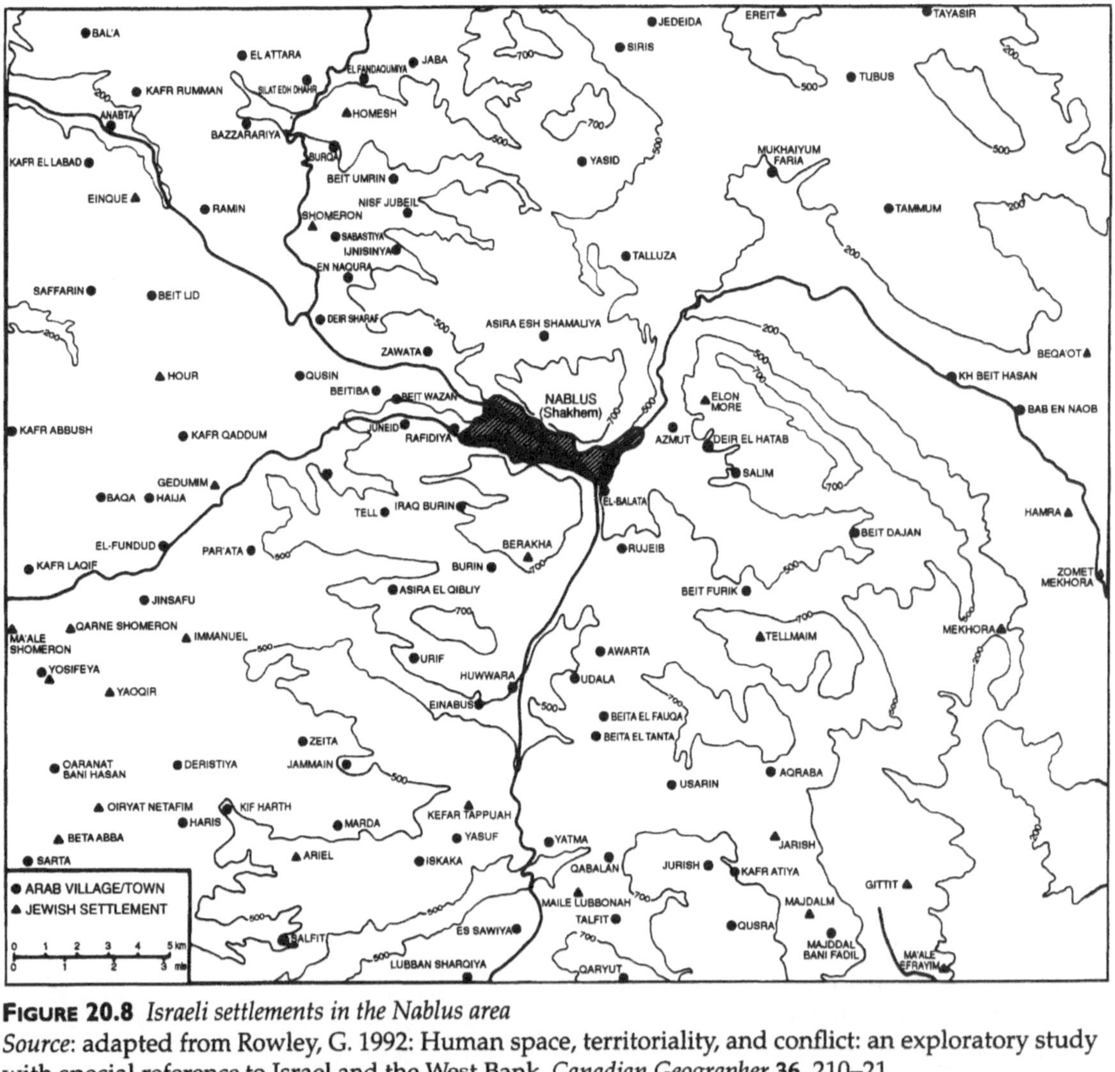

FIGURE 20.8 *Israeli settlements in the Nablus area*
Source: adapted from Rowley, G. 1992: Human space, territoriality, and conflict: an exploratory study with special reference to Israel and the West Bank. *Canadian Geographer* **36**, 210–21

FIGURE 20.9 *The destruction of the Palestinian cultural landscape*
Source: Reprinted from Falah, G. 1996: The 1948 Israeli–Palestinian war and its aftermath: the transformation and de-signification of Palestinian cultural landscape. *Annals of the Association of American Geographers* **86**, 256–85

guards. The settlers have privileged access to a wide range of scarce local resources and this is beginning to have environmental consequences. In particular, they extract ground water for irrigation and swimming pools in an area where the aquifers are already over-exploited. Water is even exported westwards into Israel. As a result, Arab consumers and farmers are short of water for drinking and farming.

The likelihood of a fully-fledged Palestinian state eventually emerging from the occupied territories of the West Bank and Gaza means that ownership and control of the landscape will become a major political issue. The creation of a robust Palestinian identity will at least partly rest on the new state's ability to control access to its natural resources, especially water and land.

IMAGINATIVE AND REPRESENTATIONAL LANDSCAPES

> Yet nature is made better by no mean
> But nature makes that mean: so, over that art,
> Which you say adds to nature, is an art
> That nature makes . . . An art
> Which does mend nature, change it rather, but
> The art itself is nature.
> (William Shakespeare, *The Winter's Tale*, Act IV, Scene III, 89–96)

In a sense Edward Said's argument about orientalism is no more than a subset of a broader argument one could mount about the role of the creative arts in our interpretation of the landscapes around us. Novels, poems, plays, paintings, sculpture, music, photography, film, television, radio, and now also computer games, virtual reality and heritage theme parks, all use the imaginative reconstruction of more or less realistic human environments in order to suspend disbelief and convey their own particular brand of entertainment and enlightenment. They all create alternative, simulated, *other* versions of reality; parallel universes in which constraints are relaxed. To take this logic further, we might even suggest that *reality* itself is made up of individual people's perceptions, that never match each other exactly, and that our own personal view of the world is really based upon an imaginative map that we revise and recreate throughout our lives as a result of the information that is presented to us and filtered through our own set of skills, beliefs and prejudices.

Place-specific art originated with the Dutch landscape painters of the seventeenth century. In English literature the equivalent fictionalized framings of real places emerged with the tradition of the regional novelists of Britain from the early nineteenth century onwards (Pocock 1981), especially the evocative works of Sir Walter Scott in Scotland, Thomas Hardy in Wessex, Arnold

Box 20.1 *The ways that geographers have used literary and other artistic portrayals of landscape, environment and nature*

- As a vivid portrayal of the personality of a region, or as raw material for the writing of regional geographies or urban social histories.
- To analyse the human perception of environment and the subjective experience of place, for instance by looking at the work of several writers who have described the same place.
- As symbolic representations of the inner meanings of the relationships between humans and nature.
- To provide a cultural refraction of reality, adding a further, qualitative dimension to the objective facts of science.
- As evidence of evolving individual, group and national identity.

Bennett and George Elliot in the west Midlands, the Brontës in Yorkshire and, in the twentieth century, D.H. Lawrence and Alan Sillitoe in Nottinghamshire and Catherine Cookson on Tyneside, among countless others around the world. These writers used their familiar surroundings, their geographically rooted identities, as a theatre of wider significance.

The fictionalization of real worlds as an attempt to explore the multiple meanings of society and human relationships has been extended in the twentieth century by literature and films written in the traditions of fantasy and science fiction. Usually set in the future, such art often crosses the bounds of known or predictable technology and creates alien and exotic landscapes to the very limit of the imagination, although many writers in the latter genre adhere strictly to the constraints of known scientific laws in their generation of new worlds beyond the solar system. Television series have explored the universe of possible landscapes, although the low budgets sets of *Doctor Who* tended to stretch the audience's credulity rather than their imagination, and Ridley Scott's film *Bladerunner*, about cyborgs in some dystopian future, has genuinely added to artistic interpretations of human/machine difference (Haraway 1991). Both are in a sense extensions of the explorations of the exotic which so characterized the geographical imagination in the colonial era, and which still dominates the agenda of Britain's Royal Geographical Society.

Case study: the Scottish Highlands

Scotland has been described as a heritage theme park in which the landscape has played a key role in the identification of nationhood. Certain sites such as Edinburgh Castle and Holyrood Palace are wrapped and sold to foreign tourists, but also to an extent to the Scots themselves, in an historic guise that assumes an artificial significance beyond the real facts of history. The Highland landscape has acquired the status of an icon in the eyes of visitors, who see Landseer's *Monarch of the Glen* and the heather-clad moors of several recent Hollywood films, but few remember that the romance of the empty skyline was achieved artificially by the clearance of its inhabitants in the nineteenth century by owners of large estates interested more in profits from sheep and grouse than in the welfare of their tenants.

Art may be seen as engaged in the creation of the Highland myth. Pringle (1988) has convincingly argued that Landseer's royal paintings, such as *Queen Victoria sketching at Loch Laggan* (1847) and *Royal sports on hill and loch* (1850) were part of the legitimation process of the naturalness of the politically *united kingdom*. The royal family are seen to be dominating peaceful Scottish pastoral landscapes, although their presence in the country for extended periods of holiday was in fact entirely new and there was such resentment at the profound economic and social crisis in Scotland that they had to be guarded by a small army. All of Landseer's royal pictures were engraved and sold in their thousands to a bourgeois audience (mostly in England) hungry to consume this royal mythology.

landscape@cyber.space.ok

The most challenging technological development for authors and consumers alike has been the mushrooming of personal computers in the 1980s and their linking up in the 1990s to the Internet to create powerful new possibilities for landscape invention, simulation and exploration. *Virtual reality* may be a new addition to the dictionary but we can argue that it is merely an enlargement of the imaginative possibilities offered by the novel for centuries. Most machine-based virtual reality relies upon simplified computer simulation models of the real world, and has yet to fully explore the subtle and complex cyberspace of science fiction writers such as William Gibson.

Virtual spaces are potentially significant, even in their present nascent form. They open an alternative, unlimited world to the rigid closure of our class and status system. Surfers of the Internet are anonymous (sexless, raceless, bodyless) or can assume new, unfettered identities of their choosing. They may converse intimately with other users of the World Wide Web or electronic mail without the usual constraints of distance (physical and social) and they may download data from remote sites which might not otherwise have been available to them. Nguyen and Alexander (1996) go as far as to speculate that the availability of an unregulated and uncensored Internet may eventually undermine the logic of liberal democracy as it has evolved over the last 300 years. The mediation of popular opinions through a representative in Parliament or the Congress will no longer be necessary and the Internet could become the space which most people inhabit for their important decision-making purposes and leisure time.

Jacky Tivers (1996) has shown how the landscapes portrayed in computer games have become more detailed and authentic as processing power has increased and multi-media machines with CD-rom drives have entered the market, and she also points to the vulnerability of the average 14-year-old game player. Games may reinforce stereotypes about other cultures and many pander to a male demand for combat and warfare. An implicit ideology lies behind games such as *SimCity* where the player creates an urban landscape within the constraints of market forces, trading and capital, or *Colonisation* which simulates territorial expansion.

Virtual reality headsets are now available and will eventually simulate three-dimensional worlds that are so realistic that participants will replace spectators as the boundaries between the virtual and real becomes blurred. Our bodies will be redefined, with limits set by the technology of implanted chips that will not just help to regulate a faulty heart or replace damaged hearing, but will also add new possibilities that will make us into cyborgs.

The spatial implications of the Internet are fundamental, not only because it crosses boundaries and annihilates the friction of distance but also because interaction is possible in a neutral plane between parties who might never otherwise meet due to incompatibilities of class, race or even personality. Business has not been slow to recognize the potential of such virtual spaces because there is often a large slice of their office work that could be based anywhere and linked to a central computer. Face-to-face contact remains vital for much business, however, and it is true that the development of tele-cottaging (Chapter 21), people working from home, has been slower than many enthusiasts predicted, but it seems inevitable that the Internet will eventually have a major impact upon the landscapes of the real world. The unfolding of the Information Age will see fibre-optic cables and satellite communication spreading and access to a broadband data highway will be as important as physical location. Cities will not disappear but some of their functions will be re-evaluated. Mitchell (1995, 24) foresees the emergence of the City of Bits:

> This will be a city unrooted to any definite spot on the surface of the earth, shaped by connectivity and bandwidth constraints rather than by accessibility and land values, largely asynchronous in its operation, and inhabited by disembodied and fragmented subjects who exist as collections of aliases and agents. Its places will be constructed virtually by software instead of physically from stones and timbers, and they will be connected by local linkages rather than by doors, passageways, and streets.

The technical limitations of the Internet mean that it cannot be universally available, however, at least in the short term. Poor communities and sparsely populated rural areas will not be cabled quickly and media corporations will vie for the limited number of broadcast bands available in the areas with most potential. Mitchell predicts the creation of a new underclass of digital hermits who are outcasts from cyberspace, newly marginalized in a partially wired world. This may be yet another technological revolution that has potential for all but which is dominated by the few, and those few may decide to cut themselves off in fortress-like urban enclaves where they can retreat into their electronic nirvana.

CONCLUSION

The age of telematics may mean the end of place-based meaning in postmodern societies. A real-time world will be of an altogether different geographical texture and not necessarily better for its inhabitants. The nightmares of *1984* and *Bladerunner* implied an urban wilderness more terrifying than the medieval visions of wild nature, with human identities sacrificed and surveillance so efficient that otherness became universal. To the modern, rational mind all of this may seem to be no more than fanciful speculation but we hope to have shown in this chapter that imagination has a powerful effect upon the way we live in the world and upon the way we mould it for a variety of purposes.

FURTHER READING AND REFERENCES

Short's (1991) book is a good introduction to the wilderness idea, and Derek Gregory gives a pithy summary of Said and other relevant writers.

Bowden, M.J. 1992: The invention of American tradition. *Journal of Historical Geography* **18,** 1, 3–26.

Carter, P. 1987: *The road to Botany Bay: an essay in spatial history*. London: Faber.

Cosgrove, D. 1984: *Social formation and symbolic landscape*. London: Croom Helm.

Daniels, S. 1993: *Fields of vision: landscape imagery and national identity in England and the United States*. Cambridge: Polity Press.

Falah, G. 1996: The 1948 Israeli–Palestinian war and its aftermath: the transformation and de-signification of Palestine cultural landscape. *Annals of the Association of American Geographers* **86,** 256–85.

Gregory, D. 1994: *Geographical imaginations*. Cambridge, Mass.: Blackwell.

Haraway, D. 1991: *Simians, cyborgs and women*. London: Routledge.

Jackson, R.H. 1992: The Mormon experience – the plains as Sinai, the Great Salt Lake as the Dead Sea, and the great basin as desert-cum-promised land. *Journal of Historical Geography* **18,** 1, 41–58.

Lutz, C.A. and Collins, J.L. 1993: *Reading National Geographic*. Chicago: University of Chicago Press.

Meinig, D.W. 1965: The Mormon culture region. *Annals of the Association of American Geographers* **55,** 191–220.

Mitchell, W.J. 1995: *City of bits: space, place, and the infobahn*. Cambridge, Mass.: MIT Press.

Nash, R. 1973: *Wilderness and the American mind*. rev. edn. New Haven: Yale University Press.

Nguyen, D.T. and Alexander, J. 1996: The coming of cyberspace and the end of the polity. In Shields, R. (ed.) *Cultures of Internet: virtual spaces, real histories, living bodies*. London: Sage, 99–124.

Pocock, D.C.D. (ed.) 1981: *Humanistic geography and literature: essays on the experience of place*. London: Croom Helm.

Pringle, T. 1988: The privation of history: Landseer, Victoria and the highland myth. In Cosgrove, D. and Daniels, S. (eds) *The iconography of landscape: essays on the symbolic representation, design and use of past environments*. Cambridge: Cambridge University Press, 142–61.

Said, E. 1979: *Orientalism*. New York: Random House.

Short, J. 1991: *Imagined country: environment, culture and society*. London: Routledge.

Tivers, J. 1996: Landscapes of computer games. Paper presented to the Social and Cultural Study Group Session, 'Virtual Geographies', Annual Conference of the Royal Geographical Society Institute of British Geographers, Strathclyde University.

Part 4

The global era

Introduction: The Post-Industrial World

This is a loosely defined term, used to indicate those high income economies where the proportion of the wealth created by manufacturing is declining and that devoted to services is increasing. Its outstanding examples are nations like the USA and UK which act as world financial centres, and Singapore where acting as a financing agent for much of South-east Asia is a major occupation. Nevertheless, all the nations that might be characterized as post-industrial are still producers of goods; they have escaped that phase only recently and the signs of it are still present. Since it is a transition rather than a revolution, chronologies are hard to define but 1950 might stand as the earliest time the term could be applied: the date when (for example) oil began to be a more important fuel in Western economies than coal.

The power for the Post-Industrial Economy (PIE) comes mostly from the same sources as its predecessors: from the fossil fuels, nuclear and hydro-power sources. There are some emerging emphases, however. The characteristic form in which the PIE wants energy is as electricity, plus refined oil for vehicles. Hence the popularity of nuclear energy (which can deliver only electricity) in a few countries. The costs and risks associated with it, especially after massive accidents like Chernobyl in 1986, have, however, retarded its development; in spite of a bad public image it accounts for 70 per cent of electricity generation in France and almost that in Belgium. But a 20 per cent maximum is the more usual figure. It is often said that capital invested in the nuclear industry has been at the expense of *renewable* energy sources and the politics have often been vicious. The PIE economy's awareness of environmental linkages of the production processes of a nation have, however, made the alternative energies attractive in certain places: Denmark has invested heavily in wind-farms, for instance, as have parts of upland Britain; tidal and wave power on a small scale have enthusiasts, and the holiday brochures for Greek

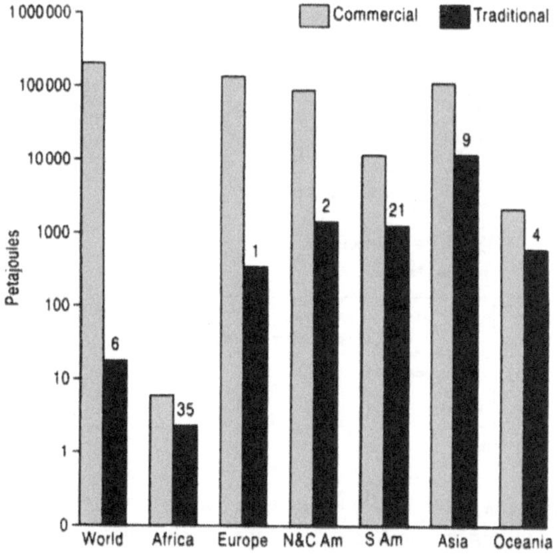

Figure P4.1 *The consumption of both commercial and traditional fuels in 1993, by major world regions. The number is the percentage of the total which is 'traditional', i.e. biofuels. The vertical axis is logarithmic.*

islands mention casually that hot water is likely to be an afternoon phenomenon because of its derivation from passive solar panels. Biomass energy is also likely to be revalued where oils and sugars can be converted to alcohol and oils for vehicle fuels. Apart from vehicles, however, technology is characterized by its miniaturization and capacity to store and process information: the palmtop computer can stand as symbol of this trend.

The PIE is not a place of shortage for perhaps 60 per cent of its inhabitants. They enjoy access to money and a reasonable degree of security. The other 40 per cent are less affluent and include an underclass of varying size, depending on state welfare policies. Where this group is large, then the cultural landscape of cities includes the shelters of the homeless, the *cardboard cities* of the underpasses and railway viaducts and concrete wastelands of public spaces. But the well-off with disposable incomes create a great cultural diversity in terms of demand for specialist shopping as well as the air-conditioned and safe malls of suburbia, for ethnic restaurants as well as world-wide hamburger chains. Each of these elements can contribute to a distinctive zone in a city and its environs and to agricultural production patterns somewhere, though not necessarily very close by since the outreach for ethnic foods or baby vegetables may be of the order of thousands of kilometres.

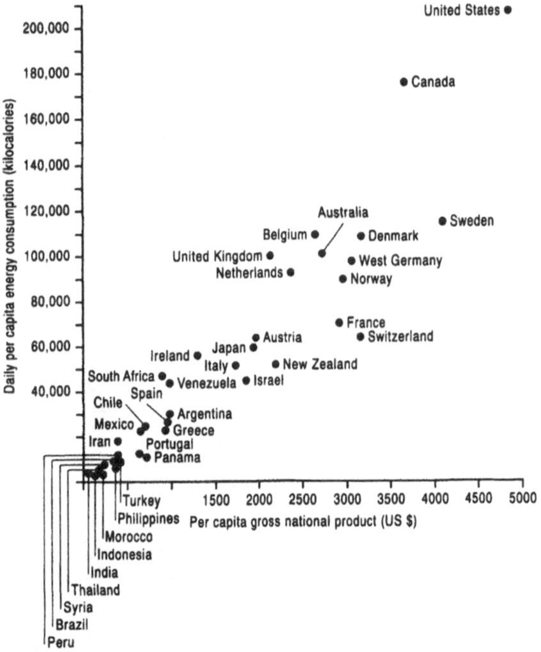

FIGURE P4.2 *Daily per capita commercial energy consumption plotted against per capita gross national product in the 1980s. This shows more or less that the richer countries consume much more energy per head of population than those which are poorer. But note that some of the poorer countries still rely strongly on biofuels, which are not counted in this reckoning, and that energy consumption is not* necessarily *an index of well-being.*

Goods are produced in large quantities for the populations of the PIEs, who are collectively known as consumers. If these goods are mass-production items such as motor cars, then they are often now produced in newly industrializing economies or even low income economies, and the production of the PIEs tends to be of short-run, specialist and tailored items. The term *segmented* is often applied, and the mode of production and consumption attracts the label *post-Fordist* to emphasize the lack of a massive investment in assembly lines. Services such as medicine, education and above all financial services seem to cater for insatiable demands.

If pollution by particles and chemicals is the characteristic downside of industrialism, the PIE is *par excellence* the site of noise. The young expose themselves to levels which affect their hearing but also cut themselves off into a private world via the Walkman, something their elders do when possible by buying a 4-wheel drive vehicle even when *off-road* means parking on the town's pavement. These are part of a segmentation of culture which means buying privacy, safety and silence and which then manifests itself in private hospital care, gated entrances to collections of houses and apartments, and looking for solitary villas for holidays: even buying an island when it is feasible. Inequality becomes then ever more manifest in PIEs since the ideological commitment to fair shares for all has not outlived even its imperfect manifestations in eastern Europe, the former USSR and centrally planned economies like Vietnam. In the landscape these inequalities can be borne by the comfortable 60 per cent provided that they are out of sight most, if not all, of the time.

21

GLOBALIZED LANDSCAPES

Globalization is about the compression of time and space horizons and the creation of a world of instantaneity and depthlessness. Global space is a space of flows, an electronic space, a decentred space, a space in which frontiers and boundaries have become permeable.

(D. Morley and K. Robins 1995: *Spaces of identity*. London: Routledge)

INTRODUCTION: WORLD SYSTEMS

How long has the world been a system of peoples interlinked by economy and politics? The answer to this question is important because the evolution of humanized landscapes has to a degree depended upon the flow of migrants, ideas and materials, and these are facilitated by the connexions forged by trade and conquest within a *world system*.

The answer given by Frank and Gills (1993) is that the world economy is about 5000 years old. They date it from 2700–2400 BC when the significant trading of surplus commodities emerged between the Indus Valley and Mesopotamia. It is no coincidence that both of these were regions of hydraulic civilization (Chapter 3) where the environment had already come under partial control through the application of irrigation water to cereals. The expanding realm of the interconnected world system thereafter was accompanied by human modification of a broader range of ecologies.

By the early Middle Ages there were eight interconnected circuits of trade that were participating in the accumulation of capital (*see* Figure 21.1). The strength of the links was dramatically demonstrated by the rapid spread of the Black Death along the established trading routes in the 1340s, as far afield as China and western Europe. An important point to note here is that Europe had no specially privileged position in this world system, which was made up of multiple nodes of power and prestige. Blaut (1993) has convincingly argued that the European miracle of rapid economic and demographic growth, coupled with technological innovation and political power focused in the nation–state, did not begin until the sixteenth century. This is also the date used by Immanuel Wallerstein (1974) and Fernand Braudel (1984) for the origins of economies of world significance.

World economies emerged in Europe after 1450, began to expand significantly only after about 1650, and became truly global after 1900. The date 1650 was significant because of the maturing of capitalism in the Netherlands and England. By then there were 20 000 trading ships in the world, two-thirds of which were owned by the Dutch, and a nascent financial system was emerging based upon promissory notes, banks and the organizational framework provided by companies such as the Dutch East India Company which traded in spices and other valuable commodities (Williams 1996).

World economies are city-centred because the city is an efficient agent of exchange and a mechanism of concentrating the wealth, skill and information necessary for mercantilist and capitalist entrepreneurship. For example, Venice was the dominant European trading city in 1500. It had begun as a defensive island settlement without any territory to speak of, but by the early sixteenth century it had acquired control over a far-flung empire in the Adriatic and east Mediterranean. Other cities rose to prominence in the sixteenth century, such as Genoa, Antwerp and Amsterdam, and later London (in the eighteenth century), so the centre of gravity of the world economy was continually shifting. It seems that no one city or state can be dominant for ever, as continues to be the case today.

The creation of an international market and international trade is a key feature of world economies, with a move away from self-sufficient regional economies producing the full range of local needs. Competition meant that the better organized or better endowed regions were most successful, capitalizing on their comparative advantage and creating specialized geographical zones. But the latter could only exist by being interdependent and complementary. The spatial structure of a world economy is made up of cores and peripheries. The core areas organize the exchange of goods, information and circulation of finance; they set prices; and they manufacture finished goods. Peripheral areas supply raw materials and act as captive markets for manufactured goods exported from the core.

In a prosperous phase of the economic cycles that seem to afflict all world economies, the core expands its influence by trading its specialized goods and services. In times of stagnation the core reduces its dependence upon the outside world for raw materials by becoming more self-sufficient. Alternatively, the countries of the core may also seek new, cheaper areas of supply and new markets in colonies under their direct control, hence stimulating imperialist expansion as a means of resolving a crisis in the world economy. Thus European countries sought to maximize opportunities by developing a portfolio of colonies, first, in tropical areas, for producing commodities such as sugar in the Caribbean, and, second, in temperate areas where markets could be developed amongst European settlers. There were two major pulses of imperialism: the period 1600–1750 and in the late nineteenth century, both coinciding with just such phases of stagnation in the international economy.

The outreach of the Europeans had effectively begun in the fifteenth century with the navigational expertise of the Portuguese and the acquisitive aggression of the Spaniards. They jostled for hegemony of the spice trade in the East Indies and eventually built empires from what had originally been a loose network of port-based trading concessions. The Dutch, the British and the

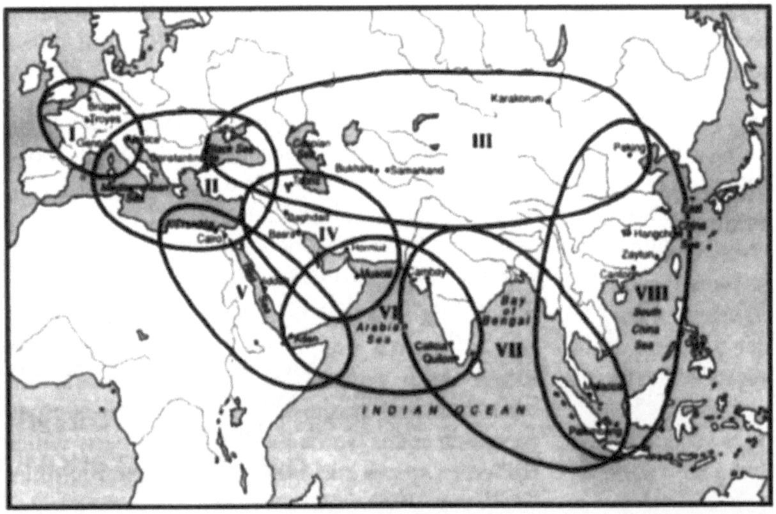

FIGURE 21.1 *The eight circuits of the thirteenth century*
Source: Abu-Lughod, J. 1993: Discontinuities and persistence: one world system or a succession of systems? In Frank, A.G. and Gillis, B.K. (eds) *The world system: five hundred years or five thousand?* London: Routledge, 278–91

French joined this expansion in the seventeenth century and together they drew upon resources varying from the products of the boreal forests (furs, wax, honey, tallow, hides, tar, timber) to the raw materials of the tropics (sugar, tobacco, cotton, indigo) (Jones 1981). This was an ecological windfall which was relatively cheap to exploit and it is not at all surprising that, as a result, Europeans sought to explore every corner of the earth in search of further wealth-creating opportunities.

World systems, environment and landscape

The environmental impact of world economies was varied. Even the early trade phase must have stimulated commodity production, as well as hunting and gathering, in the hinterland of the port cities. This would have created an incentive for landscape modifications such as woodland clearance for agriculture. Eventually in the case of the boreal forests of North America and Russia there was a depletion of certain species of fur-bearing animals as a result of trapping and the intensive cod-fishing of the Grand Banks (North Atlantic) was also not sustainable.

The imperial phase of conquest had unintended but catastrophic consequences in South America where the indigenous population had low natural immunity to common European diseases and was therefore reduced by up to 90 per cent by 1650 (Chapter 10). This inevitably led to the abandonment of the labour-intensive field systems and terraces that had been painstakingly built up over centuries and a new, alien imprint was imposed instead.

Perhaps the greatest make-over of landscape resulted from the phase of colonization. This involved the settlement of significant numbers of Europeans overseas. They transformed certain regions of the tropics with the regimented agro-ecosystems of sugar plantations in the Caribbean and tea plantations in India; and they tore at the earth to extract tin in Malaysia or copper in the Andes. But, as Crosby (1986) and Arnold (1996) observe, there were also European colonies of self-sustaining permanent settlement which did not depend solely upon trade with the metropolitan country. They relied for their success upon the manipulation of the new environment into something which could bear an approximation of the familiar agro-ecosystems of home. The tropical lowlands were, by this criterion, inhospitable and the search for *neo-Europes* was therefore confined to the temperate climes. North America, the southern cone of South America, South Africa, and the cooler fringes of Australasia experienced the recreation of European landscapes to such a degree that much of the local flora and fauna

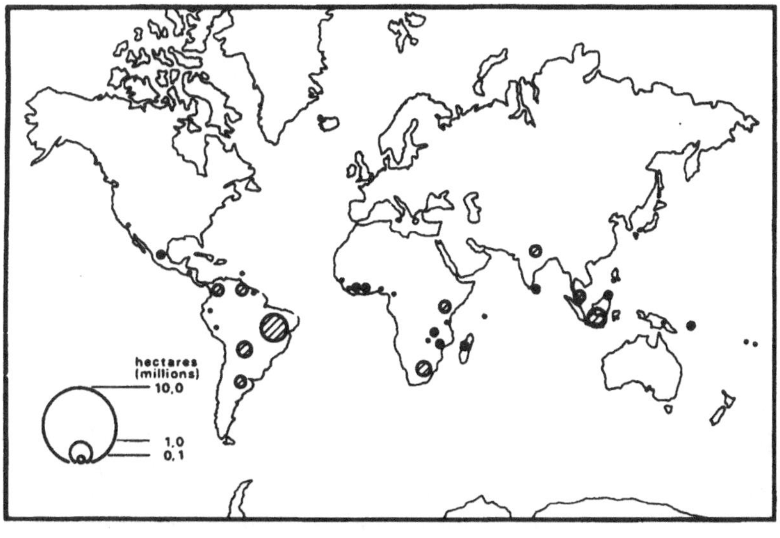

Figure 21.2 *British overseas investment in plantations, 1913*
Source: Christopher, A.J. 1985: Patterns of overseas investment in land, 1885–1913. *Transactions of the Institute of British Geographers* n.s. **10,** 452–66

followed the indigenous population into an oppressed lower profile or even extinction.

The tropical lowlands were not to escape attention, however. In the twentieth century, former colonies encouraged their populations, as part of the state-building process, to expand the ecumene of settlement and agriculture by cutting tropical rainforest and displacing any existing tribal inhabitants. In recent times this activity has been pushed up to the borders with neighbouring states on the grounds that unsettled land might invite territorial claims from rivals. The process of internal colonization is epitomized by the waves of settlers in Brazil who over the last two centuries have left the coastal belt to grow cash crops for export such as sugar (seventeenth century) and coffee (nineteenth century to present), and who in recent times have cleared the forest for cattle ranching and subsistence arable farming. Their common characteristic is the treatment of the extensive tracts of land at their disposal as a good without value, to be exploited and then discarded after a few years, the soil exhausted and worthless. The casual destruction or depletion of whole ecosystems, and resulting soil erosion and other forms of land degradation, have become a taken-for-granted outcome.

FIGURE 21.3 *The Baltic flour mill, Gateshead*
Source: P.J. Atkins

As we noted earlier, the pulses of imperial and internal colonization coincided with fluctuations in the world economy and it is possible to portray the resulting environmental and landscape change in terms of pulses or rounds of investment by traders and governments. In the most intensively settled and exploited colonies, the level of landscape transformation was extraordinary, especially where plantations (*see* Figures 21.2 and 21.5) and mines were geared to the export of materials to feed processing and manufacturing industries in the home country. The British used peninsula Malaysia in this way, ruthlessly extracting value in the form of rubber, pineapples, palm oil, coconuts and tin (*see* Figure 21.4). Such cash crop and mineral-based economies persist in poor countries down to the present day, with the power wielded now by transnational corporations but using a colonial legacy of physical infrastructure such as irrigation networks, roads, railways, and port facilities.

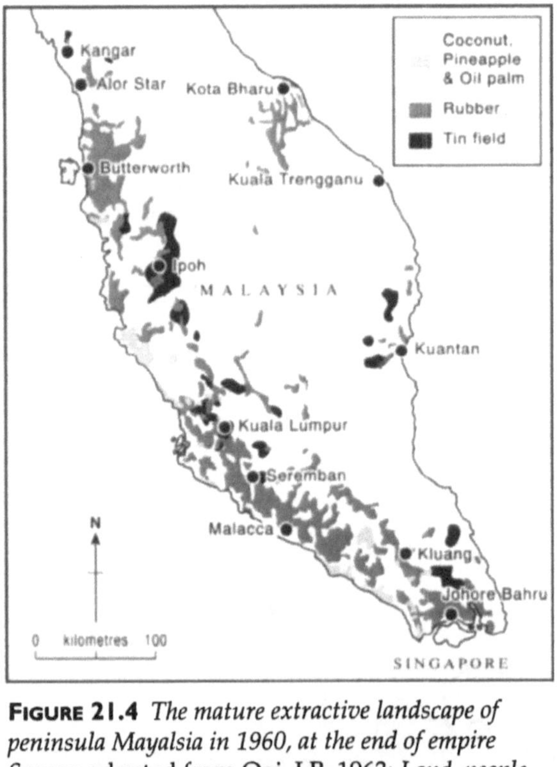

FIGURE 21.4 *The mature extractive landscape of peninsula Mayalsia in 1960, at the end of empire*
Source: adapted from Ooi, J.B. 1963: *Land, people and economy in Malaya*. London: Longman

FIGURE 21.5 *Rubber tapping in Kerala, India*
Source: P.J. Atkins

Any impression that imperial landscapes were mainly rural would be incorrect. The numbers of alien soldiers, colonizers and bureaucrats were often very small by comparison with the indigenous population, and the exercise of control was therefore usually urban-based. Towns and cities were efficient concentrators of power and much care and attention was lavished upon the choice of suitable locations and the planning of the urban environment for the greatest convenience. Pre-existing towns were either ignored or were added to with a European quarter that was often very different in ground plan. Impressive symbols (buildings, monuments) were constructed to convince the local population of the superiority and inevitability of colonial rule (*see* Figure 21.6), and much imperial townscape also had a flavour of the metropolitan country in order to create links of culture and sentiment (*see* Figure 21.7). In the post-colonial era the presence of surviving European clubs, churches, parks, and even street names may seem odd in the centre of an Asian or African city until one considers that they are merely the traces of one globalized landscape that has been replaced by another, the gleaming skyscrapers that are crowded into every central business district in the world. The landscapes of global capital have replaced those of imperial might.

FIGURE 21.6 *The Gate of India, Bombay*
Source: P.J. Atkins

A curious feature of the colonial form of imperial landscapes was the creation of reservations for displaced or endangered indigenous populations (Table 21.1). These were usually tribal peoples whose formerly extensive lands had been appropriated and who were now to be

FIGURE 21.7 *Central Tunis*
Source: P.J. Atkins

TABLE 21.1 *Native reserves in various British colonies, c. 1930*

	Total area (million ha)	Reserves (million ha)	European land (million ha)
Canada	954.3	2.1	65.8
Australia	770.4	54.1	460.7
New Zealand	26.8	1.8	15.3
South Africa	122.2	8.9	82.8
Southern Rhodesia (Zimbabwe)	38.9	8.7	13.0

Source: Christopher (1988, 162)

settled in marginal areas which had no conceivable use to the colonizers. Here the landscape was to be protected and the tribal people were prevented from participating in the modern economy. In the case of Malaysia, for instance, the inhabitants of the Malay reserves could not produce rubber (Christopher 1988).

The overall impact of the growth of population, the expansion of settlement and the intensification of the world system over the last two centuries can be seen in Table 21.2. Note in particular the changes in cropland in the Americas, although the percentage figures are somewhat misleading since they record growth from a relatively small base. Note also the destruction of forests and woodlands in Latin America, South Asia and the Middle East.

TABLE 21.2 *Estimated world land cover changes (per cent), 1700–1980*

Region	Forests and woodlands	Grassland and pasture	Croplands
Tropical Africa	–20.9	+10.1	+404.5
North Africa/ Middle East	–63.2	–5.6	+435.0
North America	–7.3	–13.7	+6666.7
Latin America	–20.3	+26.2	+1928.6
China	–57.0	–2.9	+362.1
South Asia	–46.3	–1.1	+296.2
South-east Asia	–7.1	–26.4	+1275.0
Europe	–7.8	–27.4	+104.5
Former USSR	–17.3	–0.3	+606.1
Pacific developed countries	–7.9	–4.9	+1060.0
World	–18.7	–1.0	+466.4

Source: Richards, J.F. 1990: Land transformation. In Turner, B.L., Clark, W.C., Kates, R.W., Richards, J.F., Mathews, I.T. and Meyer, W.B. (eds) *The earth as transformed by human action*. Cambridge: Cambridge University Press, 163–78.

MODERN GLOBALIZED LANDSCAPES

Drawing together post-industrial and low income economies alike today, there are certain human-dominated processes which we can characterize as global in scope. The material economy is one such. Following the example of oil and of metal ores, goods are transported very large distances: coal from Colombia to Denmark, for example; roses from Kenya to Stockholm. Such movements may be controlled by a series of companies passing on the goods from hand to hand, as it were. But, equally likely, vertical integration may have produced the transnational company (TNC) which owns both the flower farm and the flower shop. The non-material economy of information and money is the outstanding example, however. Electronic communication of data in digital form has revolutionized the speed of transmission and its penetration to almost anywhere with a telephone line or a satellite receiver.

The central phenomenon is without doubt the global corporation. Apart from size, the operations of such a unit stress the importance of speed, innovation and the irrelevance of time and space. In the early 1990s there were 37 000 of these corporations and among other things they controlled much of the direct investment in both their headquarters locations and elsewhere. This includes the developing countries, which received 25 per cent of all capital investment inflows in 1991. In general terms, services now account for about half of all such investment, rather than commodities. The concentration which is represented by the city is paralleled in the rural areas of industrial nations, where investment in agribusiness brings about hot spots of production. Hence in the European Union, 80 per cent of agricultural production comes from 20 per cent of the businesses and these are located in the regions of the Paris Basin, East Anglia, Emilia-Romagna and the southern Netherlands.

Trying to capture the essence of today's fast-changing world is difficult. Johnston *et al.* (1995) suggest that the major processes which underpin today's cultural landscapes can be crystallized out as characterized by (a) new economic processes which emphasize flexible accumulation with a polarized workforce; (b) the collapse of rural economies but the growth of mega-cities; (c) migrations of both élite workers and the very poor; and (d) religious revivals.

THE SETTING

The world is occupied as never before. Some rural places have a lower population density than say 1000 years ago, but most are more densely inhabited. Overall, the globe now has about 5.8 billion people. Average densities per continent are shown in Table 21.3 and the absolute numbers for the world in the last few decades, together with the growth rate in Table 21.4. Similarly, incomes per capita have risen, though as with population, tremendously unevenly. Nevertheless, more people than ever before have access to means of communication: 60 per cent of the global population has access to a TV set, for example. (In the early 1990s there were *c.* 600 million telephones in the world and an equal number of TV sets.) The distribution of the monetary fruits of industrialization is, as we all know, uneven and is clearly not static. This means that the cultural landscapes of some regions are undergoing very rapid transformations and that, for example, information about them given here will very likely be out of date by the time it is read. The NICs, as exemplified by Malaysia, Taiwan and Mexico, contain regions which are undergoing immense changes in both rural and urban areas as money underpins development and the population migrates towards the towns. Not all such changes may last for very long: industrialization of parts of China which are predicated on cheap labour might not survive a political change which decided against the free-market principles that encourage the driving-down of labour costs. It may well be that one of the strongest globalizing forces will be the 1990s' GATT agreements on free trade, which will locate resource use and environmental change wherever capital decides to station itself and probably reduce the role of any legislation concerned with environmental protection.

TABLE 21.3 *Regional population densities (1994) (per square kilometre)*

Africa	23
Latin America	23
North America	13
Asia	107
Europe	32
Oceania	3
World	42

Source: United Nations Demographic Yearbook

THE TECHNOLOGY

A constant theme of this book has been the key role of technology in enabling societies to

TABLE 21.4 *Regional population 1961–89 (millions)*

	1950	1960	1970	1980	1990	Annual increase (per cent)
Africa	224	282	364	476	633	2.8
Latin America	166	217	283	358	440	1.8
North America	166	199	226	252	278	1.0
Asia	1403	1703	2147	2642	3186	1.6
Europe	549	605	656	693	722	0.2
Oceania	13	16	19	23	26	1.5
World	2520	3021	3697	4444	5285	1.6

Source: United Nations Demographic Yearbook

construct their landscapes. In the post-industrial world this is no longer always so obvious: the factory chimney and the bulldozer intrude upon our senses in a way that micro-electronics may not. In fact, although many appliances have got much smaller (contrast a 1930s' television set with a pocket version from the 1990s, for instance) they may well get smaller still: from micro-electronics to nanotechnology. Prophecies for the latter are still in the realm of science fiction but include a world in which people do not *have* to do anything, which might well change landscape patterns.

The classification of the effects of the present generation of digital electronics has not yet been made. For our purposes, we might look at four particular outworkings that seem to contribute to cultural landscapes:

1. Fast air travel.
2. Technology as a substitute for actual experience.
3. Technology as a carrier of information in which symbols substitute for anything more concrete.
4. Technology as a culturally validating medium.

Fast air travel

The passenger jet itself is not entirely an electronic matter, but the control systems of modern aircraft are computer-based, as is the navigation and probably part of the air traffic control network in which it is held from start to finish of its journey. People, along with their ideas and a limited quantity of goods, can be taken almost everywhere within 24 hours. The relations between space and time have changed: absolute distance is no longer at issue: time and cost have replaced it.

Technology as a substitute for actual experience

The pocket-sized tape or disc player is used to allay some of the effects of living in towns since it shields the owner from unwanted noise and from the close proximity of many other people by creating a bubble of personal space. If we believe the more gung-ho media discussions, then virtual reality (Chapter 20) will be able to substitute for many kinds of landscape experience. One wall of a home will be able to display a mountain forest with streams, birds and all; wearing the special helmet we will be able to experience what it is like being at the top of the Jungfrau. The rather unconvincing demonstrations available on the World Wide Web are just the beginning, we are told. Less attractive in many ways is the possibility that the Internet is one more form of American cultural imperialism.

Technology as information carrier

Technology will be a carrier of information in which symbols substitute for anything more concrete. The prime example is money: the international exchanges and financing centres in London, New York, Tokyo and their ilk send vast sums to each other at the press of a key and these sums seem to bear no relation to any materials. Indeed, it might be possible that the quantity of symbolic money whirling round the exchanges of the world may exceed all the resources that it could buy. It is certainly true that much of it is simply gambling money (based for example on stock futures and derivatives) not tied even to hog futures or the possible price of copper ore in Bougainville in 12 months' time. But now and again, a piece of this money whirls off like a spark from a catherine wheel and lands in a place where it is used to extract a resource or build a dam.

Technology as a culturally validating medium

Every society has seemed to need its opinions and facts validated. In medieval Europe, such validation came from the pulpit; in the eighteenth century from books, in the first half of the twentieth from newspapers and now from the small screen. Notice how many TV ads actually portray information on a terminal within the screen and how nobody appears in a documentary programme without a PC in the background. So seeing is believing: being told on telly means it's true, especially at election time. There is an extension of this in the form of the personal telephone, when middle-rank managers attempt to exaggerate their status by being called frequently: a kind of electronic equivalent of the codpiece.

RESULTING LANDSCAPES

Most of these developments are too new to have formed distinctive cultural landscapes. Also, because they grow out of existing landscapes they may not be easily visible. One phenomenon certainly defined and intensified by the processes described above is that of the World City. This is an urban agglomeration which is not only important nationally and regionally but plays a world role as well. Hence, the great financial centres of the world form a super-league in terms of their influence. They are characterized by being the sites of the leading global markets in finance and commodities, having clusters of specialized businesses which are international in scope, concentrations of corporate headquarters, being the sites of inter-governmental organizations and being close to nodes of the global media. In spite of the ubiquity of fax, telephone and e-mail, the people in this world like to see each other and so there is an agglomerating effect. This produces the high land-price, high buildings syndrome. In turn that is intensified by the corporate status urge to have the biggest and flashiest building from the trendiest architect. The wealth from money moving spins off into smart restaurants and up-market sandwich shops together with a few boutiques selling essentials like Italian suits. The money generated spreads out into the rest of the city through, for instance, property prices and the conversion of less-intensively used sites for expensive car sales.

The number of cities that can be called world cities depends upon the method of classification. Knox (1995) uses four broad criteria to suggest that only four cities fulfil the requirements for super-league status: London, New York, Paris and Tokyo. A second division has most of the necessary functions but a smaller global share: examples are Frankfurt, Los Angeles, Rome and Zurich. Because of their low share, some conventionally included cities are very low in such a list: Singapore, Hong Kong and Vancouver, for instance. But a wider view of both global, multinational and national role might produce a different hierarchy. Here, the top tier includes only London, New York and Tokyo. A second rank

TABLE 21.5 *Post-industrial landscapes: examples of the new regions in cities, compared with a rural area*

	London	Singapore	French Alps
Technoscapes		Numerous centres of electricity based industry e.g. Jurong	Towns with hi-tech manufacturing and knowledge-based industry e.g. Grenoble
Financescapes	Clusters of corporate high-rise e.g. Wapping, the City	High-rise downtown development	Second homes of those able to afford rural retreats
Ethnoscapes	Chinatown as part of tourist fabric; Irish focus in e.g. Kilburn; South Asia in Southall	Preserved character districts e.g. Little India	Evidence of globalizing tendencies overcoming the Frenchness: pizza parlours, discos; English pubs
Mediascapes	Tourist must-see attractions like Trafalgar Square	The variety of cuisine in all regions of the state	Expectation of seeing shy/rare wildlife
Ideoscapes	Privatized public transport: bunched buses and confusing railway termini	Traffic control systems and high price of car ownership	Relaxed controls on development in rural areas compared with Switzerland or Austria
Commodityscapes	Knightsbridge and Harrods; Oxford St.	The Orchard Road shopping centres	Winter sports developments; National Parks

comprises Miami, Los Angeles, Frankfurt, Amsterdam and Singapore. The third level consists of Paris, Zurich, Madrid, Mexico City, São Paulo, Seoul and Sydney.

The working-out of all this diversity and change on to the visual landscape has provoked a number of new classification systems for the urban fabric of large cities which replace the older concentric and nodal models. Instead, terms like *Technoscapes, Financescapes, Ethnoscapes, Mediascapes, Ideoscapes* and *Commodityscapes* are produced (Table 21.5), which impose a finer grid on the structures than the other models. The contents of each need some definition in each case and it may not always be possible immediately to make maps that clearly differentiate: the HQ of Globalgreed Finance may look very similar to that of Worldtrash Media Corporation, just across the road; they both probably retained the same architect during the 5-year apogee of his popularity.

A rash of Sunday supplement articles in the 1980s predicted an opposite trend. The computer plus an Internet connexion would, it was claimed, lead to home-working on a massive scale (Chapter 20). This has to some extent occurred but it is mostly the farming out of routine communications rather than the white-hot deals or the dazzling piece of architecture which takes place. Its visibility is, for obvious reasons, muted but the data from the re-population of the countryside in Western nations is probably a reasonable index of it. Of course, it is seen as especially suitable for women; men are probably less keen because of the toddler effect ('daddy, come and . . .') and the absence of the expenses-paid business lunch. So its expansion is probably subject to quite a low ceiling. The rural areas which are peripheral to the main agribusiness centres in high income economies are, however, potential sites of countervailing forces. They provide a locale for non-industrial food (organically reared meat, fruit and vegetables for example) and for a more conservation-minded attitude to rural resource management.

It is much easier to discuss the world city than the dispersed rural network. The latter is, however, being more extensively employed not to produce but to inform, particularly by way of delivering formal education. So two world cities are allied to a rural region in this account. The lesson of the cities in landscape terms, however, is that rapid change is inevitable. So any account in print is out of date before it is even written, let alone printed.

CASE STUDY: ROADSIDE ICONS OF THE GLOBAL ERA

The petrol (gas to Americans) filling station is a symbol of the twentieth century. It is testimony to the universality of the motor vehicle throughout the developed Western world, and increasingly in the South also. We have become so used to the convenience of the regularly spaced service stations that no urban landscape would be complete without one, and they are tolerated even in picturesque rural locations where planning permission for industrial or retailing premises is at a premium. In short, we are now as dependent upon fuel for the internal combustion engine as we are upon food and drink.

Jakle and Sculle (1994) argue that petrol station architecture is distinctive and has acted as a form of advertising for the different oil companies, a method of imprinting brand loyalty and therefore creating loyal customers. Yet such roadside landscapes have also become generic elements of the globalized culture that is now spreading around the world (*see* Figure 21.8). A gas station in

FIGURE 21.8 *A filling station in 1990s' Britain*
Source: P.J. Atkins

Bangladesh or Tunisia is likely to look like one in America or France. They are now part of the language that everyone understands. Even the former distinctly male-orientation of utilitarian concrete forecourt, the greasy overalls of the attendant, the strong smells and the partially dismantled cars, has been toned down by the petrol retailers who have changed the atmosphere to make it as non-threatening as possible.

According to Relph (1987) the roadside strips in America are the epitome of the free market corporate economy. The neon signs, advertising hoardings, and the customized architecture are chaotic in their dazzle of colour, bright lights and confusion of company logos but the underlying theme is a uniform one of attracting passing trade with the comforting familiarity of replicated mini-landscapes. Since the 1970s the stylistic variety of architecture has been reduced and the franchising of well-known brands of fast food has meant that the golden arches signifying McDonald's hamburgers are now as common as Shell or Texaco filling stations (*see* Figure 21.9). The roadside has become a corporate landscape of the most bankable kind.

FIGURE 21.9 *The roadside landscape of corporate fast food*
Source: P.J. Atkins

CONCLUSION

The homogenizing influence of modernity is one of the most powerful forces in cultural history. In recent decades it has spilled over from the West into the less developed world and now touches the lives of billions of people, at least those who live an urban existence. This has been experienced in the physical outcomes of architecture and goods such as Coca Cola but less tangible traditional social values are also succumbing to the lure of Western culture through the influence of films, television, literature and, most recently, the Internet. Further developments along these lines seem inevitable.

FURTHER READING AND REFERENCES

The books of Wallerstein (1974) and Braudel (1984) are classics on world systems. Williams (1996) and Christopher (1985, 1988) are good on imperial expansion. Arnold (1996) has written a simple introductory text.

Arnold, D. 1996: *The problem of nature: environment, culture and European expansion.* Oxford: Blackwell.

Blaut, J.M. 1993: *The colonizer's model of the world: geographical diffusionism and Eurocentric history*. New York: Guilford Press.

Braudel, F. 1984: *Civilization and capitalism 15th–18th century, vol. III: the perspective of the world.* London: Collins, Chs 1 and 6.

Christopher, A.J. 1985: Patterns of overseas investment in land, 1885–1913. *Transactions of the Institute of British Geographers* N.S. **10,** 452–66.

Christopher, A.J. 1988: *The British Empire at its zenith.* London: Croom Helm.

Crosby, A.W. 1986: *Ecological imperialism: the biological expansion of Europe, 900–1900.* Cambridge: Cambridge University Press.

Jakle, J.A. and Sculle, K.A. 1994: *The gas station in America.* Baltimore: Johns Hopkins University Press.

Johnston, R.J., Taylor, P.J. and Watts, M.J. (eds) 1995: *Geographies of global change: remapping the world in the late twentieth century.* Oxford: Blackwell.

Jones, E.L. 1981: *The European miracle: environments, economies, and geopolitics in the history of Europe and Asia.* Cambridge: Cambridge University Press.

Knox, P. 1995: *Urban social geography: an introduction.* 3rd edn, Harlow: Longman.

Relph, E. 1987: *The modern urban landscape.* Baltimore: Johns Hopkins University Press.

Wallerstein, I. 1974: *The modern world system: capitalist agriculture and the origins of the European world-economy in the sixteenth century*. New York: Academic Press.

Williams, M. 1996: European expansion and land cover transformation. In Douglas, I., Huggett, R. and Robinson, M. (eds) *Companion encyclopaedia of geography: the environment and humankind*. London: Routledge, 182–205.

22

CONSERVATION

What would the world be, once bereft
Of wet and wildness? Let them be left,
O let them be left, wildness and wet;
Long live the weeds and the wilderness yet.

(Gerard Manley Hopkins 1881: *Inversnaid*)

LANDSCAPES OF PROTECTION

This book has shown time and time again that most cultural landscapes are influenced by production: of goods, services and knowledge in particular. The use of resources and the environmental consequences, together with the basic (and not-so-basic) needs of those doing the production, result in the landscapes we describe and explain. Yet examples have surfaced from time to time of priority being given to nature rather than to immediate human needs. Admittedly, the end-product may be human pleasure, but it has been recognized that only by allowing the process of nature to remain undisturbed by human societies can that end be achieved. The restraints practised by hunter–gatherers, the setting-up of gardens, the creation of picturesque landscapes, the North American cult of wilderness are all examples mentioned earlier in our text (Chapters 1, 18 and 20).

Reasons for protection

No single reason exists for the protection of what is perceived as the natural world. The variety of human cultures has produced a diversity of reasons and not all the reasons (like the divine right of kings to hunt certain animals) nor the systems of protection have endured, since other pressures for change may have prevailed. We ought to add, however, that all the protected areas now depend upon the integrity of human societies for their functioning: a 5000-megaton nuclear exchange would probably destroy much of nature as well as of humanity via the ensuing nuclear winter.

SACREDNESS

The best examples of this quality are in the sacred groves of classical Greece. Here, most communities designated an area as sacred to the special *genius loci*, the spirit of the place. That meant, for example, that the area was closed to grazing animals, that the trees (for it was usually a grove of trees in a valley) might not be felled and that fire was forbidden, except for ritual purposes. In the grove, the deity usually had a shrine, so that the area was not totally natural, but normal economic actions were definitely suspended, no doubt under threat of divine penalty for transgression. An extension of this thinking is seen in the way religious institutions often surround themselves with a protected landscape. This may often be near-natural and the intention is that it should be a place of spiritual rather than material harvest. The best examples in Europe can now be seen in the areas surrounding monasteries of the Orthodox tradition, such as Mount Athos reaching to 2033 metres above sea level on its peninsula in the Aegean. Here the forests are kept unmanaged and indeed female animals of domestic varieties are not allowed. Fire suppression means, as in many Mediterranean forests, that the vulnerability to wildfire is increased. A parallel phenomenon is seen in the hills around Kyoto in Japan, where the Buddhist monasteries own the forested terrain. Originally for meditation, it now provides a valuable reservoir of public space (and hence is not totally unmanipulated) as well as watershed protection in an area vulnerable to landslides in periods of heavy rainfall.

Pleasure

Human pleasure in the wild and untamed places of the world is often a recent phenomenon, coinciding roughly with the ability to bend much of it to human purposes when required. So the romantic movement of the nineteenth century (see Chapter 20) makes the rough places inspiring rather than terrifying (or at any rate pleasurably terrifying, like a horror movie): the Alps and the English Lakes were two places much revalued by the time the railways reached them. But the appeal of the primitive has persisted and so many such areas have been given protected status under legislation, often as National Parks. (This function may overlap with the category of wildlife protection.) The National Park is frequently supposed to be an island in time: a vignette of how it was before humans had altered the landscapes apparently permanently. The savannas of Africa, the glaciers of the Canadian Rockies, and the Biaowieza forest of eastern Poland are all thus designated. Visitors are encouraged, provided that their behaviour is deemed appropriate and that not too many facilities have to be provided. Tensions thus exist between the desire to have the awe-inspiring experience and the desire to have it in not too much discomfort. Further, ecological history often shows that the landscape is not as natural as all that and so the desirable landscapes are in some ways cultural and so their preservation may be contrary to the processes of nature which are supposed to be dominant. Yosemite NP in California, for example, is famed *inter alia* for the beauty of the meadows with oak groves along the valley floor: a landscape produced by the long-extinct Indians who fired the landscape to encourage the oaks to produce heavy yields of edible acorns.

Profit

It is often the case that the types of landscapes described as protected are under public control. They represent, as it were, a nationally (or even internationally) shared resource which somehow constitutes a shared cultural expression, rather in the manner of a Royal Family (well, sometimes) or a national bard like Shakespeare. In ages of rampant capitalism or feudalism, of course, that does not have to be so. Protected landscapes can be privately owned and the users charged to enter. (This is often done with public resources also, but largely, in order to control numbers and to provide income to offset the running costs.) The waterfall at High Force on the River Tees in County Durham, England, is an example: on the north bank of the river there is only one possible car park and only one place of entry to the path leading to the falls. Both exact an entry charge. If access to a natural landscape is difficult, then a profit-making enterprise may be allowed to monopolize the transport system: a mountain railway or a chair-lift, for example. In general, however, this practice is more commonly found with cultural monuments than with the natural world.

Biodiversity and Wildlife

The biological sciences have awakened humanity to the fact that species extinction is taking place at an unprecedented pace. Instead of one species loss every few decades, there is now one in hours or less. One way of conserving the great pool of genetic variety inherent in the millions of species of plants, animals and micro-organisms is to keep threatened species in laboratories, zoos and botanical gardens, as seeds, sperm and other tissues. But no evolution takes place in drawers and refrigerators. Hence, the protection of species in all kinds of nature reserves creates a landscape element devoted to the maintenance and indeed enhancement of biodiversity. The elements of this landscape are called by a variety of names: nature reserves, biosphere reserves and also national parks are the commonest terms. But if their purpose is to perpetuate natural diversity by preserving the web of ecosystems in which it flourishes, then human intervention may be necessary. Few such places are large enough for the ecosystem boundaries to coincide with the legal boundaries and so management is often needed to, as it were, keep the inside in and the outside out. A herbivore mammal species which is migratory may spend part of the year outside a National Park and there become a competitor for forage resource with domesticated flocks. On the other hand, a nature reserve in a small basin might well be susceptible to the inwash and concentration of biocides from surrounding agricultural land. Only in the very largest reserves is it possible to maintain a totally 'hands-off' policy and to do that may well mean restrictions on the number of humans who can visit: Antarctica is a good example.

Heritage and History

Landscape can be a powerful medium of manipulation, both in the process of its evolution and in the present as a database of historical images, myths, traditions, relicts. The selection or sometimes invention of these elements for conservation/preservation is ideologically and very often politically charged. Heritage has been called fake history (Hewison 1987) but it is often of more significance than the real past in shaping the future (Chapter 17). National symbols such as the Tower of London, Stonehenge, Edinburgh Castle, and Anne Hathaway's Cottage have come to represent in some sense *Britishness*, if only to foreign tourists. To a degree, this is innocuous but consider the potent and potentially dangerous power of imagery and symbolism in the hands of nationalists who often use them as an excuse for intolerance and belligerence. Continued Serbian military occupation of Kosovo, for example, with its largely Albanian population, is justified because the historic heartland of Serb culture is said to lie there.

The identification and preservation of valued landscapes, by the listing of buildings, and the delimitation of Conservation Areas and Sites of Special Scientific Interest, are of course a matter of opinion, and there is power in the exercise of the selection process: UNESCO recognize certain landscape types in their World Heritage List. These are:

1. Designed landscapes, for instance, historic monuments, parks and gardens.
2. Organically evolved:
   (a) relict landscapes, usually archaeological sites but also the remains of industrial and mining operations, abandoned irrigation works;
   (b) continuing landscapes. These are difficult to define but a good example is the rice terraces of the Ifugao, Philippines, which are of ancient origin but are still worked and preserved in their mature form.
3. Associative landscapes, especially natural features which have some important religious, artistic or cultural associations. The Tongariro National Park in New Zealand is listed under this heading because of a mountain range sacred to the indigenous Maori people.

Source: von Droste *et al.* (1995)

Just as important as this physical establishment of a hierarchy of objects and areas to be revered is the writing of history itself, which inevitably is partial. Much of the significance of history lies in what it leaves out and the groups whose histories are ignored or marginalized.

Even an apparently innocent guide book may have an agenda of selection and representation which is not immediately apparent. Roland Barthes made this clear in his study of the French *Guides Bleus* which purvey a bourgeois ideology and a puritanical appreciation of nature through fresh air and rugged landscapes (Duncan and Duncan 1992). They appreciate church architecture but ignore the achievements of the working-class and ethnic minorities.

In another sense, the protection of the past can be seen as akin to that of biodiversity. The visible signs of the past, no less than old maps and documents, are part of the palimpsest and so part of the cultural (rather than biological) diversity of a society. Most people's surroundings are the richer for containing a visible historical depth and some action is therefore often necessary to protect those parts of it more imperilled by age and decay. Old buildings, for example, can be adapted to new uses: a castle may house students who are less worried about the cold than their elders; a timber-framed town house looks good as the offices of an old-established law firm. Where economic change is rapid, then those who remember earlier times are reluctant to see all their personal history redeveloped or reclaimed as playing fields. So the gear and buildings at the head of a mine become a museum, with visits down the shaft to the original working faces, as with the copper mines at Falun in Dalarna (Sweden). When, for instance, the centre of a town has a harmonious collection of buildings of various ages, then public sentiment usually prevents the construction of the very modern in case it destroys the look of the whole. Fights between architects or between developers and citizen groups are common.

The past may be quarried because there seems to be little future. A nation insecure about its trading position with regard to manufactures may try to balance the books with a vigorous tourist industry based on its long and variegated past. In addition, the reconstruction of the past may well be part of that use of history which seeks to create a particular national identity. So in the case of Great Britain, the tourist is encouraged to think of the industrial past as one of working-class solidarity, of the middle classes focused on village cricket, warm beer and the Church of England, and of the upper classes in luxurious homes surrounded by landscape parks and fat cattle organically reared. These images are fostered by industrial museums which in general issue the visitor with a pair of rosy-tinted spectacles on entrance and rarely allude to the lock-outs and the wage-cuts, the industrial diseases and the rates of accidental death. For the middle class, there are periodic calls for a return to family (even Victorian) values, which presuppose a stability and harmony which no serious study has ever shown to exist, and for the upper echelons, bodies like the National Trust lovingly cherish the material exhibits of a class of people who shirked no effort to make money. Study the iconography of Lord Armstrong's house (Cragside) in Northumberland as an example: most wars in the mid-nineteenth to mid-twentieth centuries were supplied on both sides by his warships. So neither visitors from abroad nor the natives are shown anything other than a rosy and sanitized picture of the past, the better to disturb neither with uncomfortable ideas.

CASE STUDY: THE ALAMO

In 1836 a small group of Texans made an armed stand against the Mexican dictator Santa Anna and were massacred in the grounds of a mission at San Antonio known popularly as the Alamo (De Oliver 1996). Their sacrifice in the cause of the independence of the Republic of Texas (1836–45) has been commemorated in the creation of a monument managed as a tourist site.

It was not until 1905 that the Alamo buildings and grounds were acquired as a memorial. The Daughters of the Republic of Texas became its official custodians and there began a struggle between different cultural factions on how the restoration should proceed. The debate was between the Spanish Texans (Latinas or Tejanas) who wanted to recreate the spirit of the original Spanish mission and the Anglo-Texans who preferred the theme of Anglo soldiers' martyrdom. The latter group prevailed and tourists today, although they are in a town with a Latino majority, visit a shrine to the Anglo perspective of American history. This mirrors an economy and society in which Latinos are a subordinate group in terms of employment and the status hierarchy.

Not only was the Alamo appropriated as a factional and therefore oppressive symbol, but the facts of history were modified at the convenience of the restorers. Most of the mission buildings were demolished and the militarily insignificant chapel was made the centre piece. The façade of this was adorned with a bogus roof line, or parapet, that owes more to Protestant Dutch or German architecture than to the Spanish original.

The Alamo is unusual in the modern heritage industry as having no entrance fee and no commercial activity within its grounds. Its place is assured, however, as the key location within a commodified tourist landscape of San Antonio which has been established and exploited mainly with Anglo-American capital and where consumption is dominated by the spending power of the same group.

In all respects, then, we may identify the Alamo as a landscape created by and for one group in society as a means of reproducing and enhancing their power and prestige. This is, in the words of De Oliver, 'historical preservation filtered through a racist lens'.

CASE STUDY: THE UNITED KINGDOM TODAY

The past may not always be such a different country, therefore. It may be a mirror to our present concerns; more precisely perhaps it is a set of filtering lenses and the light that enters is carefully screened for correctness. History is not only incomplete because it cannot be otherwise but its incompleteness may be determined by one of today's purposes.

A place like the United Kingdom today is important in these kinds of studies for a number of reasons. The first is the role of the countryside and nature in the national culture. Running through the English-dominated culture of the twentieth century is a powerful nostalgia for a rural England, often evoked as a national icon in times of war. The resource consists of a land of high population densities where intensive agriculture has been the norm, contrasted to almost unpopulated uplands, wet and cold, where the human-produced landscape is in fact perceived as truly natural. Nearly all the land is privately owned and so much of the protection has to be achieved through persuasion and co-operation, though since 1949 there has been a series of legislative Acts in the service of nature and land conservation. An outline of the bodies involved and the main areas of their concerns for England and Wales is shown in Table 22.1.

TABLE 22.1 *Main scales and responsibilities for landscapes of protection in England and Wales*

	International	**National government**	**Local government**	**NGOs**	**Private, trusts, etc.**
Nature conservation	International wetland sites (Ramsar Convention)	National Nature Reserves; Sites of Special Scientific Interest (SSSIs)	Local nature reserves	County wildlife and conservation trusts	Protection by landowners, including sporting uses
Landscape protection		National Parks; Areas of Outstanding Natural Beauty (AONBs)	Areas of Great Landscape Value in structure plans	National Trust	Protection from change by landowners, especially if paid under e.g. set-aside
Landscape restoration			Reclamation of waste land; open-air museums	National Trust restores historic gardens and parks	
Urban fabric and built environment	World Heritage Sites (UNESCO)	Listed buildings in public and private ownership; Ancient Monuments	As owners of buildings, etc.	National Trust	Private owners

Note: Scotland and Northern Ireland are different: there is no legislation which would allow the designation of National Parks, for example.

These categories may be less separate than they look: National Parks get central funding but are in the charge of authorities whose membership is largely local, for example.

All the landscapes thus protected are human-made, with the possible exception of a few rocky islets with protected sea-bird populations. This means that a cultural decision to freeze a landscape in time or to manage it for some pre-determined end has to be taken. In the case of a nature reserve that holds the vegetation and animal communities of undrained fen (Chapter 9), then an artificially high water table has to be maintained by means of channels and sluices and indeed the reserve may be literally higher than the surrounding land because the peat has not shrunk. In the case of the moorlands, which have been created out of woodland by millennia of burning and grazing, a section of public perception demands that afforestation be kept to a minimum, especially where this consists of coniferous species. Similarly, there has been considerable concern over the loss of hedgerows in lowland agricultural regions. In part, this is due to their function as the habitats of bird and plant species but even more so as an element of traditional English lowland countryside. In fact, hedges are largely a creation of the Enclosure movement (Chapter 7) and the current situation is something like a return to early modern or even late medieval conditions: even more traditional, as it were.

ANOTHER COUNTRY

The wildlife of east and central Africa seems to belong to another world. The tropical savannas are familiar from a thousand TV programmes. They and the travel brochures portray a world separate from most human activity, impinged upon only by hovering tourists who observe, photograph and go away. The indigenous population seems to feature little except perhaps as *extras* who are as exotic as the fauna.

The savannas of Africa and their interspersed wetlands have a high biodiversity and a high biomass of visible fauna. The vegetation mosaic and its seasonal succession of utilization by herbivores allows for some very large herds of, for example, wildebeest as well as many species of birds and reptiles. The whole supports populations of high-profile predators, especially the big cats such as lion, leopard and cheetah. From colonial days, governments have allotted special status to many of the savanna areas, calling them game reserves and national parks and prohibiting the more usual kinds of development. The earliest motivations were centred upon the provision of a supply of animals to be killed on safari but after independence, the protection of the animals and their habitats as foundations of the tourist trade became paramount. The parks became big earners of foreign currency (though much of the profit was repatriated to the home countries of the tour operators) and the problems of management were mostly those of poachers after meat and ivory, encroachment upon park lands by agriculturalists, and competition with pastoralists for grasslands. Pastoralists were also unfriendly to mammal predators such as the lion. The Idi Amin regime in Uganda ruthlessly exploited the parks for ivory, so that elephant and rhino populations were drastically reduced.

Only in the 1970s was it recognized that both colonial and independent governments had set up park and reserve systems without much regard for the local people. Having no stake in the animals' future, these populations were uninterested in their conservation. Species whose migration routes went outside the parks were subject to intense human predation and the hostility to wild life outside the parks meant that increasing populations were crowded in the safe areas: more than one part of savanna Africa has experienced elephant populations so dense that they destroy both their own habitats as well as those of other species. By contrast, low elephant populations are unable to destroy enough trees to maintain essential elements of openness in the savanna vegetation. Most development plans for popular African wildlife concentrations, such as the Ngorongoro in Tanzania, now explicitly include the Masai pastoralists and their lifeways, though problems of culture conflict between the remnants of colonial attitudes, government desire for foreign currencies and the profits to be made from tourism, are all common. However, proposals which include zoning of the landscape but in which not all the zones are exclusive to wildlife, pastoralism or catchment protection, may have a future in which the abundant wildlife is included.

Conclusion

The conventional wisdom to date is that wildlife deserves its special landscapes. Its future is more assured, runs the argument, when undomesticated species are given special treatment in nature reserves, wildlife sanctuaries, National Parks and many other designated pieces of terrain. The success of such policies has been variable but there is no doubt that in crowded countries many more species would have become extinct had not such set-aside been practised.

Perhaps, however, it is time to move on from this way of viewing wildlife. Most wildlife exists outside such areas anyway, and it may be that the future of the non-human species of the planet is more secure if they co-exist alongside the humans whenever possible. They are not, then, exotic and *other*, so much as companions in the world. The world is not *our* oyster, we share it *with* the oysters. The challenge, then, is about changing human values rather than elaborating new laws and regulations.

Further reading and references

Conservation and the associated discourses of environmentalism and heritage have generated a vast literature in the last 20 years. For further discussion of the issues involved see pp. 183–85 and 312–17 of Simmons (1996).

Droste, B. von, Plachter, H. and Rösler, M. (eds) 1995: *Cultural landscapes of universal value: components of a global strategy*. Jena: Gustav Fischer.

Duncan, J.S. and Duncan, N.G. 1992: Ideology and bliss: Roland Barthes and the secret histories of landscape. In Barnes, T.J. and Duncan, J.S. (eds) *Writing worlds: discourse, text and metaphor in the representation of landscape*. London: Routledge.

Hewison, R. 1987: *The heritage industry: Britain in a climate of decline*. London: Methuen.

Oliver, M. de 1996: Historical preservation and identity: the Alamo and the production of a consumer landscape. *Antipode* **28**, 1, 1–23.

Simmons, I.G. 1996: *Changing the face of the earth: culture, environment, history*. 2nd edn, Oxford: Blackwell.

23

CONCLUSION

THE PAST, PRESENT AND FUTURE OF THE STUDY OF PEOPLE, LAND AND TIME

> We are the absolute masters of what the earth produces. We enjoy the mountains and the plains, the rivers are ours. We sow the seed and plant the trees. We fertilize the earth, ... we stop, direct, and turn the rivers, in short by our hands we endeavour, by our various operations in this world, to make, as it were another nature.
>
> (Cicero, *De re deorum* II.39, 45, 53)

THE HISTORIOGRAPHY OF LANDSCAPE STUDY: FOUNDATIONS

Landscape is a very complex and ambiguous word. It has the ability to absorb different meanings according to the culture, place and time. To the Anglo-Saxons it meant a tract of land, but Dutch artists in the seventeenth century invented the notion of a *landschap* in a picture frame. By the late nineteenth century this concern with visual appearance or landscape as a way of seeing the world had permeated much of European thought, and was reinvigorated by the German geographical tradition of *Landschaft*.

Landschaft had several interpretations. For Otto Schluter (1872–1959) the cultural landscape (*Kulturlandschaft*) was a visible material product, with little room for the invisible forces of production, but in the hands of other German writers it acquired the spiritual overtones of the *blood and soil* of the fatherland which partly at least legitimized the aggressive nationalism of 1930s' fascism (Livingstone 1992).

In France Paul Vidal de la Blache was interested in peasant landscapes. For him the essential strength of French culture lay in its rural community life, and his descriptions of the local texture of landscape in regions such as Alsace-Lorraine were masterpieces of the evocation of the human–environment relations. He identified regional landscapes, *pays*, that had distinctiveness or *personality* in terms of their customs, way of life, and local products such as wines, cheeses and architectural styles. The mutual moulding of humans and land over the centuries had led to a deep and lasting connexion that is best appreciated at the slow pace of the country dweller. Unfortunately this hymn to uniqueness had few verses and no tune for the urban and industrial era which swept across Europe in the nineteenth century. The Vidalian vision at around the time of the First World War was already being overtaken by the forces of globalization and subsequent regional geography had to begin from a different set of premises.

In Britain the concern for regions was strong in the late nineteenth century and up to the 1950s. This was partly an awareness of the utility of ordered spatial knowledge for any imperial and trading power but the work of Geddes, Herbertson, Fleure, Unstead and Roxby was far more than a matter of a descriptive inventory of factual material. Livingstone (1992) quotes from a 1916 paper of Herbertson which shows an awareness of the humanistic aspects of landscape that prefigured the later work of scholars such as Estyn Evans:

> There is a *genius loci* as well as a *Zeitgeist* – a spirit of place as well as of time. No social psychology, and no regional psychology is possible without a loving familiarity with the region ... No simple chronicle suffices for the historian, no superficial geographical inventory suffices for the geographer.

Carl Sauer in America was also interested in folk life and, like Vidal, neglected the modern era in favour of historical origins, but his inspiration was German rather than French. His work was anchored in the place-specific outcomes of cultural history that one might call a landscape. He saw the culture area as an organic whole, represented visibly by the works of humans: fields, villages, towns. For Sauer (1925): 'The cultural landscape is fashioned out of a natural landscape by a culture group. Culture is the agent, the natural area is the medium, the cultural landscape is the result.' This superorganic view of cultures and of cultural landscapes as identifiable objects is very much an American one and found little currency in Europe, where cultures have been seen rather as products of individual human actions (Shurmer-Smith and Hannam 1994).

Sauer was more concerned with cultural phenomena than with the causalism of many of his contemporaries, and it was partly his influence that inspired a human geography that owed much to materialist anthropology, with the mapping of cultural signatures (so-called cultural *spoor*) such as house types and covered bridges by Fred Kniffen, Wilbur Zelinsky and their graduate students (Hugill and Foote 1994). Unfortunately the extremes of this approach drew derision from later generations who had no respect for such chimney pot geography and, despite the valiant efforts of scholars such as Wagner, Mikesell and Meinig, cultural geography was effectively eclipsed by the 1960s due to its anti-theoretical philosophy, its preference for particular descriptions over general statements, and its lack of accommodation with the inexorable forces of urbanization and industrialization (Price and Lewis 1993).

A different perspective of the human–nature intersection, but nevertheless a branch from the same intellectual trunk, was the sub-discipline of *cultural ecology* (Turner 1989). First popularized by Harlan Barrows in the 1920s, this majors on the detailed interconnexions between humans and their environment in the sense of resource management. The strategies of diet, technology, reproduction, settlement and system maintenance are studied, and particular interest is taken in changes through time and variabilities across space (Butzer 1989). In the 1960s and 1970s cultural ecology drew inspiration from the systems theoretical approach that was then current and from the early stages of the computer revolution, which assisted with the modelling of energy and information pathways. Adaptation to risk in the form of environmental hazards was one theme, with related insights into human perceptions of their resource constraints. In sum, cultural ecology sees humans as part of the ecosystem and subjects their activities to a biological frame of thinking. This works well in archaeological and pre-industrial contexts, but is less convincing in advanced urban–industrial societies.

The latest interpretation of human–environment relations to emerge in the last 20 years has been *environmental history* (White 1985; Bailes 1988). It has been called 'one of the most exciting things to happen in American history this century' (Williams 1994) and certainly seems to have caught the imagination on both sides of the Atlantic through the writings of scholars such as Donald Worster, William Cronon and Simon Schama. In essence, environmental history is a response to criticisms that both history and human geography have taken too little interest in the physical environment. It looks at 'the role of nature in human life' or 'nature as a historical actor', for instance, the variety of institutional or socio-structural responses to resource scarcities and ecological crises. Worster (1984) identifies three areas of study:

1. The analysis of environmental change.
2. Human responses in terms of material culture, social relations, modes of production, and the regulation of human–environment relations.
3. Human perceptions of nature through laws, myths, and ethics (Williams 1994).

OTHER LANDSCAPE TRADITIONS

> Landscape is a medium of exchange between the human and the natural, the self and the other. As such, it is like money: good for nothing in itself, but expressive of a potentially limitless reserve of value.
> (W.J.T. Mitchell 1994: Imperial landscape. In Mitchell, W.J.T. (ed.) *Landscape and power.* Chicago: University of Chicago Press, 5–34)

From a different intellectual tradition from Sauer, but with similar results in terms of a richly detailed interpretation, the *local historian* W.G. Hoskins introduced notions of the historically grounded cultural landscape to a British (mainly English) audience in the 1950s (Phythian-Adams 1992). His scholarly use of both archival evidence and fieldwork observation, coupled with a palpable enthusiasm for the villages and fields of rural England, brought him both academic respectability and a large popular audience (Hoskins 1955). Hoskins' landscapes were built from local details and generalizations were few, but he resisted the morphological approach of cultural geography:

> For my own part I am not much interested in surface impressions. The three visible dimensions of a building or a landscape are not enough: they may entrance for the moment but they make no abiding impression on the mind. One needs the fourth dimension of time to give depth to the scene: one wants to know as much as possible about the past life of a place, about its human associations, and to feel the long continuity of human life on that spot before it can make its full impression on the mind.
>
> (Hoskins 1949)

Significantly, he continued the anti-modern theme of Vidal de la Blache and Sauer (Meinig 1979; Matless 1993) and thus cemented the notion for many that an authentic cultural landscape is timeless and unchanging. This backward-looking view remains the basis of rural planning regimes in many countries and the protective scheduling of landscapes by UNESCO. For Hoskins (1955): '[Since] the later years of the nineteenth century . . . and especially since the year 1914, every single change in the English landscape has either uglified it or destroyed its meaning, or both.' Hoskins was an historian by inclination but his desire to experience the landscape through the soles of his boots was widely shared by historical geographers and landscape archaeologists. Scholars such as Thorpe, Beresford, Hurst, and Taylor have contributed much in this area, as have many of their colleagues in Western Europe. Particularly prominent has been the Hodder & Stoughton series of volumes on the English landscape and the publications of the Permanent European Conference for the Study of the Rural Landscape (Baker 1988). A major Swedish interdisciplinary study is the most recent manifestation of the continuing strength of this school of thought (Berglund 1991).

In America the banner of landscape studies has been carried forward in J.B. Jackson's magazine *Landscape* (1951–), which has provided a popular outlet for a wide variety of writing about cultural landscapes, without the overt ideological baggage of *National Geographic* (Lutz & Collins 1993) or the social snobbery of British equivalents like *Country Life* or *The Illustrated London News*. Jackson saw the landscape as an outward expression of the inner essentials of human life, and an integrated unity of community and environment. He was particularly concerned with the living and working elements of both countryside and city, especially the house which he saw as the 'most reliable indication of . . . identity'.

A few geographers have clung on to notions of landscape because it gives them a focus around which they can unify. Both the physical and the human sides of the subject have something to contribute to its study, and in recent decades there has been interest in cognate disciplines such as archaeology and local history. In the minds of most geographers, however, there is an association between landscape and the old regional and cultural geographies which were superseded in the 1960s by a succession of process- and structuralist-based interpretations of space and place. A few historical geographers such as Sir Clifford Darby continued to work on landscape themes in the 1950s and 1960s, but their whiggish enthusiasm for improvements such as 'clearing the wood', 'draining the marsh' and 'reclaiming the heath' drew fire for its over-optimistic developmentalism (Coones 1992).

Other criticisms have been sharp and damaging. In the 1960s cultural landscape writing was found wanting by *positivists*, who characterized it as unscientific because of the absence of hard spatial data and theoretical frameworks. Later a naïve superficiality was identified by *structuralist* scholars who thought that a concentration on material artefacts and landscape forms ignored the crucially important social, economic and political processes that were responsible for their formation.

Really it was environmental geography rather than cultural or historical geography which kept the landscape fire burning. Probably the strongest continuity of work in our area throughout the 1950s and 1960s was in the United States. Following the theme of human impacts on the landscape pioneered by Marsh and Sauer, was an influential group of scholars whose work is perhaps best exemplified by the book *Man's role in changing the face of the earth*, edited by W.L. Thomas and published in 1956. There is much excellent work in this volume and others of the same period such as Clarence Glacken's magisterial book *Traces on the Rhodian shore*, which looks at human–environmental inter-relationships from the Ancient Greeks onwards.

Contemporary workers such as Billy Lee Turner have kept this tradition alive. Working with archaeologists in central America and elsewhere, Turner has helped to reconstruct the varied and changing implications of the human impress. European palaeoecologists have been similarly concerned, using techniques such as pollen and macro fossil analysis to chart successive societies' modification of flora and fauna, and the equally important impact of climate and sea-level changes. The study of such environmental changes during the Holocene has become a growth industry in its own right.

CHARISMA AND CONTEXT

> Every landscape is . . . a synthesis of charisma and context, a text which may be read to reveal the force of dominant ideas and prevailing practices, as well as the idiosyncrasies of a particular author.
>
> (D. Ley and J.S. Duncan 1993: Epilogue. In Duncan, J.S. and Ley, D. (eds) *Place/culture/representation*. London: Routledge.

Donald Meinig (1979) has coined the term *ordinary landscapes* to convey the notion of the continuous creation and alteration which arise more from 'the unconscious processes of daily living as from calculated landscape design'. The routines of ordinary people are of greatest significance here, as influenced by the tides of social history. Such landscapes may be invisible to us because we see them every day and take them for granted. Interpretations are therefore rarely straightforward, even if one accepts Meinig's contention that 'all landscapes are symbolic, as expressions of cultural values, social behaviour, and individual actions'. However, Peirce Lewis (in Meinig 1979) has helpfully offered seven axioms for reading the landscape:

1. Landscape is a clue to culture and 'the ordinary run-of-the-mill things that humans have created and put upon the earth [provide] strong evidence of the kind of people we are . . . In other words, the culture of any nation is unintentionally reflected in its ordinary vernacular landscape'.
2. 'Nearly all items in human landscapes reflect culture in some way. There are almost no exceptions.'
3. 'Common landscapes, however important they are, are by their nature hard to study by conventional academic means'.
4. 'In trying to unravel the meaning of contemporary landscapes and what they have to "say" about us . . . , history matters.'
5. 'Elements of a cultural landscape make little cultural sense if they are studies outside their geographic . . . context.'
6. 'Most cultural landscapes are intimately related to physical environment. Thus, the reading of the cultural landscape also

presupposes some basic knowledge of physical landscape.'

7. 'Most objects in the landscape, although they convey all kinds of "messages", do not convey those messages in any obvious way. The landscape does not speak to us very clearly.'

A breakthrough in the study of ordinary landscapes came with the subjective, qualitative work of a number of *humanistic* geographers in the 1970s. They objected to the treatment of people by spatial scientists as faceless statistics in mechanistic models, without any obvious interest in human individuality, creativity, relationships and power. This new group drew strength from psychology, literary criticism and art history (Penning-Rowsell and Lowenthal 1986; Pocock 1981). They saw landscapes through very different eyes and used the softer source material of literature and environmental perception to reveal the essences of human life rather than to formulate laws of human behaviour. Some have even documented *soundscapes* and *smellscapes* as means of writing sensuous geographies of their surroundings (Rodaway 1994).

In its most recent incarnation, the study of landscape has ridden in on the tidal wave of enthusiasm that has accompanied the cultural turn in social science. In particular, there has been a very encouraging trend in the new cultural geography to look beneath the surface appearances of shape and form, increasingly at signs, symbols, and other forms of intelligible meaning in society and landscape (Jackson 1989). An initial source of inspiration for this was the work of English-speaking Marxist cultural historians such as Raymond Williams and John Berger (Daniels 1989), but more recently French theorists have been very influential: post-structuralists such as Barthes, Bourdieu, and Foucault, and writers on postmodernism like Baudrillard, Lyotard and others.

Early, influential works in the new cultural geography were Denis Cosgrove's *Social formation and symbolic landscape* (1984) and his volume co-edited with Stephen Daniels entitled *The iconography of landscape* (1988). They introduced notions of visual metaphors or ways of seeing the landscape, including icons, spectacle, and theatre (Demeritt 1994). Significant themes have included consumption, representation, imagination, identities, authority and surveillance, conflict and resistance, and the related discourses of gender and citizenship are never far removed. Rather than being an updated version of Sauer's work, this has instead drawn heavily upon the intellectual field known as *cultural studies* and upon broader social and artistic theory. It had a revived interest in the visual, but less in terms of material objects than in the *gaze* of the observer, with all the implications this has for individual and group identities and representations, and also for the power structures which set the context of inter-personal relationships and the relationships between humans and their environment.

The future

The future of landscape studies is difficult to predict, but on the basis of recent trends it seems to lie in two directions. In the social sciences and humanities, the explosion of the new cultural studies has breathed fresh life into the landscape theme, with different methodologies and a new philosophical stance. There seems to be enough energy in this approach for much significant further work. Leading on the side of science are a growing band of environmental historians who have attracted much attention in recent years with their interest in the background to the relationships between humans and nature. Technological developments have aided with investigative methods but again there is a strong theoretical underpinning to this branch of current landscape work.

The common characteristic of both strands is a willingness to cross the boundaries of the classical disciplines such as geography, history, biology, archaeology, sociology and anthropology in order to gain insights about the interactions between humans and their environment. The future seems to lie with such broad-minded, integrative work.

References and further reading

Readers who wish to delve deeper into this fascinating area of scholarship are recommended to browse

Meinig (1979), Cosgrove (1985), Daniels (1989) and Olwig (1993, 1996).

Bailes, K.E. (ed.) 1988: *Environmental history: critical issues in comparative perspective*. Lanham, MD: University Press of America.

Baker, A.R.H. 1988: Historical geography and the study of the European rural landscape. *Geografiska Annaler* **70B,** 5–16.

Berglund, B.E. 1991: The cultural landscape during 6000 years in southern Sweden. *Ecological Bulletins* **41.**

Butzer, K.W. 1989: Cultural ecology. In Gaile, G.L. and Willmott, C.J. (eds) *Geography in America*. Columbus, Ohio: Merrill.

Coones, P. 1992: The unity of landscape. In Macinnes, L. and Wickham-Jones, C.R. (eds) *All natural things: archaeology and the green debate*. Oxford: Oxbow, 22–40.

Cosgrove, D. 1985: Prospect, perspective, and the evolution of the landscape idea. *Transactions of the Institute of British Geographers* n.s. **10,** 45–62.

Daniels, S. 1989: Marxism, culture, and the duplicity of landscape. In Peet, R. and Thrift, N.J. (eds) *New models in geography: the political-economy perspective*. London: Unwin Hyman, 196–220.

Demeritt, D. 1994: The nature of metaphors in cultural-geography and environmental history, *Progress in Human Geography* **18,** 2, 163–85.

Hoskins, W.G. 1949: *Midland England: a survey of the country between the Chilterns and the Trent*. London: Batsford.

Hoskins, W.G. 1955: *The making of the English landscape*. London: Hodder & Stoughton.

Hugill, P.J. and Foote, K.E. 1994: Re-reading cultural geography. In Foote, K.E., Hugill, P.J., Mathewson, K. and Smith, J.M. (eds) *Re-reading cultural geography*. Austin: University of Texas Press, 9–23.

Jackson, P. 1989: *Maps of meaning: an introduction to cultural geography*. London: Unwin Hyman.

Livingstone, D.N. 1992: *The geographical tradition: episodes in the history of a contested enterprise*. Oxford: Blackwell.

Lutz, C.A. and Collins, J.L. 1993: *Reading National Geographic*. Chicago: University of Chicago Press.

Matless, D. 1993: One man's England: W.G. Hoskins and the English culture of landscape. *Rural History* **4,** 187–207.

Meinig, D.W. (ed.) 1979: *The interpretation of ordinary landscapes*. New York: Oxford University Press.

Merchant, C. 1980: *The death of nature: women, ecology and the scientific revolution*. San Francisco: Harper & Row.

Olwig, K.R. 1993: Sexual cosmology: nation and landscape at the conceptual interstices of nature and culture, or: what does the landscape really mean? In Bender, B. (ed.) *Landscape: politics and perspectives*. Oxford: Berg, 307–43.

Olwig, K.R. 1996: Recovering the substantive nature of landscape. *Annals of the Association of American Geographers* **86,** 630–53.

Penning-Rowsell, E.C. and Lowenthal, D. (eds) 1986: *Landscape meanings and values*. London: Allen and Unwin.

Phythian-Adams, C. 1992: Hoskins' England: a local historian of genius and the realisation of his theme. *Local Historian* **22,** 4, 170–83.

Pocock, D.C.D. (ed.) 1981: *Humanistic geography and literature: essays on the experience of place*. London: Croom Helm.

Porteous, J.D. 1985: Smellscape, *Progress in Human Geography* **9,** 356–78.

Price, M. and Lewis, M. 1993: The reinvention of cultural geography, *Annals of the Association of American Geographers* **83,** 1–17.

Rodaway, P. 1994: *Sensuous geographies: body, sense and place*. London: Routledge.

Sauer, C.O. 1925: The morphology of landscape. *University of California Publications in Geography* **2,** 2, 19–54.

Shurmer-Smith, P. and Hanam, K. 1994: *Worlds of desire, realms of power: a cultural geography*. London: Edward Arnold.

Turner, B.L. 1989: The specialist-synthesis approach to the revival of geography: the case of cultural ecology. *Annals of the Association of American Geography* **79,** 95.

White, R. 1985: American environmental history: the development of a new historical field. *Pacific Historical Review* **54,** 297–335.

Williams, M. 1994: The relations of environmental history and historical geography. *Journal of Historical Geography* **20,** 1, 3–21.

Worster, D. 1984: History as natural history: an essay on theory and method. *Pacific Historical Review* **53,** 1–19.

INDEX

Acorns, 68, 95
Africa, 3, 5, 6, 14, 16, 21–2, 34, 41, 46, 67, 84–5, 99, 100, 102–3, 110, 113, 119, 128, 165, 169, 171, 177, 179, 208, 231, 235, 245, 259–61, 268, 272
Agriculture, xv, 1–3, 6–7, 13–25, 27, 28–34, 37–8, 40–51, 54, 57–60, 63–9, 71, 73, 76, 84–91, 93, 97–8, 102, 104–8, 110–13, 114–16, 117–18, 120, 122–3, 125, 127, 129, 131, 133–6, 138, 140–5, 148–51, 159, 162, 165–74, 178, 184, 192, 226, 237, 239, 254, 256–8, 260, 269, 271–2
The Alamo, 270–1
Alaska, 4, 179–80
Alberta, 10, 11
Alder, 8, 9, 71, 82
Allotments, 207, 236–7
Almond, 16, 167
Amaranth, 16, 115
Amazon, 16, 41, 93
The Americas, 21, 34, 110, 118–19, 165, 260
Amsterdam, 256, 264
Ancient landscape, 70, 82, 91, 142
Andes, 1, 13, 21, 257
Angkor Wat, 2
Anglo Saxons, 68, 82, 94–5, 137, 274
Antarctic, 269
Apple, 16, 65, 71, 167, 171
Apricot, 16, 167
Arabs, 22, 34, 37, 51, 104, 134, 247–9
Architecture, 18, 55, 59–60, 78, 91, 120, 132, 134, 150, 153, 155, 199, 203–4, 206, 208–10, 212, 215, 217, 224, 246, 263–5, 269–70, 274
Arctic, 1
Arid and semi-arid lands, xiv, 28, 30–1, 34–5, 51, 85, 119–21, 129, 170–1
Asia, 3, 13–14, 16, 21–2, 34, 38–9, 41, 45–6, 47, 99, 100, 119–21, 139, 165, 169, 177, 208, 215, 245, 253, 259–61, 264
Assarting, 96, 137
Aswan Dam, 108, 179
Aubergine, 16, 167
Australia, xv, 4, 16, 119, 169–70, 222, 240, 257, 260
Avocado, 16, 57, 167

Bajos, 56, 58
Baltic, 72, 91
Bananas, xiii, 16, 167
Bangladesh, 39, 104, 265
Barley, 13, 15–17, 21, 28, 33, 37, 66, 113, 123, 167, 170
Beans, 16, 19, 21, 57, 115, 117, 167
Beaver, 7, 10, 11
Beech, 10, 65, 68, 71, 95
Beet, 16, 167, 170
Belgium, 233, 253
Belize, 56, 61, 111
Benjamin, Walter, 208–10, 217
Berger, John, 240, 278
Biodiversity, 70, 99, 109, 171–2, 268–9, 272
Biotechnology, 15, 173–4
Birch, 11, 71, 82
Black Death, 67, 85, 136–7, 255
Bocage, 70, 82
Boserup, Ester, 44–5, 51
Bovines, 68, 167
Brazil, 14, 21, 95, 99, 102, 128, 258
Breckland, 73, 125
Bridgeman, Charles, 158, 226
Britain, xiv, 21, 44, 68–74, 82, 85–90, 124, 131, 139, 146, 153, 157, 158, 167, 172–3, 182, 187, 199, 203, 207, 224–7, 233, 236–7, 243, 249–50, 255, 270, 274
Brittany, 2, 82
Bronze Age, 48, 50, 71, 77, 91, 124, 126, 131, 219
Buckinghamshire, 71, 87, 226
Burial, 28, 29, 60, 61, 126, 181, 219

Cabbage, 16, 21, 167
Cacao, 16, 57, 60
Calcium, 42, 170
California, 128, 198, 214–15, 217, 268
Camels, 16, 40–1

Canals, 22, 25, 27, 30, 31, 34, 36–7, 38, 58, 59, 69, 72, 107, 108, 112, 115, 182, 186, 205, 227, 229
Cañihua, 16, 21
Capability Brown, 70, 226–7
Capitalism, 21, 98, 131–2, 135–6, 139, 146, 148, 150, 158, 188, 194, 195, 198, 203, 205, 208, 213–14, 217, 255, 268
Caribbean, 14, 256
Carrot, 16, 167
Cassava, 16, 167, 169
Castles, 134–6, 232
Cathedrals, 2, 135–6, 155, 162, 225
Cattle, 8–9, 13, 16–18, 21, 41, 42, 65–6, 68, 71, 97, 99, 123, 128, 167, 244, 258, 270
Cauliflower, 16, 167
Central America, 16, 22, 53, 55, 100, 102, 277
Cereals, 13, 15–18, 29, 37, 40, 51, 66, 129, 167, 255
Chalk, 66, 124, 138
Charcoal, 5, 7–9, 51, 70, 103, 149–50, 184, 190, 227
Cheese, 21, 78, 148, 151, 274
Chernobyl, 182, 253
Cherry, 16, 71
Chestnut, 16, 71, 167
Chicago, 197, 203
Chicken, 16, 21, 167
Childe, Gordon, 15
Chile, xv, 14, 122
China, 13–14, 16, 21–2, 33, 34, 46–7, 60, 83, 102, 104, 116–18, 119–20, 133, 135, 168, 177, 187, 255, 260–1
Chinampas, 69, 112, 114–16
Christianity, 13, 21, 37, 154, 239, 243
Citrus, 16, 21, 37, 122
Climate, xiv–xv, xvii, 1, 4–6, 8–9, 15, 17, 21, 28, 33, 40, 43, 47, 49–51, 54–6, 61, 63, 69, 78, 82, 85, 93, 108, 114, 119–20, 126–9, 136–7, 162, 167, 168, 170–1, 180, 223, 228–9, 277
Cloves, 16, 22
Coal, xv, 1, 70, 73, 153, 161, 176–7, 181, 186–7, 188–90, 191–5, 227, 253, 260
Coca Cola, 214, 265
Coconut, 16, 167, 258
Coffee, 16, 22, 99, 123, 167, 258
Collapse of civilization, 51, 53–63
Colombia, 99, 112, 260
Colonialism, xvi, 21, 49, 84, 91, 99, 103, 119, 132, 187, 222, 228, 233, 245, 250, 256–60, 272
Common Agricultural Policy, 109, 167
Common fields, 68, 82, 112, 113, 141, 151, 172
Computers, 181, 205, 214, 250–2, 254, 262, 264–5
Conifers, 4, 73, 102, 227, 272
Conservation, 97, 109, 127, 129, 172, 190, 241, 264, 267–73
Constable, John, 225, 240
Copper, 29, 50, 189–90, 257, 262, 270
Coppice, 68, 71, 73
Cotton, 16, 22, 30, 123, 167, 170, 187, 203, 257
Cree Indians, 10
Cucumber, 16, 167
Cultural ecology, 275
Cultural geography, 222, 275–8
Cultural landscape, xiv–xvii, 3, 5–7, 22, 37, 77, 162, 245, 254, 261–3, 267, 274–7
Cultural studies, 237, 247, 278
Cyberspace, 250–2

Dams, 34, 36, 38, 51, 59, 69, 108, 109, 168, 177–9, 222, 262
Darby, Sir Clifford, 93, 277
Dark Ages, 96, 134, 142, 172
Deciduous trees, 5, 7, 9, 11, 60, 100, 126
Deconstruction, 220
Deer, 1, 4, 8–10, 16, 18, 60, 65, 70–1, 94, 127
Defence, 27, 29, 32, 60, 108, 134, 144, 152–4, 232, 246–7, 256
Deforestation, 34, 49, 63, 93–6, 99–102, 128, 162
Demesne, 65, 133
Denmark, 82, 125–6, 182, 253, 260
Desert, 29, 33, 51, 53, 84, 101, 119–20, 123, 170–1, 234, 239–40, 244–5
Deserted medieval villages, 29, 85, 144
Desertification, 41, 51, 63, 128, 129, 170–1
Diachronic change, xv
Dickens, Charles, 197, 200
Diet, 6, 13, 18–19, 21, 57, 246, 275
Digging stick, 1, 25, 42
Diffusion, 13, 15, 21, 25, 120, 202
Discourse, 109–10, 220, 273, 278
Disease, 43, 60, 66, 69, 110, 122, 152, 168, 171, 200, 206, 209, 216, 222, 225, 257, 270
Disney, Walt, 214–17
Dog, 3, 16, 17
Domesday Book, 66, 68, 94
Domestication, 13–21, 25, 27, 46, 157, 165, 173, 269, 273
Dorset, 125, 232
Drainage, 31, 58, 69, 104–14, 118, 120, 126, 127, 165, 170, 172, 223, 239
Drained fields, 69, 108, 112, 118
Dual economy, 144, 149, 150–1
Ducks, 11, 16, 66, 167
Durham, 73, 134, 144, 192, 193, 268
Dyke-pond system, 104, 116–18

East Anglia, 82, 261
Ecosystems, 5, 6–7, 9, 11, 13, 22, 25, 37, 41, 43, 63, 67, 93, 97, 99, 103, 112, 117–18, 128, 129, 131, 166, 174, 234, 245, 257–8, 269, 275

Egypt, 2, 20, 25, 30, 33–4, 37, 50, 53, 179, 244, 246
Electricity, xv, 162, 176–82, 253, 263
Elk, 10
Elm, 9, 71, 73
Emmer, 17, 33
Enclosure, 82, 85–91, 92, 125, 128, 143, 145, 165, 172, 236, 272
Energy, xv, 1, 3, 6, 18, 27, 38, 107, 129, 150–1, 161–3, 166, 171, 176–85, 193–5, 253–4, 275
England, xv–xvi, 5, 7–9, 44, 45, 73, 77, 82–3, 85, 87, 90, 92, 94, 97, 103, 108–10, 128, 134, 137–40, 150, 172, 187, 191–3, 236–7, 250, 255, 268, 270–1, 276
Enlightenment, 72, 91, 224, 239
Environmental degradation, xvii, 33, 40–1, 43, 45, 47–51, 53–4, 59, 64, 67, 103, 168, 170–1, 258
Environmental history, xvi, 15, 275, 278
Essex, 68, 71, 150–1, 236, 240
Europe, xvi, xvii, 2, 3, 5–7, 11, 14, 16, 20–2, 25, 43–4, 47, 64, 68, 78–82, 85, 91, 94–7, 100, 102–3, 108, 119, 129, 131, 133–6, 140–3, 146, 148–9, 152–4, 165, 169, 177, 182, 184–6, 205, 208, 214, 217, 221–4, 231, 235, 239, 242–5, 254–7, 259–62, 267, 274–7
Ethiopia, 13, 14
Euphrates, River, 20, 27, 30, 33
European Community, 44, 169
European Union, 109, 167, 260
Exchange value, 135, 148

Factories, 150–1, 161–2, 167, 174, 176, 179, 184–9, 191, 193–5, 199–200, 203, 204–5, 262
Falklands/Malvinas, xiv
Fallow, 30, 33, 37, 43, 44, 47, 60, 69, 101–2, 141, 143
Famine, 35, 57, 67, 85, 136, 246
Fenland, 66, 104, 108–10, 118, 272
Fertile Crescent, 13, 27
Fertilizer, 44, 46, 58–9, 67, 109, 116–17, 127, 166, 168, 172, 174
Feudal system, 65, 85, 135, 137
Feudalism, 131, 133, 135–7, 139, 145–6, 148
Field system, 47, 77, 82, 84, 91, 110–14, 118, 121, 131, 141–2, 146, 172, 257
Filling station, 264–5
Fire, 1, 3–5, 7–11, 71, 93, 154, 157, 165, 190–1, 196, 197, 267–8
Fishing, 2, 10, 15, 65–6, 69, 90, 117, 257
Flax, 16, 65, 167
Flint, 5, 7, 9, 41, 57, 125
Florence, 72, 153
Fordism, 184, 193–5
Foucault, Michel, 205, 278
France, 37, 70, 72, 73, 78–9, 82–4, 91, 93, 108, 126, 150, 157, 182, 222, 226, 233, 243, 253, 257, 263, 265, 269, 274–5, 278
Frontier, 242–4

Gardens and gardening, 21, 37, 43, 58, 65, 69, 115, 157–8, 207, 223–4, 226–7, 235–7, 267–9, 271
Garlic, 16, 167
Genes, 13, 15, 17, 19–21, 87, 172, 173–4, 268
Germany, 22, 72, 82, 93, 126, 150, 180, 187, 224, 228, 243, 270, 274–5
Global era and globalization, xv, 139, 166, 185, 198, 202, 208, 214, 216, 232, 253–65, 274
Global warming, xiv, 64, 128, 187
Goats, 15, 16, 18, 21, 37, 41, 51, 167
Goose, 16, 65–6
Grain, 17–18, 21–2, 33, 44, 66, 84–5, 93, 124, 129, 136–7, 140, 148–9
Grape vine, 16, 37, 51, 84, 167
Grassland, 5, 7, 42, 108, 128–9, 165, 260, 272
Great Lakes, 10, 98
Greece, 47–51, 53, 96, 171, 224, 254–5, 267, 277
Green Revolution, 45, 167
Green villages, 144, 194
Grouse, 10, 127, 250
Guatemala, 56, 112
Guava, 16, 57
Guinea pig, 16, 21

Hardy, Thomas, 119, 125, 240, 249
Hare, 10, 65
Haussman, Georges-Eugène, 209
Hawaii, xvi
Hazel, 8, 9, 71, 167
Heath, 70, 78, 82, 124–6, 129, 131, 167, 277
Heather, 9, 124, 125, 126–8, 250
Hedges, 10, 65, 68, 70, 73, 82, 87–90, 141, 142, 157, 166, 172–3, 231, 272
Hemp, 16, 65, 167
Heritage, 134, 136, 189–90, 232, 240, 249–50, 269–71, 273
Hill stations, 222–4
Himalaya, 1, 46
Hoe, 22, 25, 44, 67, 69
Hollywood, 214–15, 250
Holocene, xiv, 4, 15, 49, 277
Homo, genus, 3, 5
Hong Kong, 107, 263
Horse, 16, 66, 128, 200, 225, 236
Hoskins, W.G., 276
Humanistic goegraphy, xvii, 275, 278
Humus, 8, 9, 42
Hunter-gathering, 1, 3–11, 15, 19, 25, 27, 267
Hwang-Ho, 34, 46–7

Hydraulic civilization, 31–4, 255
Hydro-power, 161–2, 177–9, 253

Iconography, 220, 270, 278
Imaginative landscapes, 249–50
Imperialism, 21, 50, 84, 187, 222, 245, 256–9, 262, 275
India, 13–14, 16, 22, 35, 38–9, 44, 45, 47, 60, 102–3, 133–4, 177, 187, 222–4, 257
Indigo, 16, 250
Indonesia, 99, 102, 169
Indus, 21, 33, 34, 255
Industrial Revolution, 150–1, 177, 186, 188–90, 199, 204, 225
Industry, xv–xvi, 1, 6–7, 38, 54, 64, 67–8, 70, 79, 98, 102, 106–7, 125, 131, 135, 145–6, 149–51, 158–9, 161–3, 168, 177, 179, 184–96, 198–9, 203–6, 214, 225, 227, 229, 232, 237, 239, 246, 254, 258, 260–1, 263–4, 269–70, 274–5
Infield-outfield system, 58, 142–3
Intensification, 11, 18, 25, 38–9, 43–5, 47–51, 57, 59–60, 95, 104, 111, 135–6, 167, 174, 260
Internet, 202, 214, 251, 262, 264–5
Inuit, xv, 1
Iran, 28, 29, 34, 40
Iraq, 28, 33, 34, 104, 234
Ireland, 45, 77, 263, 271
Iron, 70, 73, 96, 135, 149–50, 158, 184, 187, 189–90, 192–5, 199
Iron Age, 34, 219, 232
Irrigation, 2, 22, 25, 27, 28–40, 44, 46, 59, 69, 108, 112, 114, 115, 119–20, 123, 126, 128, 131, 165, 167, 168–70, 179, 228–9, 235, 249, 255, 258, 269
Islam, 2, 152
Italy, 51, 84, 149–50, 155, 157, 224, 226–7
Ivory, 22, 272

Japan, xvi, 14, 46, 47, 99, 214, 228, 234, 268
Java, 47, 169
Jericho, 17, 27, 28, 239
Jordan, 18, 247
Jute, 16, 167

Kandy, 221–2
Korea, 47, 208, 228–30, 239
Kulturlandschaft, xvii, 274

Labour service, 27, 34, 133, 136–7, 170
Lake District, 225, 240–1, 268
Land reform, 83–5
Landscape parks and gardens, 158, 223, 226–7, 235, 237, 239, 270
Landscapes of protection, 267
Landschaft, 274
Landschap, 274
Latin America, 121, 122–3, 152, 208, 260–1
Leisure, xiii, 44, 135, 207, 215–17, 223, 231–2, 235–7, 251
Lentil, 16, 167
Lettuce, xiii, 16
Levant, 18, 40
Limestone, 48, 50, 55–6, 59, 124, 169, 172, 192
Little Ice Age, xiv
Livestock, 21–2, 25, 65–6, 68, 85, 87, 106, 125, 128, 136–7, 139, 149, 165–6, 171–2
Llama, 1, 16, 21
Logging, 98, 99, 103
London, 136, 149, 151, 157–8, 191, 194, 197, 200, 202, 205, 212, 235–6, 256, 262–3, 269
Los Angeles, 180, 197, 202, 212, 214–16, 263–4
Lower Swansea Valley, 189–91
Lowland zone, England, 77, 82, 124–5, 150, 219, 272
Low Countries, 82, 149

McDonalds 214, 265
Machiavelli, xiv
Maize, 1, 15–16, 19–21, 42, 57–8, 61, 69, 97, 113, 115, 123, 129, 167, 169
Malaysia, 21, 47, 99, 257–8, 260–1
Mali, 171, 177
Malta, 50, 121, 153–4, 221
Malthus, Thomas, 43–4, 136
Manchester, 186, 197, 203, 205
Mango, 16, 167
Manioc, 16, 21, 42, 69, 113
Manor, medieval, 64–7, 76, 85, 96, 133, 136, 141, 145, 150–1
Manufacturing, 203, 210, 213, 253, 256, 258, 263, 270
Manure, 42, 44, 46, 65–6, 69, 114, 117, 143, 172
Markets, 27, 37, 44, 63–4, 67, 87, 117, 131, 135–8, 144, 148–50, 155, 157, 166, 170, 204
Marsh, 29, 66, 71, 94, 104, 105, 106, 108, 109, 111, 165, 167, 219, 229, 239, 277
Maya, 55–62, 110
Meadow, 8, 11, 64–5, 68, 108, 141, 225, 268
Mediterranean, 13–14, 21, 25, 37, 47–51, 53, 119, 122, 124, 128, 133, 144, 155, 179, 224, 256, 268
Melon, 16, 167
Meso-America, 14, 18–20, 33, 63
Mesolithic, xvi, 7–9, 25, 95, 124, 126
Mesopotamia, 20, 27, 28–9, 31–4, 104, 255
Metallurgy, 29, 189
Metaphor, 219–20, 245, 278
Metro Centre, Gateshead, 212, 216
Mexico, 13, 14, 18, 20, 56, 111, 114–16, 120, 261, 264
Microclimate, 114, 120
Microliths, 7, 8
Middle East, 13–18, 20–1, 25, 27, 28, 34, 120, 155, 239, 260–1

Milk, 21, 68, 87, 167
Millet, 16, 21, 167
Mining, 127, 145, 150, 159, 161, 177, 184–7, 191–3, 244, 269
Modern era, xv, 48, 76, 79, 108, 109, 119, 129, 131–2, 137, 139, 158, 161–3, 165–6, 172, 174, 176, 185, 196–208, 210, 212–13, 222, 234–6, 239–41, 252, 260, 264, 275
Modernism, 91, 119, 131, 162, 205, 207–8, 210
Modernity, 198–9, 203, 205, 208, 213–14, 265
Modernization, 108, 124, 179
Monetization, 136, 140
Moor, 70, 126–8, 131, 239, 250, 272
Mormons, 243–5
Mulberry, 16, 116–17

National Parks, 268–9, 271–3
Natural gas, 34, 51, 120
Neolithic, 8–9, 15–17, 28, 43, 77, 96, 125, 133, 239
Netherlands, 105–7, 172, 230, 255, 261
Newcastle-upon-Tyne, 152–3, 193
New England, 10, 11
Newfoundland, 2
New Guinea, 14, 176
New York, 183, 194, 202, 262–3
New Zealand, xv, 18, 44, 260, 269
NICs (Newly Industrializing Countries), 180, 195, 261
Nile, River, 20, 33–4, 179
Nitrogen, 17, 42, 57, 66, 69, 109
Nomadic pastoralism, 18, 40–1, 119, 171
Norfolk, 68, 108, 125
Normans, 95, 137, 152
North America, xvi, 4, 10–11, 14, 16, 22, 44, 84, 91, 99, 100, 102, 103, 113, 119, 128, 169, 208, 214, 217, 231, 257, 260–1, 267
North Sea, xv, 109
Northumberland, 73, 94, 270
Nottinghamshire, 227, 250
Nuclear energy, 161–2, 176–7, 181–2, 253
Nucleated villages, 82, 87, 97, 143
Nutrients, 42–3, 59, 66, 114, 117, 166, 168, 170

Oak, 8, 9, 10, 11, 65, 68, 70–4, 82, 93, 227, 268
Oats, 66, 113, 167
Oaxaca, 18, 120
Obsidian, 28, 57, 60
Oceania, 100, 102, 103, 110, 165, 177, 261
Oil, xv, 1, 41, 72, 129, 161, 169, 176, 179–80, 253, 260, 264
Olive, 16, 37, 51, 84, 122, 167
Onion, 16, 167
Open fields, 64, 66, 82–3, 85–8, 90, 141–3, 146, 165, 172
Ordinary landscapes, 91, 220, 277–8
'Other' landscapes, 222, 239–52

Pacific, 14, 16, 22, 112, 208, 260
Paddy, 46, 116
Pakistan, 21, 45
Palestine, 17–18, 27, 51, 235, 246–9
Palimpsest, xvii, 219, 269
Palladio, 91, 224
Papaya, 16, 167
Paris, 197, 200, 208–10, 217, 263–4
Parliament, 73, 82, 86–7, 90, 108, 236, 251
Partible inheritance, 142, 150–1
Pasture, 37, 64–6, 68, 70, 85, 96, 127, 128, 138–9, 141, 151, 162, 170–1, 227, 231, 260
Pays, 78–9, 274
Peach, 16, 167
Peanut, 16, 117
Pear, 16, 167
Pearl River (Zhuijang), 104, 116–18
Peat, 9, 66, 93, 104, 106, 109, 126, 272
Pepper, 16, 21, 115, 167
Peru, 14, 33, 113, 120, 122–3
Petén, 56–7, 59–61
Philippines, 43, 120, 269
pH level, 114, 117
Phosphates, 42, 172
Photosynthesis, xv, 1, 161, 162
Picturesque, 74, 225, 232, 241, 267
Pig, 16–18, 21, 37, 65, 68, 95, 117, 168–9, 214
Pineapple, 16, 167, 258
Pistachio, 16, 167
Place names, 68, 94–5
Planned landscape, 82, 88, 91
Plantations, 21, 257–8
Pleistocene, 3–6, 11, 15
Plotlands, 235–6
Plough, 25, 29, 34, 42, 44, 46, 47, 49, 65, 67, 69, 85, 91, 110, 120, 123, 127, 129, 141–2, 149, 165, 167, 219
Plum, 16, 167
Poland, 150, 268
Pollards, 67, 71, 82
Pollen, 5, 7, 60, 93–4, 277
Pollution, 63–4, 152, 166, 172, 183, 187, 195, 199, 202, 225, 254
Pomegranate, 16, 167
Population, 9–11, 15, 18, 22, 25, 27, 29, 32–4, 38, 40–9, 50–1, 54–61, 67, 85–6, 90, 96, 99, 104, 108, 110, 115, 119, 122, 127, 129, 131, 136–7, 140, 143–4, 150–3, 161–3, 169–71, 179–80, 186–7, 190, 199, 205; 207, 260–1, 271
Post-Fordism, 184, 195–6, 254
Post-industrial world, xv, 195–6, 234, 253, 260, 262–3

Postmodernism, 197, 212–17, 222, 252, 278
Potato, 16, 21, 90, 113, 123, 167
Pottery, 28, 33, 57, 60
Pre-industrial era, xv, 1–2, 22, 34–7, 43, 54–5, 60, 63, 67–8, 76, 110, 119–21, 131, 136, 151–7, 165, 182, 197, 199–200, 232, 240, 246, 275
Privatization, 64, 84, 87, 91, 145, 188, 263
Proto-industrialization, 159–51, 184–5, 195
Pulses, 17, 20, 167
Pyramid, 2, 29, 50, 54–5, 57

Qanat, 34, 69
Quarries, 50, 77, 127, 186–7
Quinoa, 16, 21, 113, 123

Railways, 70, 98, 161–2, 186–7, 192, 200, 205, 237, 254, 258, 263, 268
Raised fields, 58–9, 110–16
Rape, 16, 167
Reaves, 77, 91
Religion, xiii, 15, 74, 198, 220, 228
Remote sensing, satellite, 56, 59, 93, 95, 110, 128, 183
Renaissance, 153–5
Repton, Humphry, 158, 226–7
Reservoirs, 34, 59, 119, 169, 179, 228
Rice, 1, 16, 21–2, 40, 45–7, 51, 67, 116, 120–1, 167, 169, 228, 269
Ridge and furrow, 108, 112, 141–2, 165
Riots, 86, 109, 209
Risk, 40–3, 59, 64, 67, 87, 108, 119, 138, 140, 142, 168, 199, 245, 253, 275
Roadside landscapes, 264–5
Rockies, 98, 268
Romans, 22, 34, 37, 48, 50–1, 82, 84–5, 93, 96, 108, 135, 152, 212
Rome, 50–1, 85, 153, 155, 263
Rubber, 16, 21, 167, 258, 260
Rushes, 9, 65–6, 104
Russia, 102, 146, 214, 257
Rye, 1, 66, 167

Sacred landscapes, 73, 221, 233, 239, 267–8
Sahel, 35, 128, 171
Said, Edward, 222, 245–6, 249, 252
Salinization, 30, 33, 170
Salt, 2, 30, 33, 50, 60, 66, 70, 104, 114, 135, 149, 170, 171, 191
Sauer, Carl, 13, 25, 275–8
Scandinavia, xv, 1, 73
Science fiction, 250, 262
Science parks, 195, 214
Scotland, 45, 73, 90–1, 126–8, 134, 143, 225, 249, 250, 271
Semiology/semiotics, 220
Serf, 65, 133, 146
Sesame, 16, 30, 123, 167
Settlement, 6–7, 9–10, 18, 28–30, 32, 33, 41, 56, 58, 60, 66, 82, 90, 93, 95, 108, 119–20, 129, 134–5, 143–6, 148–9, 171, 179, 192–3, 195, 223, 233, 235, 242, 244, 246–7, 256–8, 260, 275
Sheep, 13, 15–18, 21, 33, 37, 41, 51, 65–6, 85–6, 91, 97, 127, 128, 144, 167, 182, 250
Shifting cultivation, 40–3, 57, 69, 102
Shopping centres and malls, 135, 204, 206, 208, 215–17, 254, 263
Singapore, xvi, 253, 263–4
Sisal, 22, 167
Skeleton, 18, 27, 60
Slaves, 22, 54, 59, 133, 226
Soil, xv, xvii, 9, 20, 25, 30, 32–7, 41–9, 50, 54, 56, 58, 59, 61, 64, 66, 69, 82, 85, 98, 108, 111–16, 119–21, 122–4, 126–9, 136–8, 140–1, 143, 165–72, 174, 189, 228–9, 244, 258
Soil erosion, 35, 41, 43, 47, 48, 49–51, 59, 61, 63, 85, 97, 99, 107, 113, 114, 120, 123, 168–71, 176, 228, 258
Solar energy, xv, 1, 114, 129, 161–2, 176, 254
Somerset, 71, 94
Sorghum, 16, 37, 123, 167
South America, 13, 16, 19, 21–2, 100, 103, 112, 122, 169, 177, 257
Spain, 2, 21, 34–8, 91, 110, 123, 128, 150, 157, 215, 270
Squash, 16, 19, 21, 57
Sri Lanka, 35, 221–2
The State, 30–4, 84–5, 103, 131, 167, 180, 205–8, 216, 224, 254, 256
Steel, 187, 189, 194, 203
Stinting, 65, 69, 149
Stonehenge, 54, 269
Stourhead, 226
Strip fields, 64, 82, 87, 133, 140–1
Suburbs, 57, 156–9, 188, 195, 200, 203–4, 207, 209, 214–15, 227, 235, 237, 254
Sudan, 123, 171
Suffolk, 150–1
Sugar, 16, 37, 99, 116–17, 166–7, 169–70, 256–8
Sumeria, 28–9, 30–4
Sunflower, 16, 167
Sunken fields, 69, 110–11
Surplus, 2, 27, 54, 64, 85, 129, 131, 148, 174, 255
Surveying, land, 84, 87–8, 91, 94–5, 229
Sustainability, 25, 49, 63–76, 118, 123, 131
Sweden, 82, 270, 276
Swidden, 40, 41–3, 51, 58, 59
Switzerland, 72, 263
Syria, 17, 18, 29, 170, 247

Taiwan, xvi, 261
Tamilnadu, 22, 35
Tanzania, 22–4, 113, 171, 272
Taro, 16, 112, 167
Taxation, 32, 37, 40, 54, 85, 94, 148, 151, 202
Tea, 16, 44, 257
Technology, xiii, xv–xvii, 7, 15, 20, 22, 25, 30, 32–7, 38, 44, 47, 54, 56, 83, 96, 99, 102, 104–8, 111, 119, 123, 131, 134, 141, 149–51, 153, 162, 165–6, 168, 170, 174, 176–7, 179, 182, 184–5, 188, 192, 194–5, 198–203, 213, 219, 228, 232, 239, 250–2, 254–5, 261–4, 275, 278
Telematics, 251
Television, 162, 206, 213, 214, 216, 220, 222, 243, 246, 250, 261–2, 265, 272
Temples, 22, 29–30, 32, 33, 50, 55–7, 61, 134–5, 179, 208, 226–7
Terraces, 34, 35, 46–9, 51, 59, 69, 120–4, 129, 165, 169–70, 228, 257, 269
Textiles, 28, 57, 85, 90, 149–51, 184–5, 187, 195
Thailand, 47, 102
Tigris, River, 20, 27, 30
Tikal, 57–61
Tobacco, 16, 167, 215, 257
Tokyo, 262–3
Tomato, 16, 115, 167, 173–4
Tourism, 54, 126, 134, 179, 189, 214, 225, 232, 250, 263, 269–70, 272
Trade, 7, 11, 20–2, 25, 28, 32, 44, 48–9, 51, 56–7, 60, 70, 90, 108, 131–2, 134–40, 148–9, 153, 155, 166–7, 185, 191–2, 255–8, 261, 265, 272
Transport, 21, 25, 44, 59, 70, 91, 98, 148, 150–1, 156, 166, 176, 180, 185–6, 188, 192, 195, 200–2, 205, 207, 214, 224, 246, 263, 268
Tropical rainforest, 6, 99, 103, 118, 165, 258
Tunisia, 34, 84, 104, 108, 265
Turkey, 8, 28–30, 34
Turkeys, 10, 16, 167
Tyneside, 70, 152–3, 192–3, 250

UNESCO, 179, 269, 271, 276
United Kingdom, 60, 250, 271–2
United States of America, 60, 91, 93, 97–9, 102–3, 108, 124, 128–9, 131, 158, 166, 187, 192, 199, 207, 215, 217, 233, 235, 240, 242–4, 246, 262, 265, 268, 270, 275, 276, 277
Urbanism, 27, 28–9, 31–4, 104, 152, 158, 196–7, 199, 204, 206, 214, 232
Urbanization, 28–34, 98, 114, 132, 137, 148–9, 158, 187, 199, 202, 204, 208, 275
Uruk, 28–9, 33
Use value, 135, 203
USSR, former, 83, 100, 102–3, 208, 214, 224, 228, 239, 254, 260

Vavilov, Nikolai, 13, 15
Venice, 72, 107–8, 153, 155, 256
Victorian period, 45, 194, 199–200, 205–8, 225, 227, 270
Vietnam, 47, 102, 234, 254
Virtual reality and virtual space, 250–2, 262

Wales, 73, 94, 128, 150, 172, 189, 225, 232, 236, 271
Walnut, 16, 67
War, xiv, 29, 122, 209–10, 228–9, 231–4, 246–7
Waste, 64–6, 86, 97, 104, 137, 151
Water harvesting, 34–5, 51, 59, 69, 123–4
Wear, River, 187, 192
Weeds, 17, 19, 21, 42, 43, 44, 58, 61, 66, 69, 112, 114, 142, 149, 168
Wessex, 71, 240, 249
West Bank, 235, 247–9
West End of London, 156–8
Wetlands, 56, 58–9, 104–18, 120, 169, 271–2
Wheat, 13, 15–17, 20, 28, 33, 37, 66, 123, 129, 167, 171
Wilderness, xvii, 33, 85, 91, 93, 97, 124–5, 126, 239–40, 243–4, 252, 267
Willow, 66, 71, 115
Windmills, 106–7, 109, 182
Wind power, 182–3
Wine, 78, 274
Woodland, 8–11, 25, 41, 64–5, 68–74, 76, 78, 82, 93–103, 104, 125, 126, 137–8, 143, 150, 165, 227, 235–6, 239, 257, 260, 272, 277
Wood pasture, 68, 70, 139, 171
Wool, 68, 85, 87, 137, 149, 151, 184, 187
Wordsworth, William, 240–1
World systems and world economies, 132, 255–60, 265
Writing, xvii, 14, 29, 34, 55, 57, 61

Yak, 1, 16
Yam, 16, 167
Yemen, 123–4, 129
Yorkshire, 73, 94, 126, 150, 187, 250
Yugoslavia, former, 214, 231

Zagros Mountains, 18, 29, 40
Zaire, 99, 177